Emad Elba

Strategies for protection and sustainable environmental management of the High Aswan Dam reservoir in Egypt considering climate change

disserta Verlag

Elba, Emad: **Strategies for protection and sustainable environmental management of the High Aswan Dam reservoir in Egypt considering climate change**, Hamburg, disserta Verlag, 2017

Buch-ISBN: 978-3-95935-354-0
Druck/Herstellung: disserta Verlag, Hamburg, 2017
Covermotiv: © Emad Elba
Covergestaltung: Annelie Lamers

Bibliografische Information der Deutschen Nationalbibliothek:
Die Deutsche Nationalbibliothek verzeichnet diese Publikation in der Deutschen Nationalbibliografie; detaillierte bibliografische Daten sind im Internet über http://dnb.d-nb.de abrufbar.

Zugl.: Dissertation, Leuphana Universität Lüneburg / Fakultät Nachhaltigkeit, Institut für Ökologie, 2016

Das Werk einschließlich aller seiner Teile ist urheberrechtlich geschützt. Jede Verwertung außerhalb der Grenzen des Urheberrechtsgesetzes ist ohne Zustimmung des Verlages unzulässig und strafbar. Dies gilt insbesondere für Vervielfältigungen, Übersetzungen, Mikroverfilmungen und die Einspeicherung und Bearbeitung in elektronischen Systemen.

Die Wiedergabe von Gebrauchsnamen, Handelsnamen, Warenbezeichnungen usw. in diesem Werk berechtigt auch ohne besondere Kennzeichnung nicht zu der Annahme, dass solche Namen im Sinne der Warenzeichen- und Markenschutz-Gesetzgebung als frei zu betrachten wären und daher von jedermann benutzt werden dürften.

Die Informationen in diesem Werk wurden mit Sorgfalt erarbeitet. Dennoch können Fehler nicht vollständig ausgeschlossen werden und die Diplomica Verlag GmbH, die Autoren oder Übersetzer übernehmen keine juristische Verantwortung oder irgendeine Haftung für evtl. verbliebene fehlerhafte Angaben und deren Folgen.

Alle Rechte vorbehalten

© disserta Verlag, Imprint der Diplomica Verlag GmbH
Hermannstal 119k, 22119 Hamburg
http://www.disserta-verlag.de, Hamburg 2017
Printed in Germany

Strategies for Protection and Sustainable Environmental Management of the High Aswan Dam Reservoir in Egypt Considering Climate Change

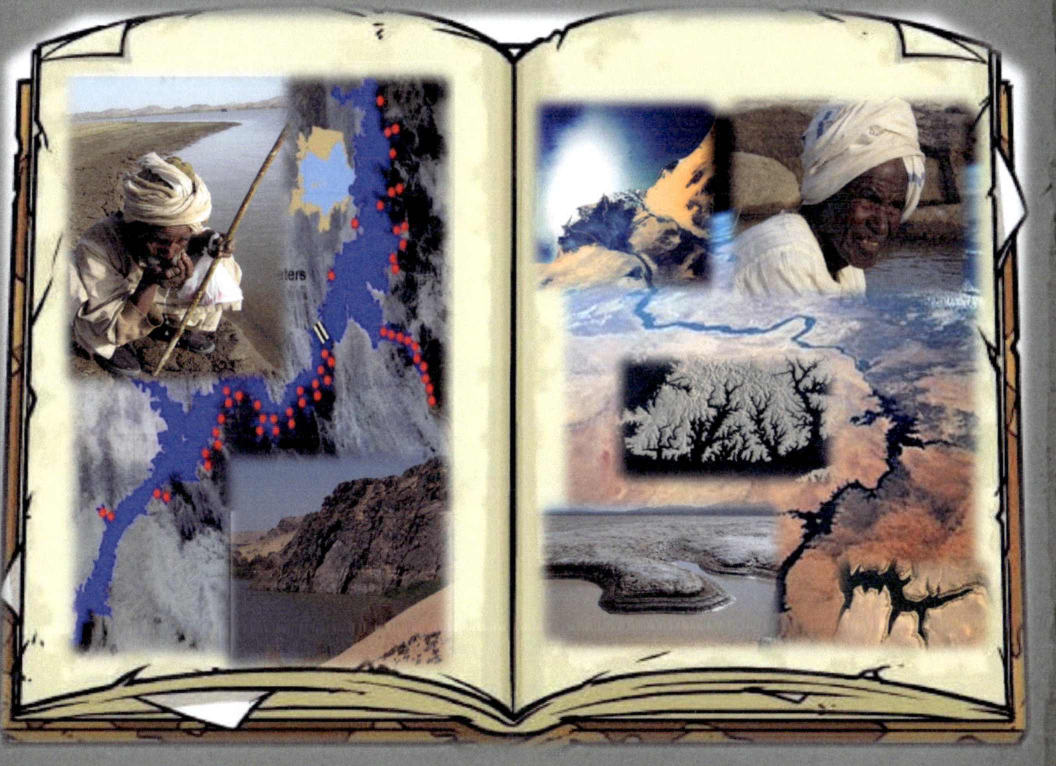

Emad Elba

Strategien zum Schutz und nachhaltigen Umweltmanagement des Assuan-Hochdamm Reservoirs in Ägypten unter Berücksichtigung des Klimawandels

Dissertationsschrift ist

der Fakultät Nachhaltigkeit
der Leuphana Universität Lüneburg

eingereicht zur Erlangung des akademischen Grades eines
Doktors der Naturwissenschaften
(Dr. rer. nat.)

vorgelegte Dissertation

von

Emad Eldin Mohamed Mahmoud Hafez Elba

geboren am 02.07.1967 in Alexandria, Ägypten

Mai 2015

Strategies for protection and sustainable environmental management of the High Aswan Dam reservoir in Egypt considering climate change

This thesis is submitted to

the Faculty of Sustainability
of Leuphana University Lüneburg

to earn the academic degree of
Doctor of Natural Science
(Dr. rer. nat.)

Dissertation submitted

by

Emad Eldin Mohamed Mahmoud Hafez Elba

Born on 02.07.1967 in Alexandria, Egypt

May 2015

Submitted on: 05.05.2015

Doctor advisor and reviewer: Prof. Dr. Brigitte Urban
Leuphana University Lüneburg, Faculty of Sustainability Sciences,
Institute of Ecology, Subject Area Landscape Change
Lüneburg, Germany

Second reviewer: Prof. Dr. Bernd Ettmer
University of Applied Science Magdeburg-Stendal,
Department of Water and Recycling Management
Magdeburg-Stendal, Germany

Third reviewer: PD Dr. Jordi Delgado
University of A Coruña,
Department of Construction Technology
A Coruña, Spain

Date of disputation: 24.09.2015

Edicated to

my great country, Egypt

the memory of:
my beloved father Mohamed Elba
my beloved wife Naglaa Helmy

my kind mother Fatma Abd Elsalam

my kind wife Dalia Farghaly

my lovely children
Mona and Abdelrahman

Acknowledgements

All praise is due to **Allah**, the lord of all creatures. It is HE who provided me with the resources and the ability that made this research possible.

When it was about time to write this acknowledgement I realized I had many people to thank and the question of where and with whom to start came up. But I decided to start from the beginning of my PhD journey.

I would like to express my deepest gratitude and appreciation to my supervisor, **Prof. Dr. Brigitte Urban**, for her excellent scientific advice and guidance, constructive suggestions, endless caring, extraordinary patience, and providing me with an excellent atmosphere to perform this research. During the work period, I found her a true academic. So, many thanks to her for everything she did to bring me to this level, and I will always remember her with respect.

I am deeply grateful to **Prof. Dr Bernd Ettmer** from University of Applied Sciences Magdeburg-Stendal for his precious time reading my thesis and for his constructive comments, technical guidance, and providing me with new ideas. I also want to express my sincere thanks to **Prof. Dr. Jordi Delgado** from the University of A Coruña, Spain for his precious time reading my thesis and accepting to be an advisor of the examination board.

I am greatly thankful for the staff of the Leuphana Graduate school in particular **Mrs. Susanne Wenzel**. The Graduate school granted me with fincial support to share my research in several international conferences and exchange experiences and knowledge. I am deeply thankful to the staff of Leuphana Writing Center **Dr. Micha Edlich and Dr. Isabell May** for their support and English revision.

This research work could not be performed without extensive field data collection in the Nile Basin. Hence, I would like to thank all those who have participated one way or another to make the data collection campaign a success. Most of all, I would like to extend my gratitude to all my colleagues in **Nile Water Sector, Ministry of Water Resources and Irrigation** for their continuous involvement. Special thanks to **Eng. Sameh Saif, Eng. Emad Ali, Dr. Aref Gharieb,** and **Dr. Yasser H. Elwan.**

I never lost contact with my home department at Nile water Sector. I would like to greatly thank the following individuals for pointing to resources and secrets of the Nile Basin hydrolgy. The love I have to the Nile is because of them. I would like to express my deepest gratitude and appreciation to the father of the Nile who passed away **Eng. Mohamadeen Amin Mohamadeen**, the best hydrologist in Nile Basin**,** and to **Eng. Ahmed Fahmy**, my first teacher and to supervisor **Eng. Ahmed Bahaa Eldin** who was the first teaching me a lot of keys about Nile Hydrology, the memory of **Prof. Dr. Ahmed Rageb** who guided me *during the preparation of the primary research plan of this work. Special thanks to* **Dr. Mohamed Abd Elatty**, **Dr. Mohamed Elshamy** *and* **Dr. Walid Omar.** *for providing me with many research results, data and knowledge.*

Special thanks go to my friends in Egypt, Germany, Ugand and Sudan for their kind inspiration, supports and most important for their friendship, specially Prof. Dr. **Fatma Abd Elrahman**, *Dr.* **Wael Roshdy**, *Dr. med.* **Mahmoud Algazzar**, *Frau* **Margot Seeber**, **Frank Dziembowski**, **Jan Schulze**, *Dr. med.* **Ahmed-Maher Mouhandes** *and Dr. med.* **Ralf Schumacher**.

Finally, I would like to thank my **mother**, *for bringing me up appreciating science and for coping with my long absence away from her. Special thanks go to my uncle* **Eng. Mahmoud Zagzoug, Dr Salah Sallam**, *and to the* **Elba family, Farghaly family, Abd Elsalam** *family and* **Sallam** *family. My deep gratitude is due to my dear wife* **Eng. Dalia Farghaly** *for her continuous support and perseverance during these years. I would like also to express my thanks to my children,* **Mona and Abd Elrahman** *who, despite their young ages, understood that I needed to be the absent present father while working on this research.*

I send it with love and gratitude.

Suderburg, Germany,

Emad Elba

May 2015

Abstract

For a broad variety of reasons, scientists in Egypt are particularly interested in the sustainable management of water and land resources. Because global climate change will, for example, adversely affect its water and land resources as well as its coastline and agriculture, Egypt is likely to become one of the most vulnerable countries in the world in the next several decades. Global warming and the subsequent rise in sea level will lead to the loss of a considerable and, in terms of economic production, very productive areas in the northern coastal zones and the delta of the River Nile. These developments will affect the lives and livelihoods of approximately 6 to 7 million people, who will have to be relocated. Many climate scenarios also predict that climate change will reduce rainfall in the Nile basin as well as the flow of the Nile River in general and in the High Aswan Dam Reservoir (HADR) in particular. Higher temperatures will cause higher evaporation as well as evapotranspiration from agricultural lands, which, in turn, will result in lower crop yields. A considerable percentage of the 18 billion cubic meter (km^3), the mean annual evaporation loss at the HADR during the past two decades, has been lost at the reservoir's 100 embayments, which are called khors and cover half of the surface area of the HADR. These losses are expected to increase by approximately 3% to 10% by the year 2100 due to global warming.

This research project aims to identify and assess several adaptation tools in order to mitigate the severe impact of climate change in Egypt in general and the area around the HADR in particular. By decreasing evaporation losses from the HADR, these tools will protect and shore up Egypt's limited water resources. In addition, they can also be used to optimize land resources and, more specifically, to identify new settlement sites for the population relocated from the northern coastal zones. The following five major steps were taken to achieve these objectives: First, databases were built to compile information on the characteristics of the HADR. Second, a digital elevation model for the bed of the HADR and a mathematical model to describe the reservoir's hydrological characteristics were developed. Third, several options for reducing evaporation losses were identified and assessed. Fourth, several options for settlement areas were evaluated. Fifth, measures to protect these new settlement areas and to achieve sustainable development in the region were identified and assessed.

Information on the following aspects were collected and used in developing the databases: meteorological data; hydrological data, especially on the discharge and water levels in the reservoir; seepage data; topographic data, including old topographic maps; the bathymetric survey of the HADR; satellite images and aerial photos; and bed soil sedimentary data. The databases show the following: the mean air temperature has increased by about 1°C per decade, and the mean evaporation rate by about 1.0 mm/day per decade, which represents an increase of 10% in terms of mean evaporation loss; the most frequent estimated surface water level was 178 m AMSL; the mean annual seepage losses were just 55 million cubic meter (hm^3), accounts for 0.1% of Egypt's share from the HADR, and can be neglected; the HADR undergoes high silt and clay sedimentary in Nasser Lake, whereas the sand sedimentary increases in Lake Nubia to 100% by end of the HADR. A Digital Elevation Model for the HADR, which describes the current hydrological characteristics of the HADR, was generated using the bathymetric survey of the HADR and satellite images. The DEM was used to calculate, the HADR's hydrological characteristics (surface area and water volume) at different water levels as well as the mean depth, surface area, and water volume of the khors. Mathematical equations were developed and evaluated to compute the surface area and water volume at different water levels. The DEM was also used to evaluate khors, several sections along and across the HADR, and changes in terms of bed altitude due to sediment and erosion events. These and other calculations provided the basis for a broad variety of scenarios that consider the impact of climate change on the region surrounding the HADR as well as the effectiveness of adaptation tools that could be used to mitigate the adverse effect of global warming.

Three main scenarios were investigated to reduce evaporation losses. Scenario 1 investigated the possibility of reducing evaporation losses by eliminating one or more khors. Eliminating Khor Kalabsha, Khor El-Alaky, Khor Genina, Khor Sara, and some small khors will reduce evaporation losses by approximately two km^3, 0.5 km^3, 0.4 km^3, 0.1 km^3 and three km^3 respectively by 2100. Scenario 2 investigated the opportunity of controlling water levels by building a dam at the El-Madik reach. The study assessed the effect of lowering the water level downstream of this dam and raising the water level upstream. Reducing the water level by five meters at an altitude of 178 m AMSL downstream of the proposed dam will, for instance, reduce evaporation losses by approximately 0.6 km^3 by the end of the twenty-first century. Scenario 3 investigated the consequence of lowering the lakebed level by removing the sediment from the bed of the HADR. By 2100, removing the sediments from Lake Nasser will reduce evaporation losses by approximately 0.2 km^3, whereas removing them from the entire HADR will reduce the evaporation losses by approximately 1.1 km^3.

This research project also considered additional tools that could be combined with the measures described above to reduce evaporation losses. A combination of the following measures is considered to be the most effective and sustainable: building a dam at the Madik section, filling Khor Kalabsha, removing sediments from the HADR, as well as using floating objects on Khor El-Alaky and other small khors. This approach is expected to save approximately 5.0 km^3 of water under the current climatic conditions and 5.55 km^3 by the end of this century. Building a new dam at El-Madik is not only likely reducing evaporation losses, but it would also increase energy resources and protect the area around Khor

Kalabsha against flooding hazards. Furthermore, the dam would be beneficial to the riparian people around Khor Kalabsha, who would have more access to agricultural lands. Covering part of Khor El-Alaky or other small khors using, for example, palm fronds, protects the balance of the fragile ecosystem and may involve economic benefits for local firms producing these biodegradable covers.

Finally, this project examines possibilities for new settlements, especially for climate change migrants from the Mediterranean coast. These plans include settlements in a total area of 1200 km^2 that would include what is now Khor Kalabsha and the surrounding areas up to 200 m AMSL. It is recommended to use this area for agricultural purposes. This kind of development is likely to involve many major infrastructure projects, for example the construction of an irrigation and drainage system. New irrigation and cultivation techniques and measures are recommended to minimize water uses, to maximize crop production, and to protect the environment. Furthermore, this study recommends to provide clean energy resources, for example electricity from hydropower stations. This resource, in addition to solar power and wind energy are likely helping smaller communities to establish new factories, which, in turn, may offer job opportunities for climate migrants.

Zusammenfassung

Aus einer Vielzahl von Gründen sind ägyptische Wissenschaftler an einer nachhaltigen Bewirtschaftung der Wasser- und Landressourcen interessiert. Aufgrund des globalen Klimawandels ist Ägypten in den nächsten Jahrzehnten wahrscheinlich eines der am meisten gefährdeten Länder auf der Welt, da sich der Klimawandel nachteilig auf Ägyptens Wasser und Landressourcen auswirkt. Insbesondere betroffen sind die nördlichen Küstenbereiche und die Landwirtschaft. Die globale Erwärmung und der daraus resultierende Anstieg des Meeresspiegels führen in Bezug auf die Produktion zu einem erheblichen Verlust wirtschaftlich nutzbarer Bereiche in den nördlichen Küstengebieten und dem Nildelta. Diese Entwicklungen werden das Leben und die Lebensgrundlage der etwa 6 bis 7 Millionen hier lebenden Menschen beeinflussen. Viele Klimaszenarien sagen vorher, dass sich die Niederschlagsmenge im Nil-Becken durch den Klimawandel, dem Nil im Allgemeinen und im Hohen Assuan Stausee (High Aswan Dam Reservoir - HADR) erheblich reduzieren wird. Höhere Temperaturen führen zu höherer Verdunstung und dem Verlust von landwirtschaftlichen Flächen. Dies führt wiederum zu einer Verminderung von früher erzielten Ernteerträgen. Ein erheblicher Prozentsatz des 18 Milliarde (10^9) Kubikmeter (km^3) betragenden mittleren jährlichen Verdunstungsverlustes im HADR in den vergangenen Jahrzehnten ging in den 100 Einbuchtungen des Reservoirs, Khors genannt, verloren. Das betrifft etwa die Hälfte der Fläche der HADR. Diese Verluste werden sich aufgrund der globalen Erwärmung bis zum Jahr 2100 voraussichtlich um rund 3% auf 10% erhöhen. Dieses Forschungsprojekt zielt darauf ab, die schweren Folgen des Klimawandels in Ägypten im Allgemeinen und insbesondere im Bereich um den HADR zu erkennen und Wege zu schaffen, diese zu mildern. Durch die Verringerung der Verdunstungsverluste aus dem HADR, werden die begrenzten Wasserressourcen Ägyptens geschützt und gesichert. Darüber hinaus können die gewonnenen Erkenntnisse auch verwendet werden, um Bodenressourcen zu optimieren und um neue Siedlungsplätze für die Bevölkerung aus den nördlichen Küstenregionen zu schaffen. Die folgenden fünf Hauptschritte wurden unternommen, um diese Ziele zu erreichen: Erstens wurden Datenbanken erstellt, um Informationen über die Gegebenheiten des HADR zusammenzustellen. Zweitens wurden ein digitales Höhenmodell für das Bett des HADR und ein mathematisches Modell zur Beschreibung der hydrologischen Eigenschaften des Reservoirs entwickelt. Drittens wurden verschiedene Optionen zur Verringerung der Verdunstungsverluste aufgezeigt und ausgewertet. Viertens wurden verschiedene Optionen für neue Siedlungsgebiete bewertet. Fünftens wurden Maßnahmen benannt um diese neuen Siedlungsgebiete zu schützen und eine nachhaltige Entwicklung in der Region zu erreichen. Informationen zu folgenden Aspekten wurden gesammelt und bei der Entwicklung der Datenbanken verwendet: meteorologische Daten, hydrologische Daten, insbesondere vom Abfluss- und Wasserpegel im Reservoir, Versickerungsdaten, topographische Daten, einschließlich alter topographischer Karten, Tiefenmessungen des HADR, Satellitenbilder und Luftaufnahmen, sowie Boden Sedimentablagerungen. Die Datenbanken zeigen: die mittlere Lufttemperatur erhöht sich um ca. 1°C pro Jahrzehnt und die mittlere Verdunstungsrate um etwa 1,0 mm/Tag pro Jahrzehnt, was eine Steigerung von 10% in Bezug auf den mittleren Verdunstungsverlust darstellt. Der häufigste geschätzte Oberflächenwasserstand beträgt 178 m über Normal-Null (üNN), die mittleren jährlichen Versickerungsverluste betrugen lediglich 55 Million (10^6) Kubikmeter hm^3, davon entfallen 0,1% auf den ägyptischen Teil des HADR und können vernachlässigt werden. Das HADR enthält einen hohen Anteil an Schlamm- und Tonsedimenten im Nasser-See, während die Sandsedimente im Lake Nubia am Ende des HADR zu 100% zunehmen.

Ein digitales Geländemodell (DGM) für den HADR, welches die gegenwärtigen hydrologischen Eigenschaften des HADR beschreibt, wurde mittels der Tiefenmessungen des HADR und Satellitenbildern erzeugt. Das DGM wurde verwendet, um zu berechnen, wie HADR die hydrologischen Eigenschaften (Oberfläche und Wassermenge) bei verschiedenen Wasserständen sowie die mittlere Tiefe, Oberfläche und Wasservolumen der Khors verändert. Mathematische Gleichungen wurden entwickelt und ausgewertet, um den Oberflächenbereich und die Wassermenge bei unterschiedlichen Wasserständen zu berechnen. Das DGM diente auch zur Bewertung der Khors an verschiedenen Abschnitten längs und quer zum HADR in Bezug auf Änderungen der Flussbetthöhe durch Sediment und Erosionsereignisse. Diese und andere Berechnungen bilden die Grundlage für eine Vielzahl von Szenarien, die die Auswirkungen des Klimawandels auf die Region um das HADR sowie die Wirksamkeit der entwickelten Tools berücksichtigen, die verwendet werden könnten, um die negativen Auswirkungen der globalen Erwärmung zu mildern. Drei Hauptszenarien wurden untersucht, um die Verdunstungsverluste zu verringern. Szenario 1 untersucht die Möglichkeit einer Verringerung der Verdunstungsverluste durch die Eliminierung eines oder mehrerer Khors. Die Eliminierung von Khor Kalabsha, Khor El-Alaky, Khor Genina, Khor Sara und einiger kleinerer Khors würde die Verdunstungsverluste um etwa zwei km^3, 0,5 km^3, 0,4 km^3, 0,1 km^3 und drei jeweils von 2100 verringern. Szenario 2 untersucht die Möglichkeit, den Wasserspiegel durch den Bau eines Staudamms in der El-Madik-Region zu kontrollieren. Die Studie untersuchte den Effekt der Absenkung des Wasserstandes flussabwärts des Dammes und die Erhöhung des Wasserstandes flussaufwärts. Die Reduzierung des Wasserstandes um fünf Meter auf einer Höhe von 178 m üNN hinter dem Damm wird vorgeschlagen, um die Verdunstungsverluste um ca. 0,6 km^3 bis zum Ende des einundzwanzigsten Jahrhunderts zu reduzieren. Szenario 3 untersucht die Folge der Absenkung des Seebodens, indem das Sediment vom Boden des HADR entfernt wird. Bis 2100, wenn die Sedimente aus dem Nasser-See entfernt worden sind, werden die Verdunstungsverluste um etwa 0,2 km^3 zu reduziert, während die Reduzierung der Verdunstungsverluste bei einer Entfernung aus dem gesamten HADR rund 1,1 km³ betragen. Dieses Forschungsprojekt bietet auch zusätzliche Tools, die mit den oben beschriebenen Maßnahmen zur Reduzierung des Verdunstungsverlustes kombiniert werden können. Auch eine Kombination der folgenden Maßnahmen gilt als effektiv und nachhaltig: Der Bau eines Staudamms in der El-Madik-Region, die Auffüllung des Khor Kalabsha, die Entfernung der Sedimente aus dem HADR sowie der Einsatz von treibenden Objekten auf Khor El-Alaky und anderen kleinen Khors. Dieser Ansatz wird voraussichtlich rund 5,0 km³ Wasser unter den derzeitigen klimatischen Bedingungen und 5,55 km³ bis zum Ende dieses Jahrhunderts speichern. Der Bau eines neuen Staudamms in der El-Madik-Region würde nicht nur die Verdunstungsverluste reduzieren, sondern auch die Energieressourcen erhöhen und dem Bereich um Khor Kalabsha Schutz vor Überschwemmungsgefahren bieten. Darüber hinaus würde der Damm den Uferbewohnern um Khor Kalabsha von Vorteil sein, denen vermehrt landwirtschaftliche Nutzflächen zur Verfügung stehen würden. Die Abdeckung eines Teil von Khor El-Alaky oder anderen kleinen Khors, zum Beispiel durch Palmwedel, würde das Gleichgewicht der empfindlichen Ökosysteme schützen und den lokalen Firmen wirtschaftliche Vorteile bringen, die diese biologisch abbaubaren Deckungen produzieren. Schließlich untersucht das Projekt Möglichkeiten für neue Siedlungen, vor allem die Klimamigranten aus dem Gebiet der Mittelmeerküste. Diese Pläne sehen Siedlungen mit einer 1200 km² umfassenden Gesamtfläche auf dem jetzigen Gebiet des Khor Kalabsha und der Umgebung bis zu 200 m üNN. Es wird empfohlen, diesen Bereich für landwirtschaftliche Zwecke zu nutzen. Hierfür bedarf es vieler großer Infrastrukturprojekte, zum Beispiel den Bau eines Be- und Entwässerungssystems. Neue Bewässerungs und Anbautechniken werden empfohlen, um den Wasserverbrauch zu optimieren, den Bodenertrag zu maximieren und die Umwelt zu schützen. Darüber hinaus empfiehlt die Studie, saubere Energieressourcen, zum Beispiel Strom aus Wasserkraftwerken bereitzustellen. Diese Ressourcen, zusätzlich zur Nutzung von Solar- und Windenergie werden kleineren Gemeinden helfen, neue Fabriken zu etablieren, diese wiederum können Beschäftigungsmöglichkeiten für Klimamigranten bieten.

Table of Content

ACKNOWLEDGEMENTS V
ABSTRACT VII
ZUSAMMENFASSUNG X
TABLE OF CONTENT XIII
LIST OF FIGURES XIX
LIST OF TABLES XXVII
ABBREVIATIONS XXXI

CHAPTER 1
PROBLEM STATEMENT

1.1 Background 1
1.1.1 Water resources management in Egypt 1
1.1.2 The effects of climate change on water and land resources in Egypt 3
1.2 Previous studies 4
1.3 Significance of this study 7
1.4 Aims and objectives of this research 8
1.4.1 Research objectives 9
1.4.2 Research questions 10
1.5 Limitations 10
1.6 Structure of the thesis 11

CHAPTER 2
LITERATURE REVIEW

2.1 Climatic change 13
2.1.1 Observed changes in the global climate 14
2.1.1.1 Atmosphere 14
2.1.1.2 Mountain glaciers 15
2.1.1.3 Sea level rise 16
2.1.2 Climate prospects in the future 18
2.1.2.1 Greenhouse gas emission scenarios 18
2.1.2.2 Climate models 19
2.1.2.3 Projections of future change 21
2.2 Water losses from open water bodies 22

2.2.1	Seepage losses	22
2.2.2	Absorption losses	23
2.2.3	Evaporation losses	23
2.2.3.1	Factors affecting the evaporation losses	24
2.2.3.2	Empirical formulas for estimating evaporation rate	25
2.2.3.3	Measurement of evaporation	30
2.2.3.4	Estimating evaporation from satellite data	35
2.3	Methods for reducing evaporation	36
2.3.1	Biological methods	36
2.3.1.1	Floating plants	36
2.3.1.2	Wind breakers	37
2.3.1.3	Palm fronds	38
2.3.2	Physical shade structures	39
2.3.2.1	Suspended covers	39
2.3.2.2	Floating covers and floating objects	40
2.3.3	Chemical layers	44
2.3.4	Design features	44
2.3.4.1	Deeper storages	44
2.3.4.2	Reduction of exposed water surface	45
2.3.4.3	Underground storages of water:	45
2.4	Sedimentation in large reservoirs	45
2.4.1	The problem of reservoir sedimentation	46
2.4.2	Measuring sedimentation rate	47
2.4.3	Procedure against sediment problems	48
2.4.3.1	Minimize sediment entering reservoir	49
2.4.3.2	Minimize deposition of sediment in reservoir	49
2.4.3.3	Lost storage replacement	50
2.4.3.4	Remove sediment from reservoir	51
2.4.3.5	Evaluation of sedimentation management methods	53
2.5	Summary	53

CHAPTER 3
PHYSIOGRAPHIC, DEMOGRAPHIC, AND HISTORICAL BACKGROUND

3.1	Nile Basin	57
3.1.1	General description	59
3.1.2	Hydrology of the Nile Basin	61
3.1.2.1	The Equatorial Lakes Plateau	62
3.1.2.2	Sudd and central Sudan basin	65
3.1.2.3	White Nile	66
3.1.2.4	Ethiopian plateau basin	66
3.1.2.5	The main Nile	68

3.1.3	Nile Basin climate	70
3.1.4	Nile water agreement	74
3.2	Egypt	74
3.2.1	Geographic situation	75
3.2.2	Geology (Geomorphological Features)	75
3.2.2.1	Western Desert	75
3.2.2.2	Western Desert	78
3.2.2.3	Eastern Desert	78
3.2.2.4	Nile Valley	79
3.2.2.5	The Nile Delta	79
3.2.2.6	Sinai Peninsula	80
3.2.2.7	The Fayum and Wad Rayan depressions	80
3.2.3	Alluvial Soils of Egypt	80
3.2.4	Demographic data	82
3.2.5	Climate	83
3.2.5.1	Temperature	84
3.2.5.2	Relative humidity	85
3.2.5.3	Rainfall	86
3.2.5.4	Wind	87
3.2.5.5	Evaporation rates	88
3.2.5.6	Climate change's effect on Egypt	89
3.2.6	Flora and Fauna	89
3.2.6.1	Flora	89
3.2.6.2	Fauna	90
3.2.7	Land uses	91
3.2.8	Water resources	92
3.2.8.1	The supply system	93
3.2.8.2	Different water uses	95
3.3	The High Aswan Dam (HAD)	96
3.3.1	Preliminary investigations about building the HAD	97
3.3.2	Location	98
3.3.3	The design of the HAD	98
3.3.4	Construction	103
3.3.5	Function and operation	106
3.3.6	Safety and monitoring:	108
3.4	The High Aswan Dam Reservoir (HADR)	108
3.4.1	Geography	108
3.4.2	HADR morphology and storage capacity	109
3.4.3	Geology of the HADR area	112
3.4.4	Climate	113
3.4.5	Hydrology of the HADR	115
3.4.6	Sedimentation in the HADR	117
3.5	Summary	120

CHAPTER 4
MATERIAL AND METHODS

4.1   Materials 123
4.1.1   Meteorological data 124
4.1.1.1 Meteorological stations 124
4.1.2   Hydrological data 127
4.1.3   Seepage data 127
4.1.4   Topographic data 129
4.1.4.1 Old topographic maps 130
4.1.4.2 Bathymetric survey 131
4.1.4.3 Satellite images 134
4.1.4.4 Aerial photos: 136
4.1.5   Bed soil sedimentary data 137
4.2   Methods 137
4.2.1   Softwares 137
4.2.1.1 Spreadsheet-based software 137
4.2.1.2 Geographic information systems 137
4.2.1.3 SPSS statistical packages 138
4.2.1.4 ERDAS Imagine 138
4.2.1.5 ENVI EX 138
4.2.2   Creation of databases 138
4.2.2.1 HADR Meteorological database (HADRMTDB) 139
4.2.2.2 HADR Hydrological database (HADRHYDB) 141
4.2.2.3 Contour lines for the HADR 141
4.2.2.4 Spot heights for the HADR 146
4.2.2.5 HADR Seepage Data Base (HADRSPDB) 146
4.2.2.6 HADR Bed Soil Sedimentary Categories Database (HADRBSDB) 147
4.2.3   HADR model development 147
4.2.3.1 HADR Digital Elevation Model (HADRDEM) 147
4.2.3.2 HADR mathematical models 149

CHAPTER 5
RESULTS AND DISCUSSION

5.1   Produced databases 153
5.1.1   HADR Meteorological Database (HADRMTDB) 153
5.1.1.1 Air temperature 153
5.1.1.2 Relative humidity 157
5.1.1.3 Wind speed 161
5.1.1.4 Water temperature 163
5.1.1.5 Evaporation rates 167
5.1.1.6 Forecasted evaporation rates 171

5.1.2	Hydrological Database (HADRHYDB)	171
5.1.2.1	Aswan gauge station:	171
5.1.2.2	Wadi Halfa gauge station:	176
5.1.3	Topographic maps	177
5.1.3.1	Contour lines for the HADR	177
5.1.3.2	Spot heights for the HADR	178
5.1.4	HADR Seepage Database (HADRSPDB)	179
5.1.5	HADR Bed Soil Sedimentary Categories Database (HADRBSDB)	181
5.2	Modeling HADR	183
5.2.1	HADR Digital Elevation Model (HADRDEM)	183
5.2.2	HADR mathematical models	184
5.2.3	HADRDEM and mathematical model verification	185
5.3	Exploring the HADR	189
5.3.1	Exploring the HADR Khors	189
5.3.2	Exploring the HADR cross sections	192
5.3.3	Exploring the HADR sediment conditions:	193
5.4	Scenarios	201
5.4.1	Eliminating Khors	201
5.4.1.1	Eliminating Khor Kalabsha	202
5.4.1.2	Khor El-Alaky	208
5.4.1.3	Khor Genina	213
5.4.1.4	Khor Sara	218
5.4.1.5	Other small Khors	223
5.4.2	Control of water levels	224
5.4.2.1	Establishing a new dam at the El-Madik reach	225
5.4.2.2	Scenarios for investigating the optimal operational water levels upstream and downstream of proposed new dam	227
5.4.3	Lowering the lakebed level	232
5.5	Combining scenarios	235
5.5.1	Building a new dam at El-Madik section and filling Khor Kalabsha	235
5.5.2	Building new dam at El-Madik section and removing the sediments from HADR	238
5.5.3	Establishing new dam at El-Madik section simultaneously with filling Khor Kalabsha and removing the sediments from the HADR	241
5.5.4	Evaluation of the proposed measures to reduce the evaporation losses	242
5.6	Summary	245

CHAPTER 6
PROSPECTIVE PROJECTS

6.1	Identification of new measures to reduce evaporation losses	249
6.1.1	Building a new dam at the El-Madik section	249
6.1.2	Filling Khor Kalabsha	250
6.1.3	Covering El-Alaky Khor and other small khors	251

6.1.4 Lowering lakebed altitudes 252
6.1.5 Measures to reduce evaporation losses... 253
6.2 Identification of new settlement areas for potential migrants from the Mediterranean coast 254
6.2.1 Potential areas to be settled and forms of land use... 254
6.2.2 The required infrastructure for sustainable development 256
6.3 Summary 258

CHAPTER 7
CONCLUSIONS AND RECOMMENDATIONS
7.1 Summary of findings 259
7.2 Recommendations and future study 262

REFERENCES 267
APPENDIX A 285
APPENDIX B 293
APPENDIX C 299
APPENDIX D 301
APPENDIX E 305
APPENDIX F 311
APPENDIX G 321
APPENDIX H 351

List of Figures

Figure 1.1: The High Aswan Dam Reservoir (HADR) at 181.6 m AMSL overlaid a mosaic of false color Landsat TM 5 images acquired in 1996, composites of bands 5, 4, and 1 2

Figure 1.2: Result Based Management (RBM) flowchart for the impacts of this study 9

Figure 2.1: Changes in the earth temperature during the past century (IPCC, 2007) 15

Figure 2.2: Changes in length of selected glaciers (IPCC, 2001a) 16

Figure 2.3: SRES scenario families (IPCC, 2001b) 19

Figure 2.4: Layout of a Global Climate model (LNFDC, 2008) 20

Figure 2.5: Range of global temperature rise for different SERS scenarios according to different climate models (IPCC, 2001a) 20

Figure 2.6: Class A pan (FAO, 1986) 31

Figure 2.7: The Sunken Screened Pan (Das and Saikia, 2009) 32

Figure 2.8: Colorado Sunken Pan (FAO, 1986) 33

Figure 2.9: The Floating Pan (Rich, 2004) 33

Figure 2.10: Evaporimeter, Piche Type (FUESS, 2013) 34

Figure 2.11: Evaporigraph, Piche type (Fuess, 2013) 34

Figure 2.12: Elephant Butte Reservoir Temperatures in °Kelvin $\times 10^4$ (Herting et al., 2004) 35

Figure 2.13: Various biological (plant) covers (Ramey, 2004), modified by Elba 2014 37

Figure 2.14: Air flow around a windbreak showing the extent of various microclimate zones in terms of multiples of windbreak height (H) (Helfer et al., 2009) 38

Figure 2.15: Evaporation pan covered with palm fronds (Alam and Al Shaikh, 2013) 39

Figure 2.16: Suspended covers (super span type) covering the Bemm River in Australia (TechSpan, 2007) 40

Figure 2.17: Continuous Floating covers Evapcap (Craig et al., 2005, WPDC, 2010) 41

Figure 2.18: Aquacaps covers (NYLEX, 2002) 41

Figure 2.19: Aqua Armour (AQUA, 2006) 42

Figure 2.20: Hexprotect tiles (AWTT, 2012) 42

Figure 2.21: AgFloats - recycled tires on a community dam at Blyth, SA- Australia (Clarke, 2009) 43

Figure 2.22: ECC Floating ball (Stuck, 2010) 43

Figure 3.1: Location of the world's major river basins (Digout, 2001) 57

Figure 3.2: The locations of 61 river basins in Africa and the Nile Basin (Rekacewicz et al., 2005) 58

Figure 3.3: Nile Basin catchment area (NBI, 2012) 60

Figure 3.4: Schematic longitudinal profile of the River Nile (NBI, 2012) 61

Figure 3.5: Scheme of upper Nile average annual flow (NBI, 2012) 62
Figure 3.6: Main subbasin of Equatorial Lakes Plateau catchment area (Karyabwite, 2000). 63
Figure 3.7: Map of the Sudd and Central Sudan Basin (Karyabwite, 2000) 66
Figure 3.8: Map of the Ethiopian Plateau (Karyabwite, 2000) 67
Figure 3.9: Map of the Main Nile Area (Karyabwite, 2000) 69
Figure 3.10: Average Monthly Nile Flow (NBI, 2012) 69
Figure 3.11: The Average Annual Total Rainfall in mm from 1960 to 1990 (NBI, 2012) 71
Figure 3.12: The Average Annual Potential Evapotranspiration in mm/ year from 1960 to 1990 (NBI, 2012) 72
Figure 3.13: Average monthly rainfall over the upper parts of the White Nile basin (Hurst and Black, 1943, Sutcliffe and Parks, 1999) 73
Figure 3. 14: Average monthly rainfall over the middle parts of the White Nile basin (Hurst and Black, 1943, Sutcliffe and Parks, 1999) 73
Figure 3.15: Average monthly rainfall over the basins of the Blue Nile and the main Nile (Hurst and Black, 1943, Sutcliffe and Parks, 1999) 73
Figure 3.16: Digital Elevation model of Egypt 76
Figure 3.17: Geologic map of Egypt - Generalized by M., A., Hammad in 1975; after R. Said in 1962, Geological Survey of Egypt - Cartographic preparation by the Soil Survey Institute, Wageningen, the Netherlands (Panagos et al., 2011) 77
Figure 3.18: Soil map of Egypt (FAO, 2005) 81
Figure 3.19: Expected population growth in Egypt (MWRI, 2005). 83
Figure 3.20: The hyper-arid regions of North Africa (Matsuura, 2003) 83
Figure 3.21: Average annual mean of daily temperatures (°C) (EEAA, 2001) 85
Figure 3.22: Annual mean of daily relative humidity (%) (Hegazi et al, 2005, EEAA, 2001) 85
Figure 3.23: Mean annual precipitation (mm) (EEAA, 2001) 86
Figure 3.24: The predicted wind speeds of Egypt determined by mesoscale modeling (Mortensen et al., 2006) 87
Figure 3.25: Potential monthly average daily evapotranspiration rates for the different agro-climatic regions in mm (MWRI – NWRC, 2002) 88
Figure 3.26: Map of Nature Protectorates (EEAA, 2005) 90
Figure 3.27: Population growth and water availability (NWRP, 2005) 92
Figure 3.28: Water Resources in Egypt in 2010 (MWRI, 2010) 93
Figure 3.29: Water uses in Egypt in 2010 (MWRI, 2010) 95
Figure 3.30: Location of the HAD and OAD (NWRP, 2005) 98
Figure 3.31: Cross-section of the High Aswan Dam, generalized by NWRP (2005) after Abul-Atta (1978) 99
Figure 3.32: Map of HAD facilities (NWRP, 2005) 101
Figure 3.33: Cross section of the emergency spillway (NWRP, 2005) 101

List of Figures

Figure 3.34: Toshka spillway and depression (including the Toshka South Valley project) (NWRP, 2005) 102

Figure 3.35: Overview over the Toshka depression and filling sequence in times of flooding (NWRP, 2005) 103

Figure 3.36: Inflow to the HADR and water demand in Egypt (NWRP, 2005) 107

Figure 3.37: Khors' names and locations on the HADR (Zwieten et al., 2011) 109

Figure 3.38: Schematic overview of the storage zones in the HADR system (NWRP, 2005) 110

Figure 3.39: Simplified geological map of the HADR area - (Said, 1993) modified by Elba 2014 112

Figure 3.40: Distribution of Meteorological Network Stations on the HADR 114

Figure 3.41: Daily Water Levels in the HADR from August 1964 to January 2010- (NWS, 2012) 115

Figure 3.42: History of annual release of the HAD (NWS, 2012) 116

Figure 3.43: Actual Levels of HADR between August 1999 and October 2007 (NWS, 2012) 117

Figure 3.44: Minimum Water Levels of HADR between August 1999 and October 2010 generated by Elba 2014 (NWS, 2012) 117

Figure 3.45: Longitudinal section along the deepest points in HADR (HDA-MWRI, 2009) 119

Figure 4.1: Workflow of the study inputs and outputs 124

Figure 4.2: Distribution of piezometers around the HADR (HDA-MWRI, 2009) 128

Figure 4.3: Cross Section at the Adindan Area (HDA-MWRI, 2009) 129

Figure 4.4: The contour lines available from the 1:100000 topographic maps and the boundary of each map overlying a mosaic of Landsat images of the HADR with false colors bands 7,4,2. 130

Figure 4.5: Location of the sections surveyed by the MWRI mission in 2007 131

Figure 4.6: Points surveyed in 2007 on Lake Nasser and Lake Nubia 132

Figure 4.7: Path and row of the downloaded image sets from Landsat 1,2,3 MSS 135

Figure 4.8: Path and row of the downloaded image sets from Landsat 4,5 MSS 135

Figure 4.9: Path and row of the downloaded image sets from Landsat 4,5 TM 135

Figure 4.10: Path and row of the downloaded image sets from Landsat 7 ETM+ 136

Figure 4.11: Aerial Photos acquired for the Nile south Egypt before the HAD construction overlaid by the HADR current morphology at 180 m AMSL 136

Figure 4.12: General methodology flowchart for creating database and geodatabase for meteorological data. 140

Figure 4.13: Grid coverage interpolated from point coverage, (ESRI, 2011) 140

Figure 4.14: General methodology flowchart for creating hydrological database for the HADR 141

Figure 4.15: General flowchart of the used data to create the contour lines coverage 142

Figure 4.16: Methodology flow chart of the object oriented classification using ENVI EX image processing software - (ENVI EX, 2013) 143

Figure 4.17: An overall workflow of the data processing methodology applied to the pilot area. 144

Figure 4.18: Overall workflow of data processing methodology applied to Landsat images. 145

Figure 4.19: Overall workflow of data processing methodology applied to Landsat images. 145

Figure 4.20: Overall workflow of the bathymetric survey measurements processing methodology 146

Figure 4.21: Overall methodology workflow for creating digital elevation models.............. 148

Figure 4.22: Overall methodology workflow for estimating sediment and erosion events. ... 149

Figure 4.23: A scheme for estimating the reservoir area and volume from the DEM, where L is the altitude and A is the lakebed area at this altitude. 150

Figure 5.1: Monthly variation of the mean values of air temperature in °C before constructing the HAD at Aswan station from 1953 to 1963 and at Wadi Halfa station from 1959 to 1963 (based on the data measured by ARHDA – section 4.1.1.1) 154

Figure 5.2: Monthly variation of the mean values of air temperature in °C after constructing the HAD, at Aswan (1977-2010), El-Alaky (1995-2007), Abu Simbel (1995-2007), Kalabsha (2004-2008), Amada (2004-2007) and Toshka (2004-2006) (based on the data measured by HDA – section 4.1.1.1) 154

Figure 5.3: Histogram of the monthly mean air temperature in °C measured at two and four meters above surface water level at all the meteorological stations along HADR 155

Figure 5.4: Distribution of mean air temperature of the HADR 157

Figure 5.5: Monthly variation of the mean values of relative humidity before constructing the HAD at Aswan station (1953-1963) and at Wadi Halfa station (1959-1963) 158

Figure 5.6: Monthly variation of the mean values of relative humidity after constructing the HAD at different meteorological stations 158

Figure 5.7: Histogram of the monthly mean relative humidity measured at two and four meters above surface water level of all the meteorological stations along the HADR .. 159

Figure 5.8: Distribution of mean relative humidity of the HADR 160

Figure 5.9: Monthly variations of wind speed at different meteorological stations of the HADR 161

Figure 5.10: Histogram of the monthly mean wind speeds measured at two and four meters above surface water level at all the meteorological stations along the HADR in m/s ... 161

Figure 5.11: Distribution of mean wind speeds of the HADR 163

Figure 5.12: Monthly variations of water surface temperature at different meteorological stations of HADR °C 164

Figure 5.13: Histogram of the monthly mean water temperature measured at all the meteorological stations along the HADR °C 165

Figure 5.14: Distribution of mean surface water temperatures on the HADR 166

Figure 5.15: Monthly means of variations of the evaporation rate from the meteorological stations along the HADR in mm/day for the 2010s 168

Figure 5.16: Histogram of the monthly means of evaporation rate measured at all the meteorological stations along the HADR °C 168

List of Figures

Figure 5.17: Distribution of mean evaporation rate of the HADR 169

Figure 5.18: Surface water level in m AMSL measured at HAD gauge station from 1964 to 2010 172

Figure 5.19: Histogram of the mean daily surface water levels measured at Aswan gauge station in m AMSL 172

Figure 5.20: Histograms of the daily surface water levels measured at Aswan gauge station for last five decades 174

Figure 5.21: Monthly mean, maximum and minimum surface water levels at Aswan gauge station for the last five decades 175

Figure 5.22: Monthly mean surface water levels measured at Wadi Halfa and Aswan gauge stations 176

Figure 5.23: Histogram of the differences in the monthly mean surface water levels between Aswan and Wadi Halfa gauge stations 176

Figure 5.24: Contour lines extracted from satelitte images and areial photos 178

Figure 5.25: Points surveyed on Lake Nasser and Lake Nubia between 1999 and 2010 179

Figure 5.26: Observations of upstream water level and Piezometers water level at Adindan Sector 316 km upstream the HAD from 1965 to 2009 179

Figure 5.27: Distribution of bed soil sedimentary along the HADR's cross sections in 2007 181

Figure 5.28: Chart of percentages of the HADR's bed soil sediment textures in 2007 182

Figure 5.29: Digital elevation model of the entire area of the HADR in m AMSL and HADR boundary at 181.6 m. 182

Figure 5.30: The isolated HADR digital elevation model (HADRDEM) in m AMSL. 183

Figure 5. 31: Evaluation of the relationship between surface area in km^2 and water level in meter AMSL and the optimum equation 184

Figure 5.32: Evaluation of the relationship between water volume in km^3 and water level in meter AMSL and the optimum equation 185

Figure 5.33: Evaluation of the relationship between water volume in km^3 and surface area in km^2 and the optimum equation 185

Figure 5.34: The HADR altitudes (m) AMSL and the corresponding surface area calculated by the HADRDEM and by the NWS in km^2 186

Figure 5.35: Scatter diagram between the HADR surface area calculated by the HADRDEM and the HADR surface area calculated by the NWS in km^2 186

Figure 5.36: The HADR's altitudes and the correspondence water volume calculated by the HADRDEM and by the NWS in km^3 187

Figure 5.37: Scatter diagram of the HADR water volume calculated by the HADRDEM and the HADR water volume calculated by the NWS in km^3. 187

Figure 5.38: Scatter diagram of the HADR's surface area calculated by the HADRDEM in km^2 and calculated by the HADR mathematical model −Equation 5.1 in km^2 188

Figure 5.39: Scatter diagram of the HADR water volume calculated by the HADRDEM and the HADR water volume calculated by mathematical model−Equation 5.2 188

Figure 5.40: The main khors along the HADR - modified after (FAO, 2012) 189

Figure 5.41: Isolated DEMs for studied khors deduced from the digital elevation model of HADR (HADRDEM) 190

Figure 5.42: Volumes of individual HADR khors in km^3 and surface areas in km^2 192

Figure 5.43: The width of the investigated sections and main sections along the HADR 192

Figure 5.44: A longitudinal section along the deepest points of HADR from 1964 to 2010. 194

Figure 5.45: Sedimentation and erosion of lake Nubia and Lake Nasser reflected by changes of the altitudes from 1964 to 2009 195

Figure 5.46: The HADR digital elevation models (HADRDEM) for 1999, 2010 and the corresponding changes of the lakebed 196

Figure 5.47: The HADR digital elevation models (HADRDEM) for 1999, 2003, 2006, and the corresponding changes of the lakebed 197

Figure 5.48: The HADR digital elevation models (HADRDEM) for 2006, 2007, 2008, and the corresponding changes of the lakebed 198

Figure 5.49: The HADR digital elevation models (HADRDEM) for 2008, 2009, 2010, and the corresponding changes of the lakebed 199

Figure 5.50: Cross sections through the HADR at Dabarosa, Abd Elkader and Halfa Dighaim sections 200

Figure 5.51: Sedimentation accumulated at major sections in Lake Nubia from 1977 to 2010 201

Figure 5.52: DEM of Khor Kalabsha, the alternatives suggested for eliminating Khor Kalabsha and the cross section at the optimal alternative 203

Figure 5.53: Chart for assessing alternatives to eliminate Khor Kalabsha 204

Figure 5.54: The alternatives of filling and cutting off of Khor Kalabsha 204

Figure 5.55: The area after filling Khor Kalabsha using the filling material from surrounding area up to 200 m 205

Figure 5.56: The mean monthly water level in m AMSL and the expected reduction of surface area in km^2 after a potential elimination of Khor Kalabsha two decades ago 206

Figure 5.57: The effect of removing Khor Kalabsha on the HADR's hydrological characteristics 206

Figure 5.58: DEM of Khor El–Alaky, the alternatives suggested for eliminating Khor E–Alaky and the cross section at the optimal alternative 209

Figure 5.59: Chart for assessing alternatives for the elimination of Khor El-Alaky 210

Figure 5.60: The monthly mean water level in m AMSL and the expected reduction of surface area in km^2 after a potential elimination of Khor El-Alaky, calculated in the past two decades 211

Figure 5.61: The effect of eliminating Khor El-Alaky on the hydrological characteristics of the HADR 211

Figure 5.62: DEM of Khor Genina, the alternatives suggested for eliminating Khor Genina and the cross section at the optimal alternative 214

Figure 5.63: Chart for assessing alternatives for the elimination of Khor Genina 215

Figure 5.64: The mean monthly water level in m AMSL and the expected reduction of surface area in km^2 after a potential elimination of Khor Genina, from 1990 to 2010 216

Figure 5.65: The effect of eliminating Khor Genina on HADR's hydrological features....... 216

Figure 5.66: DEM of Khor Sara, the alternatives suggested for eliminating Khor Sara and the cross section at the optimal alternative 219

Figure 5.67: Chart for assessing alternatives for elimination of Khor Sara........................... 220

Figure 5.68: The monthly mean water level in m AMSL and the expected reduction of surface area in km^2 after a potential coverage of Khor Sara two decades ago 221

Figure 5.69: Cross sections at El-Madik section in 2000, 2004, 2007, and 2010................... 224

Figure 5.70: Alternatives for choosing the optimal location of the proposed new dam at El-Madik section, and the cross section of the optimal alternative...................................... 225

Figure 5.71: Evaluation of the different alternatives for dam location at El-Madik section.. 226

Figure 5.72: The HADR hydrological characteristics, and its two parts after the construction of the potential new dam 227

Figure 5.73: The optimal drop of the water level downstream the new dam at different water levels and the resulting reduction of the surface area 229

Figure 5.74: Variations of the HADR water level after a potential constration of the new dam from 1990 to 2010.. 230

Figure 5.75: The potential drops of water levels downstream of the new dam and the reduction of surface area exptected for different water levels after a potiential construction of the new dam from 1990 to 2010. 230

Figure 5.76: The optimal drop of water level in Part Two downstream of the proposed dam and the reduction of surface area after a potential fill of Khor Kalabsha 237

Figure 5.77: The optimal drop of water levels in Part Two downstream of the new dam and the reduction of surface area after a potential fill of Khor Kalabasha two decades ago 237

Figure 5.78: The expected reduced surface area after a potential execution of different investigated measures.. 244

Figure 6.1: The boundaries of Wadi El-Alaky Biosphere, HADR, and Khor El-Alaky, overlaid in a mosaic of false color Landsat TM 5 images taken in 1986, composites of bands 5, 4, and 2 252

Figure 6.2: Proposed measures for reducing evaporation losses from the HADR 253

Figure 6.3: Proposed new settlements and cultivated lands on Khor Kalabsha.................... 255

Figure 6.4: Satellite image provided by Google Earth of Khor Kalabsha and cultivated areas around it in 2012. 256

Figure 6.5: Proposed locations for the main and secondary irrigation canals and the main pump station at Khor Kalabsha. 257

List of Tables

Table 2.1: Estimates of the contribution of different sources to observed sea level rise (IPCC, 2007) 17

Table 2.2: Main climate models (GCMs) used in the report (IPCC, 2007) 21

Table 3.1: Major rivers in the world (Karyabwite, 2000) 58

Table 3.2: Nile Basin countries areas and rainfall (FAO, 1997) 59

Table 3.3: Egypt, Major Soil Groups and Non Soil Land Covers (%) (FAO, 2005) 82

Table 3.4: Monthly average annual potential evapotranspiration rates in Egypt in the different agro-climatic regions in mm/day. (MWRI – NWRC, 2002) 88

Table 3.5: Summary of Technical Data of the Project (Abul-Atta, 1978) 105

Table 3.6: The quantities of work divided between the two stages (NWS, 2012) 105

Table 3.7: Number of workers on the project (NWS, 2012) 106

Table 3.8: The Distribution of Workers (NWS, 2012) 106

Table 3.9: Original Surface area and stored volume as functions of HADR level (HDA-MWRI, 2009, NWS, 2012, Abul-Atta, 1978) 111

Table 3.10: Summary of the storage zones within the HADR (HDA-MWRI, 2009) 111

Table 3.11: Rainfall records from 1961 to 2010 at the HADR (NWS, 2012) 113

Table 4.1: Data available from each meteorological station along the HADR 125

Table 4.2: Summary of the measured parameters by the meteorological stations on HADR 126

Table 4.3: The HADR cross-sections' names, codes, West and East head geographical coordinates, and distances upstream from HAD in Kilometers (HDA-MWRI, 2009).. 133

Table 4.4: Average Coefficients of Permeability Test 147

Table 5.1: The mean, median, maximum and minimum monthly air temperatures in °C measured at two and four meters above surface water level at all the meteorological stations on the HADR 156

Table 5.2: The mean, median, maximum and minimum monthly air temperatures in °C measured at two and four meters above surface water level at different meteorological stations on the HADR 156

Table 5.3: The mean, median, maximum and minimum monthly relative humidity measured at two and four meters above surface water level at all the meteorological stations on the HADR 159

Table 5.4: The mean, median, maximum and minimum monthly relative humidity measured at two and four meters above surface water level at different meteorological stations of the HADR 160

Table 5.5: The mean, median, maximum and minimum monthly wind speeds measured at two and four meters above surface water level at all the meteorological stations on HADR 162

Table 5.6: The average, median, maximum and minimum monthly wind speed measured at two and four meters above surface water level at different meteorological stations of the HADR..... 163

Table 5.7: The mean, median, maximum and minimum monthly water temperature measured at all the meteorological stations on the HADR..... 165

Table 5.8: The mean, median, maximum and minimum monthly water temperature measured at different meteorological stations of the HADR..... 166

Table 5.9: The average, median, maximum and minimum monthly means of evaporation rates measured at all the meteorological stations along the HADR..... 170

Table 5.10: The average, median, maximum and minimum monthly means of evaporation rates measured at the various meteorological stations of the HADR for the different decades..... 170

Table 5.11: Forecasted additional Losses in the evaporation rates until end of the century based on the climate models simulated by MWRI - (LNFDC, 2008)..... 171

Table 5.12: The number, mean, median, maximum, minimum, range and standard division of the daily surface water levels measured at Aswan gauge station of each month..... 173

Table 5.13: The mean, median, maximum and minimum daily surface water levels measured at Aswan gauge station of each decade..... 174

Table 5.14: List of utilized image sets with acquisition dates and corresponding water levels of the HADR in meters AMSL..... 177

Table 5.15: Annual average water losses by seepage (hm^3) in the HADR..... 180

Table 5.16: List of the studied khors of the HADR, their surface areas in km^2 and their volume in KM3 and corresponding mean water depths in m..... 191

Table 5.17: Annual evaporation losses before and after a potential elimination of Khor Kalabsha for the past two decades and the expected reduction in evaporation losses in km^3..... 207

Table 5.18: Expected reduced evaporation losses in km^3 from the HADR after a potential elimination of Khor Kalabsha under current climatic conditions and according to the ECHAM5 and HadCM3 climate models..... 208

Table 5.19: Annual evaporation losses before and after a potential elimination of Khor El-Alaky for the past two decades and the expected reduction of evaporation losses in km^3 212

Table 5.20: Expected reduced evaporation losses in km^3 from the HADR after a potential elimination of Khor El-Alaky under the current climatic conditions and according to the ECHAM5 and HadCM3 climate models..... 213

Table 5.21: Annual evaporation losses before and after a potential elimination of Khor Genina for the past two decades and the expected reduced evaporation losses in km^3..... 217

Table 5.22: Expected reduced evaporation losses in km^3 from the HADR after a potential elimination of Khor Genina under the current climatic conditions and according to the ECHAM5 and HadCM3 climate models..... 218

Table 5.23: Annual evaporation losses before and after a potential coverage of Khor Sara for the past two decades and the expected reduction of evaporation losses in km^3..... 222

Table 5.24: Expected reduced evaporation losses in km^3 from HADR after a potential coverage of Khor Sara under the current climatic conditions and according to the ECHAM5 and HadCM3 climate models 222

Table 5.25: Expected reduced evaporation losses in km^3 from the HADR after a potential coverage of small Khors under the current climatic conditions and based on the results of ECHAM5 and HadCM3 climate models 223

Table 5.26: The optimum drop of the water level downstream the new dam at different water levels and the potential reduction in the surface area 228

Table 5.27: The mean water level measured from 1990 to 2010, calculated annual evaporation losses before and after a potential construction of the new dam, and the potential reductions of evaporation losses 231

Table 5.28: The effect of a potential removal of the sediment deposits from the Lake Nasser on the evaporation losses 233

Table 5.29: The effect of a potential removal of the sediment deposits from the HADR on evaporation losses 234

Table 5.30: The optimal drop of the water level downstream the suggested new dam at different water levels and the resulting reduction of surface area after a potential fill of Khor Kalabsha 236

Table 5.31: The optimal drop of the water levels in Part Two downstream the new dam at different water levels, and the reduction of surface area after a potential removal of 1.1 km^3 of sediment from Lake Nasser 239

Table 5.32: The optimal drop of the water levels in Part Two downstream the new dam at different water levels, and the reduction of surface area after a potential removal of 6.6 km^3 of sediments from the entire HADR 240

Table 5.33: The optimal drop of the water level in part 2 downstream the suggested dam and the reduced surface area after a potential removal of 6.6 km^3 sediment from the HADR and a potential fill of Khor Kalabsha 241

Table 5.34: Comparison between the different combinations based on the expected reduction in evaporation losses (%) after a potential execution of different measures 243

Abbreviations

AMSL	Above Mean Sea Level
ARHDA	Aswan Reservoir and High Dam Authority
AT2	Air Temperature measured at two meters
AT4	Air Temperature measured at four meters
BP	Barometric Pressure
CGCM	Canadian Global Climate Model
CSIRO	Commonwealth Scientific and Industrial Research Organization
DEM	Digital Elevation Model
DGM	Digitale Geländemodell
DGPS	Differential Global Positioning System
DS	Down Stream
ECHAM	European Centre Hamburg Model
EMA	Egyptian Meteorological Authority
ESRI	Environmental Systems Research Institute
ETM+	Enhanced Thematic Mapper Plus
EVAP	Evaporation Losses
GFDL	Geophysical Fluid Dynamics Laboratory
GISS	Goddard Institute for Space Studies
HAD	High Aswan Dam
HADCM	Hadley Centre Coupled Model
HADR	High Aswan Dam Reservoir
HADRDEM	High Aswan Dam Reservoir Digital Elevation Model
HADRHYDB	High Aswan Dam Reservoir Hydrological database
HADRMTDB	High Aswan Dam Reservoir Meteorological Database
HADRSPDB	High Aswan Dam Reservoir Seepage Database
hm^3	1000,000 m^3 - Cubic Hectometer - Million Cubic Meter
IDW	Inverse Distance Weighted

km³	1,000,000,000 m³ - Cubic Kilometer - Billion Cubic Meter
MSS	Multispectral Scanner
MWRI	Ministry of Water Resources and Irrigation
NCAR	National Centre of Atmospheric Research
NR	Net Radiation
NWS	Nile Water Sector
PJTC	Permanent Joint Technical Commission
RH2	Relative Humidity measured at two meters
RH4	Relative Humidity measured at four meters
TM	Thematic Mapper
UKMO	UK Meteorological Office
üNN	über Normal-Null
US	Up Stream
USGS	United States Geological Survey
UTM	Universal Transverse Mercator.
WD2	Wind Direction measured at two meters
WD4	Wind Direction measured at four meters
WS	Water Surface temperature
WS2	Wind Speed measured at two meters
WS4	Wind Speed measured at four meters

Chapter 1

Problem Statement

This chapter introduces the problem statement by giving a brief overview about water resources management in Egypt and the effects of climate change on water and land resources. It will state the aims and objectives of this study and describe the structure of the thesis.

1.1 Background

Only 5% of Egypt's one million km² territory are arable land, which is situated along the Nile, in the valley and delta. 95% are desert, and it has 3500 km of coastline along the Mediterranean and the Red Sea (Elsharkawy et al., 2009). Egypt is located in the hyper-arid regions of North Africa and West Asia. It is considered one of the hottest and driest countries in the world with annual rainfall rates in most parts of less than 50 mm (Nour El-Din, 2013) and limited water resources. For these reasons, it is highly vulnerable to climate change (Slingo, 2011).

1.1.1 Water resources management in Egypt

The annual water supply is estimated at about 58 billion cubic meters (km³), of which 55.5 km³ are drawn from the Nile and stored in the High Aswan Dam Reservoir (HADR). The water extracted from the HADR represents 96% of Egypt's total water resources (Allam and Allam, 2007). The remaining 4% is drawn from groundwater and rainfall. The rainfall and flashfloods amount to about 1.5 km³. On the Mediterranean coast, rainfall usually occurs during winter with a maximum annual precipitation of 150 mm (UNFCCC, 1999). Groundwater, usually extracted on the Sinai Peninsula and in the western desert, amounts to about one km³ (El-Tahlawi et al., 2008, ICID, 2002).

The HADR maximum water level is 182 Above Mean Sea Level (AMSL) with a maximum water content of about 160 km³. About 350 km of the reservoir are located in Egypt and called Lake Nasser while 150 km extend through Sudan and are called Lake Nubia. This entire area is classified as subtropical, with a hot and very dry desert climate. Rainfall is extremely scarce during the winter season and estimated to be less than one mm (Slingo, 2011). Over the last two decades, measured mean annual evaporation losses have been estimated to be around 18 km³, approximately 11% of the total volume. The HADR has many embayments, which are called khors by the local population. There are 100 khors (El-Shabrawy, 2009), leading to a shoreline of more than 12000 km, as illustrated in Figure 1.1. The area of these khors spans about 3000 km² (45% of the lake surface area). Hence, these khors are considered to be one of the major causes for the reservoir's high evaporative losses (Ebaid and Ismail, 2010).

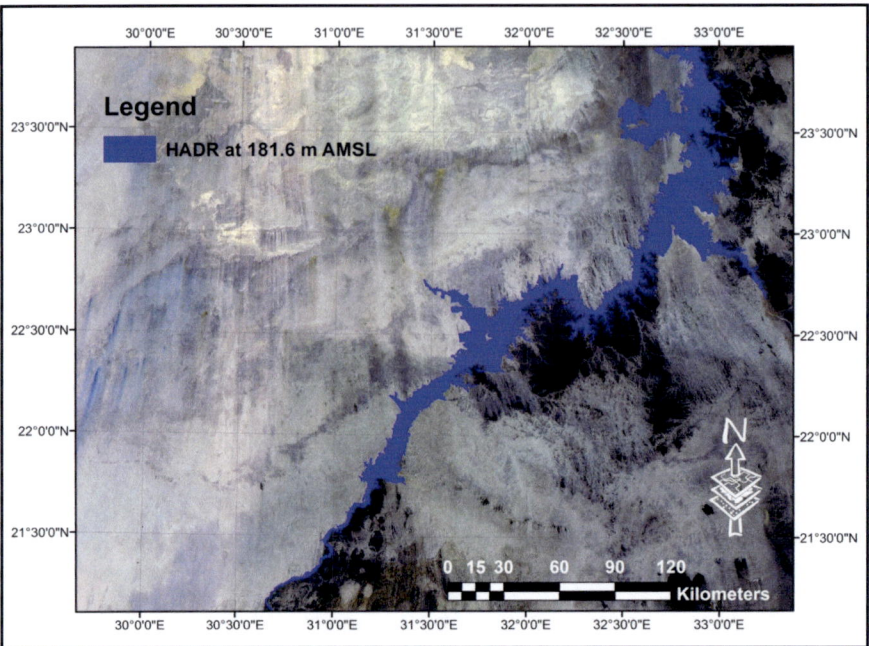

Figure 1.1: The High Aswan Dam Reservoir (HADR) at 181.6 m AMSL overlaid a mosaic of false color Landsat TM 5 images acquired in 1996, composites of bands 5, 4, and 1.

Egypt's water demands from agricultural, industrial, and municipal sectors exceeded 79 km^3 in 2010 (MWRI, 2011). The gap between supply and demand is closed by using the shallow ground water around the Nile and reusing agricultural drainage (LNDFC, 2008). However, the rapid increase in population and the development of the industrial sector further increase the gap between supply and demand (Allam and Allam, 2007). Furthermore, agricultural expansion through land reclamation is limited by water scarcity (Osman and Hanna, 1995). In addition, Egypt's agricultural production is not sufficient to meet the needs of the growing population. Egypt's population was estimated to be about 82.5 million in 2011 and predicted to grow to 135 million in 2100 (UN, 2013).

The major challenge water management experts are facing is to close the rapidly increasing gap between the limited water availability and the escalating demand for water from various economic sectors. The rate of water utilization has already reached its maximum in Egypt, and climate change will exacerbate its vulnerability (IPCC, 2008). Therefore, Egyptian water management specialists have to protect and develop this vital resource. They face a continuous challenge to achieve sustainable development. The major obstacles facing them are continued population growth, environmental degradation, and the limitations with regard to water supply (NWRP, 2005). Therefore, Egypt has to optimize its water usage and maximize its water resources. One of Egypt's foremost strategies to maximize the water resources is to reduce the evaporation losses from the HADR (EEAA 2010).

1.1.2 The effects of climate change on water and land resources in Egypt

Recent climate change is one of the major problems faced by the entire world, especially the effects of climate change on flooding and various health risks (Mynett and Vojinovic, 2009). With its limited water resources, Egypt will be significantly affected by the impact of climate change, especially in the Nile basin (Elshamy et al., 2009, Strzepek et al., 1996). Climate change affects Egypt's share of water from the Nile by causing flow fluctuations and higher evaporation losses from the HADR, which can reach up to 20 km^3 per year (Attia, 2008). These losses are expected to increase due to global warming (Bates et al., 2008). As the flow of the river Nile is likely to change considerably, it will also greatly affect water management in Egypt. Current water management strategies, used to secure the water supply in Egypt, may not be sufficient to meet these demands in the future. Both water supply and demand are expected to change in response to climate change, population and industrial growth, as well as the upstream development projects in other Nile Basin countries (LNFDC, 2008). An analysis of the results of current climate models shows that the uncertainty concerning the flow predictions that Egypt can expect is significant. This variation is caused by the widely varying predictions for changes in rainfall in various models, and trends differ from model to model, but one thing is certain: water resources in Egypt will be greatly affected by climate change.

In light of these developments, the Ministry of Water Resources and Irrigation of Egypt (MWRI) developed climatic scenarios (low, medium, high) based on the results from eleven Global Circulation Models (GCM) for the Special Report on Emissions Scenarios (SRES) B2 emission scenario (IPCC, 2007). In these models, the effects of climate change on air temperature and evaporation losses were studied by MWRI. Based on the Climate Model ECHAM5, developed by the Max Planck Institute for Meteorology in Hamburg, it is predicted that the mean annual evaporation losses will increase by about 0.47 km^3, 0.88 km^3, and 1.66 km^3 for the years 2030, 2050, and 2100, respectively. Using the Climate Models HadCM3, developed by the UK Hadley Centre for Climate Prediction and Research, it is estimated that the yearly evaporation losses will increase by about 0.52 km^3, 0.80 km^3, and 1.46 km^3 for the same years. This means that evaporation losses at the lake's surface area of 6500 km^2 will be about 3% to 10% higher by the year 2100 compared to the mean annual evaporation rates for the last 30 years (LNDFC, 2008).

In addition, Egypt's food security is threatened by climate change and climatic variability as well. According to data collected by the Egyptian Meteorological Authority (EMA) for the period 1961-2000, there is a possible warming trend of the air temperature, which means an increase in hot days. This increase in temperature will reduce crop yields (Hassanein and Meany, 2007, Tolba and Saab, 2009). Furthermore, rising temperatures reduce the organic matter in the soil, which is not renewable due to the construction of the High Aswan Dam (HAD). The HAD has blocked transportation of a large amount of the alluvial sediment that provides organic matter to the Nile Valley and delta soils (Elewa, 1985). As a result, agricultural productivity has decreased.

There is no doubt that the sea level along the northern coast of the Nile Delta will rise in the coming decades. This trend is caused by both local subsidence as well as by global sea level rise. Although it is slow, it will go on for several decades. Estimates of the rate of relative rise vary and are in the order of 30-50 cm by the year 2050 and between 50-100 cm by the year 2100, compared to year 2000 levels according to results of the fourth Assessment Report of the Intergovernmental Panel on Climate Change (IPCC, 2007). Therefore, Egypt's coastal zones, with their relatively low elevations, are vulnerable to climate change hazards because of the expected sea level rise. This rise will affect the coastal zones's water resources, agricultural resources, tourism, and human settlements (Elsharkawy et al., 2009). Besides, the Delta also suffers from land subsidence increases from west to east. The coastal land subsidence and degradation are the result of a lack of sediment, which is trapped by the HAD (Frihy, 1992). These coastal lands are the foremost arable lands, as they are used to grow 50% of the national crop production and as over 30% of the Egyptian population live there (Dasgupta et al., 2007)

To assess and mitigate the effects of climate change, the Egyptian Coastal Research Institute (CoRI) used two models to estimate land subsidence rates. Three scenarios were considered, including Intergovernmental Panel on Climate Change (IPCC) scenarios B1 and A1F1 as well as a new CoRI scenario that assumes a linear increase rate of air temperature till 2100. The models estimated the total affected area and the percentage of the Nile Delta area, including the presence of the Mohamed Ali Sea Wall constructed in 1830, which protects the lowlands at Abu-Qir Bay. A loss of about 760 km^2 represents 3% of the total fertile land in the delta, requiring the relocation of six to seven million people from the Nile Delta (CoRI, 2009). To help people adapt to these impacts, the government intends to offer supplementary arable lands and settlements to encourage people to relocate from these areas to new communities near to the valley.

1.2 Previous studies

The limited water resources in Egypt have forced experts and scientists of water resources management to conduct several studies about integrated water resources management in Egypt. Previous studies investigated increases of water resources through a variety of projects along the upper Nile, which were stopped due to the political conflict in Sudan (Awulachew, 2012). Therefore, previous scholarship in the last century focused on the demand side and investigated the opportunities of optimizing water needs in different sectors.

El-Fellaly and Saleh (2004) studied the control of water demand for major agricultural crops in Egypt, namely rice, cotton, sugarcane and tree crops. They introduced an agreement to change cropping patterns in Egypt to cope with the water situation. The provisions on the agreement included, for example, replacing sugarcane with sugar beet and modernizing existing sugar factories, particularly in Upper Egypt. In addition, it included limiting cultivated area for rice to 1.0 million feddans to satisfy national demand, to leave some potential for export, and to stop soil salinization and seawater intrusion. Furthermore, the agreement took into account the development of new crop varieties, especially genetically engineered types of rice, which lead to higher productivity and less water consumption. Lastly, it considered designing an analytical cropping pattern for each region in the country based on climatological conditions, soil characteristics and water resources availability, and it advised farmers to follow these cropping patterns.

Abdin et al. (2009) investigated the rational water use in Egypt. They discussed challenges facing water resource planning in Egypt with an emphasis on supply and demand. They stated that "most of the supply options are exhausted and cannot maintain significant enlargements." They demonstrated that the demand side management requires some potential for water saving through cost recovery as one of the financial instruments for water conservation. However, this can have negative social effects and environmental implications. They addressed some environmental problems such as soil salinization due to drainage water reuse or reduced water applications. They recommended a balanced approach that considered pricing and soil salinity to avoid the environmental problems. They raised awareness among farmers concerning salinity management and introduced new water-saving and salinity-resistant crops.

Recent scholarship focused on HADR and its high evaporation losses and investigated the opportunities of reducing these losses. Since the construction of the HAD, many studies have offered methods for estimating evaporation losses. In contrast, only few recent projects have discussed methods for reducing them. Hassan et al. (2007) investigated reducing the evaporation from Lake Nasser using new environmentally safe techniques. They confirmed the accuracy of some previous methods to estimate the average annual and monthly evaporation losses on the lake. The evaporation losses were calculated based on empirical calculations estimating the reservoir surface areas with respect to water level, an approach that has been used since 1978. They conducted some experimental work for estimation of evaporation. They found out that the annual average of the daily evaporation rate from Lake Nasser is 6.3 mm/day and that the average annual water lost by evaporation is about 12.5 km^3. They listed several methods for reducing evaporation losses: For example, they designed a system to cover a part of the lake surface using small circular foam sheets between big ones with an coverage efficiency of about 90%. This system can be adjusted so that it does not prevent the passage of sunlight to aquatic life. This system can save more than one million cubic meters (hm^3) of lost water from Lake Nasser by covering 0.50 km^2 with circular foam sheets.

Ebaid and Ismail (2010) also investigated the reduction of evaporation from Lake Nasser. The study aimed to evaluate the reduction of evaporation of Lake Nasser's waters caused by disconnecting (fully or partially) some of its khors. They integrated remote-sensing Geographic Information System (GIS) techniques and aerodynamic principles. They used Landsat 7ETM+ images to calculate the surface temperature from Landsat's thermal band in March and, applying aerodynamic principles, thereby evaporation depth and approximate evaporation volume for the entire lake. They developed a Digital Elevation Model (DEM) based on the available contour maps before the construction of the HAD to determine the depths of the khors. Consequently, they assessed the properties of some khors to identify the khors that could be disconnected. The study demonstrated evaporation losses between 2.73 mm/day at the centre of the reservoir and 9.58 mm/day at the edge, leading to about 0.86 km^3/month in March. Finally, they recommended disconnecting two khors with approximate construction heights of 8 m and 15 m to save 2.4 km^3 annually.

The irregular distribution of population over Egypt and the climate change hazards at the northern shores will force the Egyptian government to relocate its citizens and to build new communities. The Egyptian government intends to resettle approximately 1.5 million people in the Lake Nasser area by the year 2017. On the other hand, the HADR is a critical national resource and therefore an environmentally sensitive area. The comprehensive development plan for Aswan and Lake Nasser was developed in 2003 by the United Nations Development UNDP Program and the Aswan Governorate to avoid weakly planned settlements, ineffective and unsustainable desert agriculture, and negative impacts on the ecosystem of the reservoir (OWARA, 2008). Moreover, several studies were conducted about the development plans to achieve sustainability of the HADR.

Abulnaga and El-Sammany (2004) investigated de-silting Lake Nasser with slurry pipelines. The paper, analyzing previous studies, showed that up to 134 hm^3 end up in the lake annually, with 130 hm^3 deposited and four hm^3 passing through the HAD to the valley north of Aswan. The paper proposed to mine these rich sediments using modern lake mining techniques for hydraulic dredging such as jet and slurry pumps as well as floating pipelines in high-density polyethylene. Sediments would be pumped in a slurry form to engineered mud ponds and then be used for reclamation and the construction of new farms and communities. The de-silting of Nasser Lake would be executed on local, national, and international levels. Small-scale dredgers can be used for local farming. Preliminary calculations showed that a 20 megawatts pumping system can move 10 hm^3 of sediments to the shores.

OWARA completed a project by the Institute for Technologies in the Tropics at University of Applied Sciences, Cologne. In the paper on this project published in 2008, they investigated the sustainable rural infrastructure management in the Lake Nasser region. The study area was a new settlement area near the village of Kalabsha, which is characterized by a simple infrastructure. It has about 800 inhabitants of resettled farmers from the north. The project developed a plan for a future-oriented settlement and an infrastructure management plan to achieve sustainable development. The study investigated the profitable cultivation of the land by developing a concept for appropriate crops and irrigation systems. Moreover, the study investigated the optimization of the water and energy use through development of concepts for drinking water treatment and supply systems, wastewater treatment and disposal systems, as well as a suitable energy supply system. The study showed that for a sustainable development, more profitable crops and enhanced agricultural practices need to be implemented in order to lower the financial investments made by farmers. Moreover, the study developed recommendations for soil and water management and promoted organic agricultural practices in the area (OWARA, 2008).

The Near East Foundation (2010) also studied the possible human and environmental effects of the new resettlement area west of the HADR, but with a distinct emphasis on climate change, especially concerning its impact on human health. They found that the rise in temperature increases water fluctuation, pollution related to bacteria, algae intensification, and water-borne diseases. The study incorporated different stakeholders to help settlers adopt more efficient and sustainable behavioral, environmental, institutional, and technological adaptation strategies and methods that sufficiently mitigate the effects of climate change on human health. The study recommended that farmers use drip irrigation, organic fertilizers, and new heat resistant crops. Finally, the study suggested adaptation strategies to improve the resettlement and national development projects through conducting capacity-building workshops for stakeholders and policy makers. They developed conservation plans and strategies to support the conservation of the region's soil and water resources, which were then integrated into the main strategy of the High Dam Lake Development Authority.

1.3 Significance of this study

To address the problems resulting from the impact of climate change on Egypt's resources, this study aims to identify adaptive measures to reduce Egypt's vulnerability. The focus here is on water and land resources. Numerous adaptation tools can be developed for minimize water demand and optimize water supply, approaches that will require numerous studies. A broad variety of adaptive measures can also be developed to lower the risks resulting from sea level rise.

For specifically, this study examines measures that can reduce the evaporation losses from HADR and their potential as adaptive measure for increasing the water supply. It also considers how new settlement sites around the HADR can function as adaptive measures to protect and support the people living in the northern coastal zones.

Previous studies on evaporation reduction from the HADR by Hassan et al. (2007) suggested that covering a very small area is likely to save only one hm^3. Ebaid and Ismail (2010) also investigated the elimination of some khors; however, they did not take into account the effect of eliminating these khors on the hydrological characteristics of the reservoir. The study measured the evaporation losses based on a certain day and did not utilize the meteorological data measured during the past years. They used an old DEM for the reservoir before the construction of HAD due to the lack of a more recent one. Moreover, they based their model on current climate conditions and did not consider the effects resulting from climate change. Although some studies investigated the effects of sea level rise on the northern shores, they hardly considered adaptation strategies to mitigate the effects of climate change in Egypt in general and near the HADR in particular.

Therefore, this study intends to model the hydrological conditions of the HADR by creating an up-to-date high-resolution DEM, which simulates the bottom of the lake after consequent erosion and sedimentation events following the construction of the HAD. This DEM is generated using contour lines extracted from satellite images acquired after operation began at the HAD as well as spot heights from a recent bathymetric survey of the reservoir. Moreover, the lake meteorological characteristics measured by the meteorological stations in the HADR and the expected changes concerning evaporation losses due to climate change are investigated and considered. Alternatives for reducing evaporation losses are then proposed. The effects of the proposed alternatives on the reservoir's hydrological and environmental characteristics are investigated. The resulting evaporation losses saved are computed with respect to current and expected climate changes. Finally, the study investigates several opportunities for the settlement of people relocated from the northern coastal region who are vulnerable due to the sea level rise and whose existence and livelihoods will depend on a sustainable use of HADR and development of the surrounding regions.

1.4 Aims and objectives of this research

This thesis aims to identify adaptation tools in order to reduce Egypt's vulnerability due to climate change and to achieve two outcomes. Figure 1.2 shows a Result Based Management (RBM) flowchart for the study including the main outputs and the corresponding outcomes to achieve the impacts. The two expected outcomes are: i) maximizing the water resources by decreasing the evaporation losses from the HADR; ii) maximizing the land resources by identifying new settlement sites for the population relocated from the northern coastal zones. Achieving these two outcomes will require a broad variety of adaptation tools.

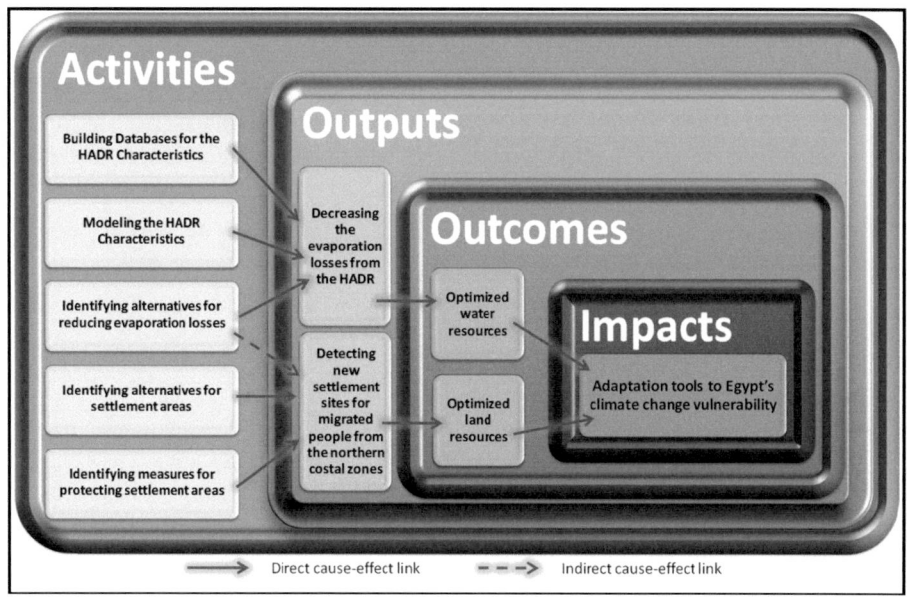

Figure 1.2: Result Based Management (RBM) flowchart for the impacts of this study.

Five main activities are required in order to accomplish the intended outputs. These activates are a) building databases for the HADR characteristics; b) modeling the HADR hydrological characteristics; c) identifying alternatives for reducing evaporation losses; d) identifying alternatives for settlement areas; e) identifying measures for protecting settlement areas. Activities a and b are associated directly with the first output, that is, "decreasing evaporation losses from the HADR." Furthermore, activities d and e are associated directly with the second output, "identifying new settlement sites for people relocated from the northern coastal zones." On the other hand, activity c is associated directly with the first output and indirectly with the second output. Identifying alternatives for less evaporation can lead indirectly to the creation of some areas that can be used for settling people from the northern coastal zones.

1.4.1 Research objectives

In this thesis, new adaptation methods to mitigate the effects of climate change on the water and land resources of Egypt, especially the increasing evaporation losses from the HADR and the lands lost due to sea level rise, are developed. Therefore, the main objectives of this study are to:

- establish databases for HADR's hydrological and meteorological characteristics,

- model the HADR's hydrological characteristics,
- analyze the evaporation losses and highlight their trends,
- identify alternatives for decreasing evaporation losses,
- investigate the effects of climate change on the environment of the HADR based on proposed alternatives,
- highlight the effects of global warming and climate change on the Egyptian shores,
- study the sediment problems and define alternatives for decreasing, removing, and utilizing them,
- identify new settlement areas for populations relocated from the northern coastal regions.

1.4.2 Research questions

This thesis will answer the following questions:
- What is the state of water resources in Egypt?
- How can the HADR secure an annual water quota of 55.5 km^3 for Egypt?
- How will climate change affect water and land in Egypt?
- What are the impacts of sea level rise on the Northern coastal zone and the Nile delta?
- Is there any evidence that future inflows in the HADR may change due to climate change?
- What is the vulnerability of Egypt's food security to climate change?
- How can water resources in Egypt be developed?
- How can water losses from water resources in Egypt be reduced?
- How can the evaporation losses from the HADR be reduced?
- How can strategies mitigate the effects of climate change?
- What are the opportunities for developing new communities around the HADR?
- How can the people's life and environment be protected in the new settlement areas?
- How can the HADR region be developed in a sustainable manner?

1.5 Limitations

This study is limited by a number of factors including climate model uncertainties, scarcity of data, and non-adequacy of some the data available. Therefore, this research will not address the following issues:

Problem statement 11

- the potential impact of climate change calculated with the help of current climate change models and the meteorological data compiled for this study, as these models were not accessible,

- the effect of the new development projects in the upper Nile basin countries, which affect the natural flow to the HADR, due to the lack of information about current and planned projects,

- the effect of water quality and system ecology on the HADR management, due to the non-adequacy of water quality data in the HADR and data of the ecology system of it and the surrounding area, the methods for optimizing the water demands downstream the HAD because it is beyond the scope of this study,

- the environmental impact assessment and the feasibility studies of the proposed measures for reducing evaporation losses due to the lack of economic, ecological and environmental data,

- the environmental impact assessment and the feasibility studies of the measures proposed for new resettlement areas due to the lack of census of the current settlers around the HADR and the people facing the hazards of sea level rise,

- the feasibility study of removing the sediment deposits from HADR and the opportunities of utilizing them are beyond the scope of this study, as they required exhaustive surveys, and comprehensive reports. This information is not available, and would require further research,

- the design and the capital costs of the proposed hydraulic structures are beyond the scope of this study, as they required hydraulic models, exhaustive surveys, and comprehensive reports. This information is not available, and would be very expensive to create it,

- the design and the capital costs of the proposed hydropower station and the expected electricity production are beyond the scope of this study.

1.6 Structure of the thesis

In order to achieve the objectives described above, this thesis is divided into the following eight chapters:

Chapter 1: Introduction

Chapter one introduces the problems that water resources management specialists in Egypt face and discusses how climate change will affect the water and land resources in Egypt. This chapter describes the aims and the significance of this study, and it includes a more detailed discussion of the research question as well as limitations.

Chapter 2: Literature Review

Chapter two surveys previous researches on HADR and related issues, for instance on the impact of climate change on Egypt. It introduces global climate change models, methods for estimating the evaporation losses, remote sensing tools, as well as techniques for decreasing evaporation of open water reservoirs. In addition, the problems of sedimentation in the reservoirs and methods of managing it are discussed.

Chapter 3: Physiographic, Demographic, and Historical Background

Chapter three provides information on the physiographic, demographic, and historical background of Egypt. It includes a description of the Nile Basin as well as an overview of Egypt's resources in general and the High Aswan Dam and its reservoir in particular.

Chapter 4: Material and Methods

This chapter introduces the materials and methods used in this study. The data used is described, the methodology is explained in detail, the required software is provided, the databases established for HADR's characteristics are shown, and the models developed for the HADR's hydrological characteristics are presented.

Chapter 5: Results and Discussion

This chapter lists the main results, including detailed descriptions and analyses of the databases produced for this study as well as the models created of the HADR and their verification. It provides also an exploration of the HADR and analyses of some case studies to decrease evaporation losses. Moreover, some case studies are combined and reanalyzed to identify optimum alternatives to mitigate the effects of climate change.

Chapter 6: Prospective Projects

This chapter proposes prospective projects for mitigating the effects of climate change. It identifies new measures to reduce evaporation loss and assesses their advantages and disadvantages. Finally, it identifies new settlement areas for people relocated from the northern coast.

Chapter 7: Conclusions and Recommendations

The last chapter includes the conclusions and recommendations of this study. It provides a summary of the study's results and recommendations and suggests possible directions for future studies.

Chapter 2

Literature Review

This chapter introduces a review of relevant literature on global climate change and selected climate models. It also discusses recent researches on water loss from open water areas in general and seepage, absorption, and evaporation in particular. In addition, this chapter summarizes studies that explain factors affecting evaporation losses as well as methods used to estimate evaporation rates, and to reduce them. Finally, the problem of sedimentation in reservoirs, the methods used to measure sedimentation rates, and procedures to reduce sedimentation, as discussed in recent scholarship, are examined.

2.1 Climatic change

Climate is usually defined as the long-term average weather (IPCC, 2001a), (Mcguffie and Sellers, 2001). The World Meteorological Organization (WMO) defines the climate of a region as the statistical description of the mean and variability of temperature, precipitation, humidity, wind, and other climatic variables over three decades or more. These quantities are most often surface variables, such as temperature, precipitation, and wind, but in a wider sense, the "climate" describes the state of the climate system (WMO, 2010).

Since the beginning of industrial revolution at the end of the 18th century, humans have started to combust fossil fuels on a large scale. This has led to an increase in the atmospheric concentration of CO_2 by 35% between 1750 and 2005 (IPCC, 2007). This increasing concentration has an effect on the radiative forcing of the earth. Atmospheric concentrations of other Greenhouse Gases (GHGs) have also increased, and the effect of these gases on the atmosphere has obviously increased as well. Due to the complexity of the climate system, a series of reactions occur of which the most important reaction is the water vapor reaction (WMO, 2010). The general temperature of the earth is expected to rise by 1.1°C - 6.4°C in 2100. This wide range is caused by various possible scenarios that involved different rates of future man-made emissions as well as by modeling uncertainties (IPCC, 2007).

Despite CO_2 and other natural GHGs (methane, nitrous oxide, etc.), there are a number of other gases that have been introduced into the atmosphere by human activities. These include Chlorofluorocarbons, which were not present in the atmosphere before the 1930s. Man-made GHGs add to the effect of increasing natural GHGs. Their total effect at the surface is often expressed in terms of the effect of an equivalent increase in CO_2 (WMO, 2010). The increasing amount of aerosols in the atmosphere plays a special role in this process. Due to that, the radiative forcing is complex and not yet recognized. The direct effect is the diffusion of part of the received solar radiation back into space. This causes a negative radiative forcing that may partly, and locally even completely, offset the enhanced Greenhouse effect. However, due to the short atmospheric lifetime of aerosols, radiative forcing is very heterogeneous in space and in time. This complicates the effect on the highly non-linear climate system. Some aerosols, such as soot, absorb solar radiation directly, leading to the local heating of the atmosphere, while others absorb and emit infrared radiation, adding to the enhanced Greenhouse effect. Aerosols may also affect the number, density and size of cloud droplets. This may change the amount and optical properties of clouds, and hence their reflection and absorption. It may also have an impact on the formation of precipitation. These are potentially important indirect effects of aerosols, resulting probably in a negative radiative forcing of as yet very uncertain magnitude (IPCC, 2001a). Further land-use change, due to urbanization, deforestation, and agricultural practices, affect the physical and biological properties of the Earth's surface. Such effects change the radiative forcing and have a potential impact on regional and global climate (IPCC, 2007).

2.1.1 Observed changes in the global climate

Since 1850, there is great evidence that the global climate is changing due to human-induced emissions of Greenhouse gasses. Such emissions have led to increasing atmospheric concentrations of these gasses, which in turn affect the global radiation balance. The general expectation will result in a warmer world and that the hydrological cycle will accelerate. On a global scale, climatic changes mean an increasing temperature, known as global warming, more precipitation and more evaporation in addition to sea level rise. However, although these general trends are recognized, the effect of the magnitude and even the direction of change on certain regions are not yet clear (IPCC, 2001a). Some of the changes are described below.

2.1.1.1 Atmosphere

Earth surface temperatures since 1865 have been reconstructed by different studies using different global temperature databases. Although the details vary, the overall picture clearly shows that global temperature has risen. This is confirmed at the continental scale for all continents (Figure 2.1). The average atmospheric water vapor content has increased since at least the 1980s over land and ocean as well as in the upper troposphere. This rise is generally consistent with the extra water vapor that warmer air can hold. Black lines show observations (dashed when spatial coverage is less than 50%), blue shades indicate the modeling range using natural forcing only, and red shades indicate the modeling range for both natural and man-made forcing (IPCC, 2007).

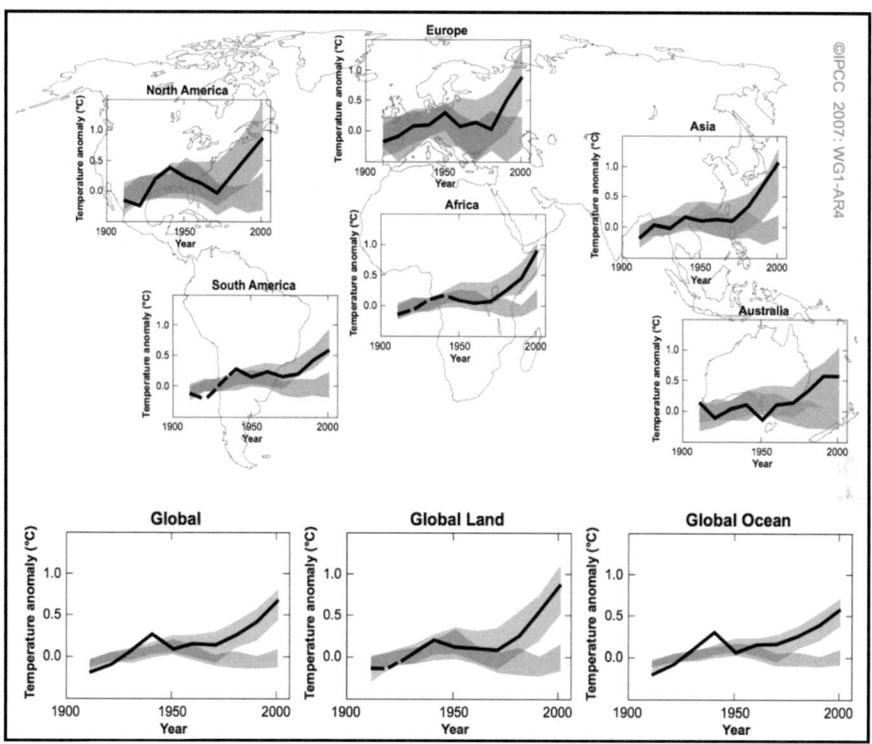

Figure 2.1: Changes in the earth temperature during the past century (IPCC, 2007)

2.1.1.2 Mountain glaciers

Evidence for increasing temperature also comes from the retreat of mountain glaciers. On the global scale, most glaciologists consider air temperature to be the most important factor reflecting glacier retreat. For the last two to three decades, the general picture is one of widespread retreat, notably in Alaska, Franz-Josef Land, Asia, the Alps, Indonesia and Africa as well as tropical and sub-tropical regions of South America. In a few regions, a significant number of glaciers are currently advancing, for example Western Norway, and New Zealand (Kohler and Maselli D., 2009). In the European Alps, there are also indications that current glacier recession is reaching levels not seen for perhaps a few thousand years as a result of the exposure of radiocarbon-dated ancient remains in high glacial saddles. Here, there is no considerable ice flow, and melting is assumed to have taken place in situ for the first time in millennia, for example the finding of the 5000-year-old Oetzal "ice man." Figure 2.2 shows the world wide glacier retreat (IPCC, 2001a).

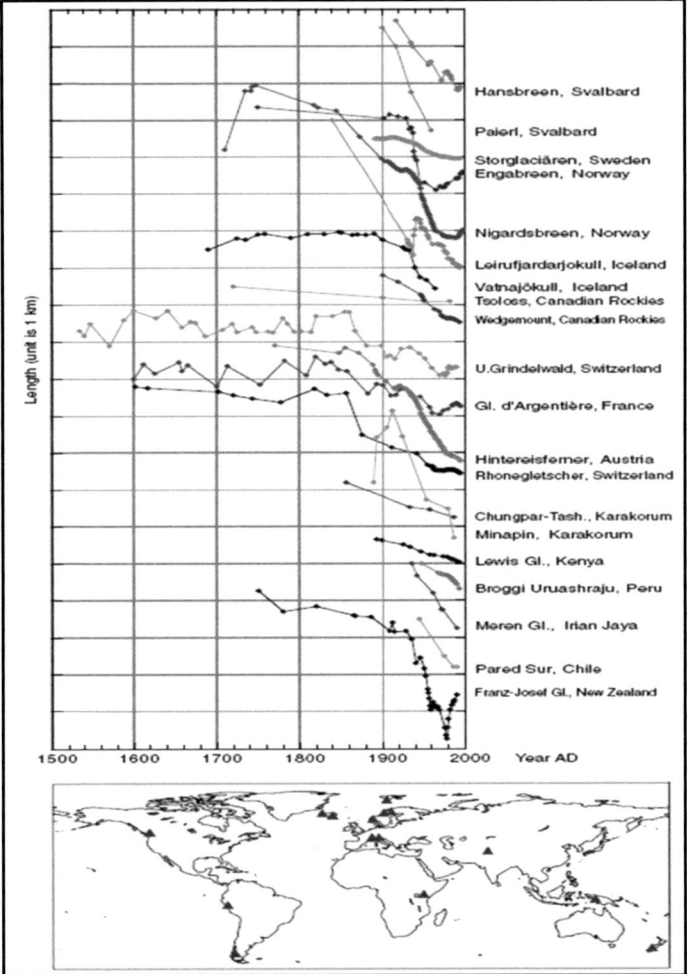

Figure 2.2: Changes in length of selected glaciers (IPCC, 2001a)

2.1.1.3 Sea level rise

Between 1961 and 2003, the global average sea level rose at an average rate of 1.8 mm per year. The rate was even higher between 1993 and 2003, namely about 3 mm per year. There is high confidence that the rate of observed sea level rise increased from the 19th to the 20th century as well. The total 20th century rise is estimated to be 0.17 m (IPCC, 2007). Several factors have caused the global sea level rise (Table 2.1).

There are four factors that mainly cause the sea level rise. First, the thermal expansion of the oceans due to heating, which has increased the volume of the oceans and consequently raised the sea level (Rayner, 2003). Second, the melting of the small glaciers affected the sea level rise, i.e. all glaciers apart from the ice sheets of Greenland and Antarctica where the melt water from the glaciers flows from the land into the ocean, thus adding to the volume of the ocean and leading to a sea level rise (Kaltenborn et al., 2010). Third, changing the mass balance, and therefore the volume of ice, of the ice sheets of Greenland and Antarctica has affected the sea level, where this may lead to a sink or a rise in sea level depending on the net effect (Sørensen et al., 2011). Finally, the melting of the underside of the floating ice shelves at Antarctica, where warming of the ocean will increase the melt rate, may lead to the acceleration of the flow of ice from the Antarctic continent into the ocean (Kaltenborn et al., 2010, Rayner, 2003, IPCC, 2007).

Table 2.1: Estimates of the contribution of different sources to observed sea level rise (IPCC, 2007)

Source of sea level rise	Rate of sea level rise (in m per century)	
	1961 - 2003	1993 – 2003
Thermal expansion	0.042 ± 0.012	0.15 ± 0.05
Glaciers and ice caps	0.050 ± 0.018	0.077 ± 0.022
Greenland ice sheets	0.05 ± 0.12	0.21 ± 0.7
Antarctic ice sheets	0.14 ± 0.41	0.21 ± 0.35
Sum of individual climate contributions	0.11 ± 0.05	0.28 ± 0.07
Observed total sea level rise	0.18 ± 005*	0.31 ± 0.07*
Difference (observed minus sum of estimated climate contributions)	0.07 ± 0.07	0.03 ± 0.10

*Data prior to 1993 was obtained from tide gauges, and after 1993 from satellite altimetry.

Furthermore, the sea surface temperature has obviously risen since the end of the 19th century (Figure 2.1). Observations since 1961 show that the mean temperature of the global ocean has raised to depths of at least 3000 m and that the ocean has been absorbing more than 80% of the heat added to the climate system. Such warming causes seawater to expand, contributing to sea level rise (Rayner, 2003, IPCC, 2007). New data since the IPPC's fourth Assessment Report (AR4) from 2007 show that losses from the ice sheets of Greenland and Antarctica have probably contributed to sea level rise between 1993 and 2003 (IPCC, 2007). The flow speed has also increased for some Greenland and Antarctic outlet glaciers, which drain ice from the interior of the ice sheets. The corresponding increased ice sheet mass loss has often been followed by thinning, reduction or loss of ice shelves or loss of floating glacier tongues. The dynamical ice loss is adequate to explain most of the Antarctic net mass loss and approximately half of the Greenland net mass loss. The remainder of the ice loss from Greenland has occurred because losses due to melting have exceeded accumulation due to snowfall (Kaltenborn et al., 2010, Sørensen et al., 2011).

2.1.2 Climate prospects in the future

The future of the global climate can be projected through predicting future GHG emissions. GHG emissions are the product of very complex dynamic systems, determined by driving forces including the demographic development, socio-economic development, and technological change. Their future forecasts are highly uncertain. Scenarios are a proper tool to analyze alternatives for driving forces and their influence on the future emission outcomes and to assess associated uncertainties as well. Such scenarios assist in climate change analysis, including climate modeling and the assessment of impacts, adaptation, and mitigation (IPCC, 2001b).

2.1.2.1 Greenhouse gas emission scenarios

Based on assumptions of the expected global driving forces, different emissions of GHGs can be expected. Different emissions of Greenhouse gases lead to different future concentrations of these gases in the atmosphere. The Intergovernmental Panel on Climate Change (IPCC) published a special report on emission scenarios in 2000 (IPCC, 2001b) and presented the Standard Reference Emission Scenarios (SRES). The SRES have been developed to explore future developments in the global environment with special reference to the production of GHGs and aerosol precursor emissions. The scenarios are based on story-lines of how the world may develop in the future. Four families of scenarios have been adopted along two axes. On one axis, the level of globalization of the solutions varies (between global and regional), while on the other axis, the solutions may come from an increase of material wealth or from sustainability (Figure 2.3). Associated with these scenarios are emissions of Greenhouse gases and concentrations of Greenhouse gases in the atmosphere. A description of each scenario is summarized below.

SRES A1:

This scenario assumes that the future world will include very rapid economic growth, low population growth and a rapid introduction of new and more efficient technology. Major underlying themes are economic and cultural convergence and capacity building, with a substantial reduction in regional differences in per capita income. In this world, people prefer personal wealth rather than environmental quality. The A1 scenario family is further differentiated into three groups that describe alternative directions of technological change in the energy system. The three A1 groups are distinguished by their technological emphasis: fossil intensive (A1FI), non-fossil energy sources (A1T), or a balance across all sources (A1B) (IPCC, 2001b).

SRES A2:

This scenario assumes that the world will become very heterogeneous. The emphasis is that of strengthening regional cultural identities, with an emphasis on family values and restricted traditions, high population growth, with less concern for rapid economic development (IPCC, 2001b).

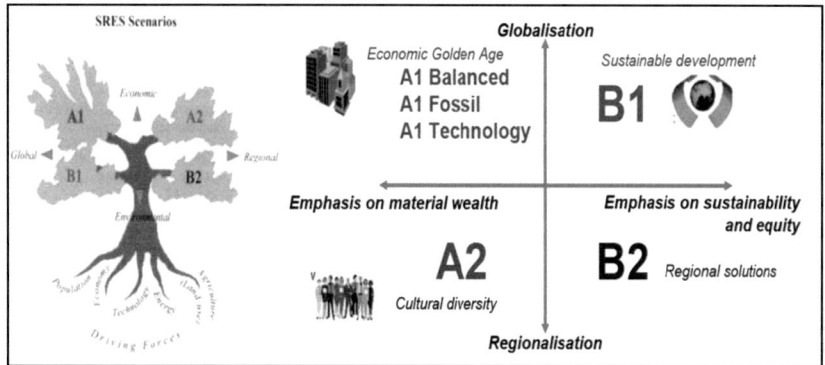

Figure 2.3: SRES scenario families (IPCC, 2001b)

SRES B1:

This scenario assumes that the world will be convergent with rapid change in economic structures, "dematerialization" and introduction of clean technologies. The emphasis is on global solutions to environmental and social sustainability, including intensive efforts for rapid technology development, dematerialization of the economy, and improving equity (IPCC, 2001b).

SRES B2:

This scenario assumes that the world will be heterogeneous as well. The emphasis is on local solutions to economic, social, and environmental sustainability with less rapid and more diverse technological changes, including an emphasis on community initiatives and social innovations to find rather local than global solutions (IPCC, 2001b).

2.1.2.2 Climate models

Numerical models are the most appropriate tools to appraise changes in climate. These models are referred to as Atmospheric Ocean General Circulation Models (AOGCM), or simply GCMs. Such models describe the earth's climate and the oceans' circulation in three dimensions. The models are based on physical laws of conservation of mass, energy and momentum. Figure 2.4 shows the general layout of such a model. These models are able to provide various weather variables such as air pressure, rainfall, temperature, wind speed, and humidity. Nowadays, the number of climate models is growing. The results of these models vary despite the fact that all of them are based on physical laws, particularly for rainfall simulations. This variation is partially due to the coarse spatial scale of the models that does not allocate an accurate demonstration of the earth's surface. Hence, the IPCC suggested the use of no less than three climate models for impact assessment. Figure 2.5 shows the average global temperature rise for different SERS scenarios and the range produced by different climate models. Numerous models are being developed over the world. Samples of these models are presented in Table 2.2 (IPCC, 2007).

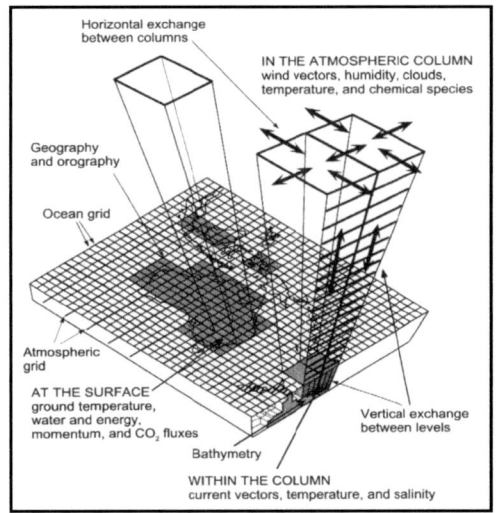

Figure 2.4: Layout of a Global Climate model (LNFDC, 2008)

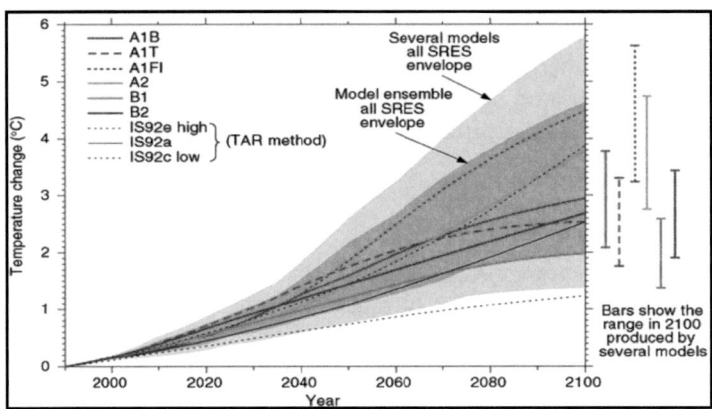

Figure 2.5: Range of global temperature rise for different SERS scenarios according to different climate models (IPCC, 2001a)

Table 2.2: Main climate models (GCMs) used in the report (IPCC, 2007)

ECHAM4	Germany - developed at the Max Planck Institute, originally derived from cycle 17 of the European Centre for Medium-Range Weather Forecasts (ECMWF) model.
HADCM3	UK - Hadley Centre Coupled Model, version 3, developed at the Hadley Centre in the United Kingdom
CSIRO	Australia, CSIRO-Mk 2.0 model developed at the Commonwealth Scientific and Industrial Research Organization
CGCM2	Canada - developed by the Canadian Centre for Climate Modeling and Analysis (CCCMA)
GFDL	USA, GFDLTR90 model, U.S. Department of Commerce/National Oceanic and Atmospheric Administration (NOAA)/Geophysical Fluid Dynamics Laboratory (GFDL),
NCAR	USA, CSM_TR98 model developed by the National Centre of Atmospheric Research
CCSR (or GISS)	USA, GISSTR95 model, National Aeronautics and Space Administration (NASA)/ Goddard Institute for Space Studies (GISS), USA
UKMO	UK: HADCM 2 model, UK-Meteorological Office -Hadley Centre for Climate Prediction and Research

2.1.2.3 Projections of future change

The following projections are summarized by the IPCC (2007) based on an analysis of GCM model results for a large number of simulations of the SRES scenarios and model experiments with GHGs and aerosol concentrations held constant after the year 2000 or 2100. Man-made warming and sea level rise would continue for centuries due to the timescales associated with climate processes and feedbacks, even if GHGs concentrations were to be stabilized. Both past and future man-made carbon dioxide emissions will continue to contribute to global warming and sea level rise for more than a millennium due to the timescales required for removal of this gas from the atmosphere (IPCC, 2007).

Temperature:

For the next two decades, the warming of the Earth's temperature or earth temperature of about 0.2°C per decade is projected for a range of SRES emission scenarios. Even if the concentrations of all GHGs and aerosols had been kept constant at year 2000 levels, a further warming of about 0.1°C per decade would be expected, mainly due to the oceans' slow response. Best-estimate projections from models indicate that decadal-average warming over each inhabited continent by 2030 is insensitive to the choice among SRES scenarios and is very likely to be at least twice as large as the corresponding model-estimated natural variability during the 20th century. Continued Greenhouse gas emissions at or above current rates would cause further warming and induce many changes in the global climate system during the 21st century that would very likely be larger than those observed during the 20th century. Advances in climate change modeling now enable best estimates and likely assessed uncertainty ranges to be given for projected warming for different emission scenarios (IPCC, 2007, IPCC, 2013). Global mean temperature is projected to likely exceed by 0.3°C to 4.8°C by the late twenty first century, relative to 1986 –2005 (IPCC, 2013).

Sea level rise

Model-based projections of global average sea level rise at the end of the 21st century are mainly based on increased ice flow from Greenland and Antarctica and loss of mass from glaciers and ice sheets at the rates observed for 1986-2005. Global mean sea level rise is likely projected to be 0.42 to 0.98 m in 2100, relative to 1986–2005, with a rate during 2081–2100 of 8 to 16 mm yr^{-1}. (IPCC, 2013)

2.2 Water losses from open water bodies

The water losses from open water bodies include evaporation, seepage and absorption losses. Estimations of total water losses from open water bodies can be determined using the water budget method or using direct measurements and calculations for evaporation, seepage and absorption losses. Direct measurement of water losses under field conditions is not feasible, at least not in the sense that one is able to measure river stage, discharge etc. Consequently, a variety of techniques has been developed for determining or estimating vapor transport from water surfaces. The most obvious approach involves the balance of a water budget. The water budget can be represented by the following equation, which accounts for the total amount of water in the watershed:

$$Inflow - Outflow = Change\ in\ Storage \qquad Equation\ 2.1$$

Assuming that the change in storage ΔS, surface inflow I, surface outflow O, subsurface seepage O_g and precipitation P can be measured, water losses L can be calculated from equation 2.2.

$$L = P + I - O - O_g - \Delta S \qquad Equation\ 2.2$$

This approach is simple in theory, but its application rarely produces reliable results since all errors in measuring inflow, outflow and change in storage are directly reflected in the computed water losses. Determining the water losses by seepage and by absorption enables estimating the evaporation losses according to Equation 2.3.

$$E = L - Q_{abs} - Q_s \qquad Equation\ 2.3$$

Where E is the evaporation losses, L is the total water losses, Q_s is the water losses by seepage and Q_{abs} is the water losses by absorption (Sokolov and Chapman, 1974).

2.2.1 Seepage losses

Water is lost from earthen-banked dams via permeation through the underlying soil. This seepage depends on several factors, including permeability of the soil and the pressure exerted on the soil by stored water. Permeability can vary over a wide range, depending on soil structure and composition. Wet areas that are observed downstream of a water storage, often thought to be natural springs, are generally seepage flows from the storage itself (MDEQ, 2000, Cedergren, 1998).

In 1856, Henri Darcy published a formula governing flow through porous media. The formula, now known as Darcy's law, was based on the study of water flow through vertical filters in laboratory experiments (MDEQ, 2000, Cedergren, 1998). The experiments indicated that the quantity of flow is expressed by the following equation:

$$Q_s = K I A = K I H L \qquad \text{Equation 2.4}$$

Where Q_s is the discharge in m³/day, A is the area of flow in m², I is the hydraulic gradient, K is the coefficient of Permeability in m/day, H is the thickness of the seepage face in m, and L is the length of seepage face in m. According to Equation 2.4, the seepage per unit area depends on two factors, namely the coefficient of permeability of the formation and the hydraulic gradient (MDEQ, 2000, Cedergren, 1998).

2.2.2 Absorption losses

Due to rises of water level in a reservoir, bank storage has been formed along the two banks of the reservoir. The bank storage changes every year until it reaches its steady state at the stabilized reservoir level. The dry rock below the flood water level is saturated with water every year. This saturation in the total balance of the reservoir will be a new loss. Once the reservoir is filled to its maximum level and the flooded rock is fully saturated, then no more water will be lost from the reservoir to bank storage as far as the water level does not decline. The losses due to saturation of rock at the different accumulation levels of the reservoir can be estimated according to the following equation:

$$Q_{abs} = N * A \left[\frac{(W_1 + W_2)}{2} - 120 \right] \qquad \text{Equation 2.5}$$

Where Q_{abs} is the absorption (saturation) losses in km³, A is the increase in reservoir surface area due to change of water level from W_1 to W_2 in km², W_1, W_2 are the maximum water levels for two successive years in meter, and N is the average porosity of the soil (Cedergren, 1998).

2.2.3 Evaporation losses

Evaporation is defined as the change of state of a liquid into a vapor at a temperature below the boiling point of the liquid. Evaporation occurs at the surface of a liquid where water molecules with the highest kinetic energies change into the gas phase. The resulting decrease in the average kinetic energy of the remaining molecules of the liquid is expressed as a decrease in the temperature of the water (Burston, 2008). Evaporation is the process whereby liquid water is converted to water vapor (vaporization) and removed from the evaporating surface (vapor removal). Water evaporates from a variety of surfaces, such as lakes, rivers, pavements, soils and wet vegetation (Allen, 1998).

Brutsaert (1982) mentioned that the two main factors influencing evaporation from open water surfaces are the supply of energy to provide the latent heat of vaporization and the ability to transport the vapor away from the evaporative surface. Solar radiation is the main source of heat energy. The ability to transport vapor away from the evaporative surface depends on the wind velocity over the surface and the specific humidity gradient in the air above it. Chow (1964) mentioned that the rate of evaporation depends on the vapor pressure of the body of water and on that of the air. This vapor pressure depends on water and air temperature, wind speed and direction, atmospheric pressure, quality of the water, and the nature and shape of the surface (Strzepek et al., 1994). The evaporative rate on a water body is affected mainly by temperature, humidity of the air, and wind conditions.

The water evaporation from large open water bodies is considered the main reason for water losses worldwide. The estimation of evaporation is difficult and not straightforward due to the numerous factors affecting the climate and consequently evaporation. The estimation methods can be empirical equations or measurement tools. Before discussing these methods, the evaporation and the factors affecting it are defined.

2.2.3.1 Factors affecting the evaporation losses

Numerous factors affect the evaporation phenomenon. The main factors are solar radiation, temperature, wind, atmospheric pressure, the nature and shape of the surface, and the quality of water (Brown, 2001, Das and Saikia, 2009). A summary of the effects of these factors is provided below.

Solar radiation

The quantity of water evaporated from a surface depends mainly on the heat quantity that the surface receives from the sun. The heat quantity received by a surface varies depending on its geographic location. This heat exchange between the atmosphere, the soil surface and the water surface is achieved through heat convection and conduction (Brown, 2001, Das and Saikia, 2009).

Temperature:

The water and air temperature significantly affect the rate of evaporation. The vapor pressure of a body of water increases with temperature because the kinetic energy of the water particles rises with increasing temperatures, leading to higher evaporation rates (Brutsaert, 1982).

Wind:

Wind is effective in removing water particles in the air and get air capable of holding more water vapor. The larger the movement of air above the water, the higher the evaporation rate. When the wind speed is high enough to remove all water particles escaping from the water surface, a further increase in velocity will not increase evaporation proportionally (Das and Saikia, 2009).

Atmospheric pressure:

The decrease in atmospheric barometric pressure with increased altitude increases the escaping rate of water particles from a free water surface because there are fewer particles in the atmosphere above the evaporating surface, hence causing higher evaporation rates (Brown, 2001, Das and Saikia, 2009).

Quality of water:

The vapor pressure of pure water under given conditions is determined by its temperature. When a solute is dissolved in water, the vapor pressure of the water is reduced. Since the rate of evaporation is proportional to the difference in vapor pressure between the water and atmosphere, lowering the vapor pressure of the water will reduce the rate of evaporation (Brutsaert, 1982, Das and Saikia, 2009). Therefore, the salt content in water affects the rate of evaporation. Experimental studies show that the rate of evaporation decreases with increasing salt content in water. In the case of sea water, the evaporation is 2% to 3% less as compared to fresh water, even though the other conditions are the same (Brutsaert, 1982, Das and Saikia, 2009).

Nature and shape of surface:

A body of water with a flat surface has greater vapor pressure than one with a concave surface, but less than one with a convex surface under the same conditions. Moreover, evaporation is a surface phenomenon, and the quantity lost through evaporation from water bodies is directly dependent on the extent of its surface exposed to the atmosphere. The higher the open surface area, the larger the evaporation losses (Brutsaert, 1982, Das and Saikia, 2009).

2.2.3.2 Empirical formulas for estimating evaporation rate

Evaporation is one of the main components in estimating the water budget of reservoirs. Accurate estimations of evaporation are essential for the solution of water management problems. Reliable evaporation data are required for planning, designing, and operating reservoirs, ponds, irrigation and drainage systems and navigation canals, especially in arid zones (Brutsaert, 1982, Strzepek et al., 1994). Evaporation from large water surfaces is a complex process depending on several physical and climatic conditions. Usually, it cannot be measured directly. A number of physical bases or empirical methods have been developed to estimate evaporation under different conditions using observed meteorological data.

The estimation of evaporation is not a simple issue as a number of factors can affect evaporation rates, notably the climate and the physiography of the water body and its surroundings. Most of the known empirical evaporation formulas were based on Dalton (1802), with modifications for factors affecting evaporation. Dalton discovered that the rate of evaporation from a free water surface depends upon the difference between the vapor pressure at saturation for the temperature of the water and the vapor pressure actually existing in the air above the water surface (Korzukhin et al., 2011). From this premise, many other investigators have derived formulas to mathematically express the rate of evaporation in terms of the meteorological elements causing or affecting it. Some of these methods are listed below.

Fitzgerald (1876 - 1887)

Fitzgerald compiled a careful and complete series of observations under both controlled laboratory conditions and natural conditions and proposed the following formula:

$$E = (0.4 + 0.199\,W)(e_s - e_d) \qquad \text{Equation 2.6}$$

Where E is the evaporation in inches per 24 hours, W is the wind speed in miles per hour, e_s is the vapor pressure of the saturated air in inches of mercury at the temperature of the water surface, and e_d is the mean vapor pressure of air in inches of mercury above the water surface. Altitude effect is not considered (Rohwer, 1931).

Russel (1888)

Russel studied evaporation measured by the Piche evaporimeter at 18 stations of the Weather Bureau spread throughout the United State, and prepared the following formula:

$$E = \frac{[(1.96\,e_w + 43.88)(e_w - e_d)]}{B} \qquad \text{Equation 2.7}$$

Where e_w is the vapor pressure in inches of mercury for the mean wet bulb temperature, e_d is the mean vapor pressure of air in inches of mercury above the water surface, and B is the mean barometric reading in inches of mercury at 32° F (Rohwer, 1931).

Bigelow (1907 - 1910)

Bigelow conducted an extensive series of experiments for the Weather Bureau. His suggested formula has the following form:

Literature Review

$$E = 0.138 \frac{e_s}{e_d} \times \frac{de}{dS} \times [1 + 0.07\,W] \qquad \text{Equation 2.8}$$

where the derivative $\frac{de}{dS}$ is the rate of change of the maximum vapor pressure with temperature, and the other symbols are designated as in the above formula (Rohwer, 1931).

Meyer and Freeman (1917)

Meyer and Freeman developed the following formula independently:

$$E = (0.5 + 0.05\,W)(e_s - e_d) \qquad \text{Equation 2.9}$$

They approached the problem of evaporation from a considerably different angle, where the indicated symbols W, e_s, and e_d, are the same as defined in above formulas (Rohwer, 1931).

Cummings and Richardson (1927)

Cummings and Richardson proposed the following formula:

$$E = \frac{(H-S-C)}{L(1+R)} \qquad \text{Equation 2.10}$$

Where H is the net radiation received in calories per sq. cm., S is the heat stored in a unit column of water in calories, C is a correction factor for interchange of heat between the wails and the water, L is the latent heat of the water in calories, and R is the Bowen's Ratio, between the heat carried away by convection and the heat carried off by vapor.

Folse (1929)

Folse published the results of detailed investigations into evaporation and runoff of the Great Lakes (Superior, Michigan, and Hurm). The suggested evaporation formula is:

$$E = (e_s - e_d)[0.319 + 0.358(W - 10.8)] \qquad \text{Equation 2.11}$$

where the indicated symbols W, e_s, and e_d, are the same as defined in the above formulas (Rohwer, 1931).

Carl Rohwer (1931)

Rohwehr published the Fort Collins experiments on evaporation, which included observations under different conditions, principally on the Colorado Sunkon Pan, the Weather Bureau Land Pan, the Floating Pan of the U. S. Geological Survey, an 85 foot tank, and a small reservoir. Out of these experiments, he developed the following formula for evaporation:

$$E = 0.771(1.465 - 0.0186\ B\ )(0.44 + 0.118W)(e_s - e_d) \qquad \text{Equation 2.12}$$

Where W, B, e_s, and e_d, are the same as defined in the formulas above (Rohwer, 1931).

Penman (1948)

Penman presented a theory and a formula for the estimation of evaporation from weather data, which allows computing the evaporation from a free water surface using readily available standard meteorological data only. Penman's equation estimates evaporation from the free surface water body by considering the energy budget balance at the water surface. The potential evaporation [mm d^{-1}] is a fairly complex function of humidity, wind speed, radiation, and temperature (Penman, 1948). He developed the following equations.

$$E_p = \frac{\Delta}{\Delta+\gamma}(R_n + A_h) + \left(\frac{\gamma}{\Delta+\gamma}\right)\left[\frac{6.63(1+0.536 U_2)D}{\lambda}\right] \qquad \text{Equation 2.13}$$

$$D = \left[\frac{e_s(T_{max})+e_s(T_{min})}{2}\right]\left[1 - \frac{RH}{100}\right] \qquad \text{Equation 2.14}$$

$$\lambda(T) = 2.501 - 0.002361 T_s \qquad \text{Equation 2.15}$$

$$e = e_s \frac{RH}{100} \qquad \text{Equation 2.16}$$

$$e_s = 0.6108 \exp\left[\frac{(17.27\ WST)}{(237.3+WST)}\right] \qquad \text{Equation 2.17}$$

$$\Delta(T) = \frac{409 e_s}{(237.3+T)^2} = \frac{2503.06\ \exp\left[\frac{12.27 T}{237.3+T}\right]}{(237.3+T)^2} \qquad \text{Equation 2.18}$$

$$\gamma = 0.0016286\ \frac{P}{\lambda} = \frac{P}{1537.675 - 145 T} \qquad \text{Equation 2.19}$$

Where R_n is the net radiation exchange (water equivalent) at the surface of the body of water [mm d^{-1}], Ah is the energy absorbed to the water body (water equivalent) [mm d^{-1}], D is the average vapor pressure deficit (e_s-e) over the estimation period [kPa], T_{max} and T_{min} are the maximum and minimum temperatures, respectively, over the period of estimation [°C], RH is the average relative humidity over the period of estimation [%], U2 is the wind speed measured at 2m elevation [m s^{-1}], λ is the latent heat of vaporization of water at temperature T (the surface temperature of the water body, °C) [MJ kg-1], e is the ambient vapor pressure of water vapor in the air [kPa], e_s is the saturated vapor pressure of water in air at temperature T [kPa], Δ is the rate of change of e_s with respect to T [kPa T^{-1}], γ is the psychrometric constant [kPa °C^{-1}], and P is the parametric pressure equals to 101.3 [kPa] at mean sea level (Penman, 1948, Maidment, 1993).

Harbeck (1962)

Harbeck's equation was developed according to the bulk aerodynamic approach. It is based on the concept of mass transfer theory, which relates evaporation to the processes that affect the removal of water vapor from the boundary layer above the air-water interface at the surface of the lake. As wind speed over the water surface increases, water vapor is removed from the system more rapidly. This causes the vapor pressure gradient above the lake to increase, thereby increasing evaporation. Therefore, evaporation can be directly related to wind speed and the vapor pressure gradient (Maidment, 1993, Harbeck, 1962, FAO, 2001). Harbeck's equation can be given as:

$$E = N\, u_2\, (e_s - e_a) \qquad \text{Equation 2.20}$$

$$e_s = 0.6108\, e^{[\frac{(17.27\,.WST)}{(237.3+WST)}]} \qquad \text{Equation 2.21}$$

$$e_{sa} = 0.6108\, e^{[\frac{(17.27\,.AT_2)}{(237.3+.AT_2)}]} \qquad \text{Equation 2.22}$$

$$e_a = RH_a \times e_{sa} \qquad \text{Equation 2.23}$$

Where E is the evaporation rate in mm/hour, N is the empirical mass transfer coefficient, u_2 is the wind speed at two meter height above water surface in m/s, e_s is the saturated vapor pressure at water surface in kPa, WST is the water surface temperature in °C, e_{sa} is the saturated vapor pressure of air in kPa, AT_2 is the air temperature at two m in °C, e_a is the actual vapor pressure of the air in kPa, and RH_a is the relative humidity at two m in percentage (Maidment, 1993, Harbeck, 1962, FAO, 2001).

Priestly-Taylor (1972)

Priestley-Taylor's equation was developed according to the energy budget approach. It is a model of the Energy Balance Approach and neglects all components except radiation, heat fluxes, and the change in energy storage. This is an applicable method for a large lake with minimal variation in daily lake level and in daily lake bottom temperature (Priestley and Taylor R. J., 1972, Maidment, 1993, FAO, 2001). The equation can be given as:

$$E = \frac{1.74 \, \Delta \, (NR - G)}{\lambda \cdot (\Delta + \gamma)} \qquad \text{Equation 2.24}$$

$$\Delta = \frac{4098 \, e_{sa}}{(AT_2 + 237.3)^2} \qquad \text{Equation 2.25}$$

$$\gamma = \frac{1.103 \, BP}{0.662 \, \Delta} \qquad \text{Equation 2.26}$$

$$G = 4.182 \times 10^{-3} \times \sum [0.5 \, dz(\, d(T_1) + d(T_2))] \qquad \text{Equation 2.27}$$

Where Δ is the slope of the saturated vapor pressure versus temperature at two m, e_{sa} is the saturated vapor pressure of air in kPa, AT_2 is the air temperature at 2 m in °C, NR is the net solar radiation in MJ/m², λ is the latent heat of vaporization in MJ/kg, WST is the water surface temperature in °C, γ is the psychometric constant in kPa, BP is the Barometric pressure in kPa, G is the heat storage (MJ/m²), dT_1 and dT_2 water temperature difference for two times steps of one hour at frequently two depths, and dz is the layer thickness (depth zone) in meters (Priestley and Taylor R. J., 1972, Maidment, 1993, FAO, 2001).

2.2.3.3 Measurement of evaporation

As mentioned before, measuring evaporation from open water bodies is not a simple matter as there are a number of factors that can affect the evaporation rates, particularly the climate and physiography of a water body and its environment. It is difficult to measure the dissipation of water vapor directly on the surface and use it as a measurement of evaporation. It is possible to estimate evaporation rates by measuring the decrease in the water level of a container, or even the water storage itself. Measuring evaporation rates from large water storages is challenging because of the complex environmental conditions (Finch and Calver, 2008, Maidment, 1993, FAO, 2001). Examples of some measurements tools are presented below.

Evaporation pan:

Pans have been used to estimate evaporation for over 200 years (Finch and Calver, 2008). The evaporation pan is the most commonly used instrument to measure evaporation. It is made of stainless steel and is filled with water that is exposed to the open air. When water evaporates, the water level falls. The differences of the water level are measured by using the suspension measuring rod suspended in the smoothing pipe. Storage of heat within the pan can be appreciable and may cause significant evaporation during the night while most crops transpire only during the daytime. There are also differences in turbulence, temperature and humidity of the air immediately above the respective surfaces. Heat transfer through the sides of the pan occurs and affects the energy balance (Finch and Calver, 2008, Das and Saikia, 2009). There are over twenty-five different designs, and the list will continue to grow (Brutsaert, 1982). Some of the widely used pans are described below.

a) The weather bureau class A land pan

The U.S. Class A pan is the most widely used evaporation pan. It is made of galvanized iron and is 120 centimeter in diameter, 25 cm deep and is set on a 15 cm high wooden grillage so as to raise the water surface a little more than 30 cm above the ground level (Figure 2.6). The water is kept about 25 mm below the edge of the pan. It is thus exposed to air on all sides. Evaporation is measured by means of a pointer gauge located in a stilling well. The equipment of a station that measures evaporation include instruments to observe associated meteorological conditions such as temperature, precipitation, wind and humidity.

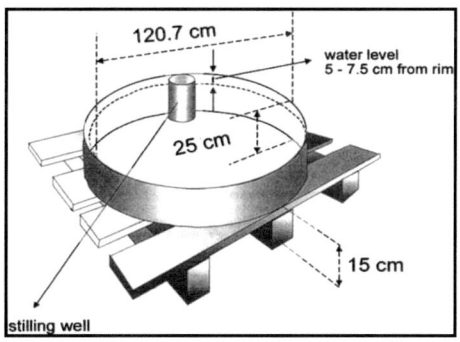

Figure 2.6: Class A pan (FAO, 1986)

The water in the Weather Bureau pan is not subject to the effect of wind to the same intensity as the open water body because of the protection given by the rim and the surrounding objects. Alternatively, the amount of water contained in the pan is small so that much more of the sun's energy heats the water surface, causing evaporation, and less energy is stored in the water. There is also substantial heating of the metal in the rim above the water. Therefore, the water in the class A pan has a higher temperature during daylight hours than the water of a nearby open water body. On the other hand, conditions are reversed at night when the class A pan cools to a lower temperature than the adjacent large body of water. The average evaporation measured from these pans is higher than from the actual evaporation from open water bodies under the same climatic conditions. Thus, the values of measured evaporation with a class A pan have to be multiplied by a coefficient for calculating probable values of evaporation from open water bodies. The coefficient values range from 0.60 in the summer to 0.82 in the winter. An average coefficient of 0.70 has been defined for estimating the annual evaporation loss (Finch and Calver, 2008, Das and Saikia, 2009, Fao, 1986).

b) The sunken screened pan

The sunken screened pan evaporimeter was developed by Sharma and Dastane in 1968. The screened evaporation pan shades water in the pan by a 6 mm mesh galvanized wire screen, which intercepts some of the heat energy, thus reducing water loss from the pan to make it similar to an open water area (Figure 2.7). Based on experiments conducted over a period of 10 years, the annual coefficient of the screened pan is found to be 1.0, but the monthly coefficient is greater than 1.0 in the winter and less than 1.0 in the summer (Das and Saikia, 2009, Dougherty, 1975).

Figure 2.7: The Sunken Screened Pan (Das and Saikia, 2009)

c) Bureau of plant industry pan

The Bureau of Plant Industry pan was the first evaporation pan to be used in Texas. It was designed by the U.S. Department of Agriculture and was installed at San Antonio in 1907. This pan is 180 cm in diameter, 60cm deep and buried in the ground within 10 cm of the top. Water is maintained almost at ground level. Evaporation is measured by a pointer gauge in a well outside the pan. The measured evaporation is about 5% higher than open water bodies' evaporation (Dougherty, 1975).

d) The colorado sunken pan

The Colorado Sunken Pan is one of the oldest evaporation pans and was developed in the late 1800's. It is a square pan, with 92 cm width and 46 to 92 cm deep. It is set in the ground so that 5 to 15 cm extends above the surface (Figure

2.8). The water surface in the pan is maintained at around the elevation of the ground surface. Only a margin of 2.5 cm upward or downward is permitted. Evaporation is measured using a pointer gauge. The pan coefficient for open water bodies varies between 0.45 and 1.10, with an average of 0.80 (FAO, 1986).

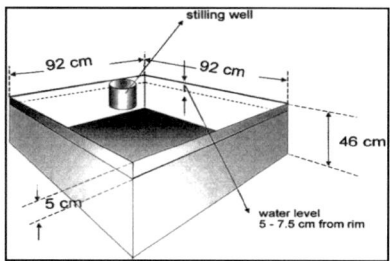

Figure 2.8: Colorado Sunken Pan (FAO, 1986)

e) The floating pan

The floating pan is widely known as the U. S. Geological Survey pan. It has been used to measure the evaporation from wide water surfaces to obtain results under conditions identical with large bodies of water. The pan is a square with a 91.5 cm width and is 45 cm deep as shown in Figure 2.9. It is mounted on a raft that floats on a moderately large body of water, equipped with baffles to avoid surging. Evaporation is measured by replacing the water lost using a special cup with a volume equal to that created by a depth of 0.25 mm over the area of the pan. The sides of the pan are usually wet due to wave action, leading to a changeable increase in evaporation. The pan's coefficient for open water body ranges from 0.70 to 0.82 with an average of 0.80 (Klink, 2006, Rich, 2004).

Figure 2.9: The Floating Pan (Rich, 2004)

Evaporimeter

Jelinek and Hann created the Evaporigraph in 1883. This instrument, sometimes called an atmometer, measures the rate of evaporation of water into the atmosphere. There are two types of this device. The first type measures the evaporation rate from a free water surface, while the second type measures it from a continuously wet porous surface. In the first type, the level of water in a tank or a pan, often sunk into the ground so that the water surface is at ground level, is measured by a micrometer gauge. The evaporation is the difference after accounting for increases due to rain and decreases due to deliberate draining. In the second type, the evaporation rate is computed according to the rate of weight loss of a wet pack of absorbent material such as Piché evaporimeter. The Piché evaporimeter uses an inverted graduated cylinder of water with a filter paper seal at the mouth as shown in Figure 2.10. Evaporation takes place from the wet filter paper and thus reduces the water in the cylinder, so that the rate of evaporation can be read directly from the graduations marking the water level (Zyl and Jager, 1987).

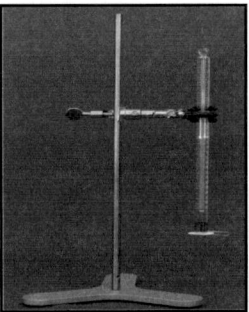

Figure 2.10: Evaporimeter, Piche Type (FUESS, 2013)

Evaporigraph, Piche type

The instrument (frontispiece) records potential evaporation by using the porous cardboard disc according to Piche, which has been universally tested for several decades as a standard evaporator. As the thin disc with its minimum of mass quickly takes on the temperature of the wet-bulb thermometer, it operates practically free of inertia, as shown in Figure 2.11 (Fuess, 2013).

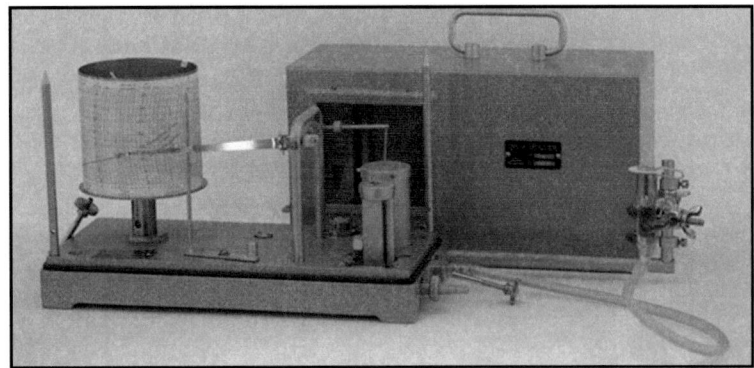

Figure 2.11: Evaporigraph, Piche type (Fuess, 2013)

2.2.3.4 Estimating evaporation from satellite data

Quantifying evaporation from water bodies can be very difficult. Reservoir evaporation measurements from an inflow-outflow water balance method or using pan measurement methods are time and laboratory intensive, particularly for large reservoirs. Additionally, the accuracy of the methods may be affected by environmental factors and some of their assumptions. Depending on the accuracy of measurements sought, the methods used to obtain evaporation can become even more complex. These methods may be simplified if less accurate results are acceptable. There is a need to create an accurate and convenient method to measure the evaporation loss that can be used for large scale reservoirs. Satellite images, either Landsat or ASTER images, hold large amounts of data that may be used to estimate evaporation as they have infrared bands. This type of data must be completely understood before it can be applied to the determination of evaporative loss from a large body of water. Therefore, some researchers have developed a remote sensing tool to estimate evaporation loss in reservoirs. The model uses the energy balance principle to measure daily evaporation losses in mm (Wang, 2008, Stuck, 2010, Shaltout and El-Housry, 1997, Herting et al., 2004).

For example, Wang et al. (2008) used the ASTER images to find the different temperatures on the surface of the Elephant Butte Reservoir as shown in Figure 2.12. In this image, darker images represent cooler temperatures while the lighter colors represent warmer temperatures. Thus, with the remotely sensed data, the temperatures across the entire lake can be determined. The temperature difference between the ASTER image and the measured temperature was 1.7°C, which is considerable. As evaporation is a function of temperature, wind, and humidity, these factors are also dependent upon each other. All of these factors play together to generate evaporation. Overall, temperature may be the major factor that influences evaporation according to the researchers. Therefore, the bulk-aerodynamic method was utilized because of its dependence on the water surface temperature estimated by ASTER image. When the calculations were executed, the estimated bulk-aerodynamic evaporation for the ASTER image was much higher than the measured one.

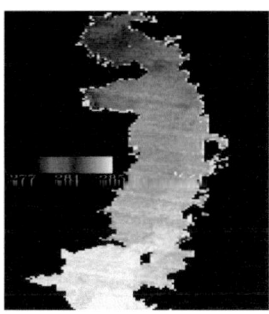

Figure 2.12: Elephant Butte Reservoir Temperatures in °Kelvin × 10^4 (Herting et al., 2004)

2.3 Methods for reducing evaporation

Over the years, a number of commercial products have also been developed to minimize water losses specially in Australia, India, China, and USA where evaporation rates are very high. Therefore, these countries have established many factories to manufacture wide range of products for controlling the evaporation losses (Sinha et al., 2006). Numerous techniques are available to reduce evaporation losses. The selection of proper technique relies on many factors, such as the water body size, characterize, purpose uses, and cost. However, the price of water is more invaluable in comparison to the technique costs, particularly in arid lands (Yao. X et al., 2010). The methods of evaporation control can be grouped under two extensive categories, namely short term measures and long term measures. They can be classified into four main classes including biological, physical, and chemical methods, as well as management methods e.g. structural and design methods (Sinha et al., 2006, Jennison, 2003). Descriptions of these methods are presented below.

2.3.1 Biological methods

Biological covers can provide a significant decrease in the volume of evaporation. Biological covers such as aquatic plants, palm fronds and nearby vegetation can reduce evaporation by reducing any or all of incident solar radiation, surface area or wind speed. These plants consume water through the transpiration process, therefore, plant transpiration must be taken into consideration. Studies in Thailand have shown that duckweed can reduce evaporation up to 10% (Jennison, 2003, Caruso, 2005). Wind breaker trees and plants on the banks such as Eucalyptus and Pine trees also consume high water by transpiration, but do not have large enough reducing effects to balance the water they transpire as fully grown trees (Wullschleger et al., 1998). Examples of the biological methods are given below.

2.3.1.1 Floating plants

Figure 2.13 shows the various biological (plant) covers. Floating aquatic plants such as water lily (Nymphaea odorata), small duckweed (Lemna minor), great duckweed (Spirodela polyrhiza) and water meal (Wolffia columbiana) enable lowering the net evaporation of water storages by forming a physical wall to the flux of vapor, thereby preventing the interaction between air and the boundary layer, which reduce incident solar energy. However, not all water plants effectively reduce evaporation rates. For example, water hyacinth and water lotus have wide extended leaves that increase the transpiration surface area and slightly reduce the surface area, resulting in higher evaporation rates (Jennison, 2003, Caruso, 2005, Ramey, 2004, Craig et al., 2005). Therefore, research on effectiveness must be applied to the different aquatic plants before introducing them to a reservoir. Moreover, some aquatic plants can affect water quality and the natural flora dynamics, and therefore reduce its suitability for recreation (Caruso, 2005, Sinha et al., 2006).

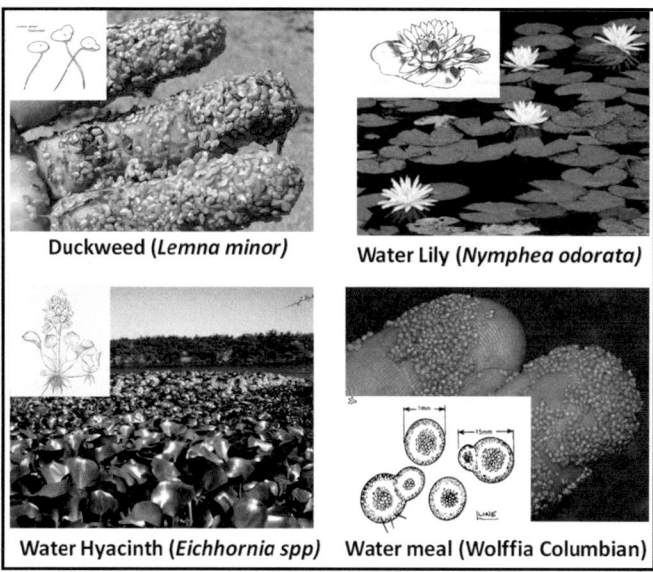

Figure 2.13: Various biological (plant) covers (Ramey, 2004), modified by Elba 2014

2.3.1.2 Wind breakers

Wind is one of the most important factors that affect the rate of evaporation loss from water surface. The greater the movement of air over the water surface, the greater the evaporation rate. Planting of trees normal to windward direction is found to be an effective measure for the slowing down of evaporation loss. Wind breaks consisting of tree 'shelterbelts' reduce the wind speed by 80% for a distance of 5 times the height of the trees (Helfer et al., 2009). However, Robert et al. (2007) found that the resulting increased turbulence and surrounding temperature in the downwind zone of the wind breaks tends to increase rather than reduce evaporation on large water surfaces. The distance downwind where wind speed is reduced is proportional to the height of the windbreak (H). If the wind is normal to the windbreak, wind speed is reduced for 3 to 5 H upwind and 20 to 30 H downwind. The area of reduced wind speed downwind is called the sheltered zone as shown in Figure 2.14. The smaller the wind angle, the lesser sheltered area is available (Robert et al., 2007, Helfer et al., 2009).

Helfer et al. (2009) showed that evaporation can be reduced by up to 35% on the windward side of the windbreak barriers and that evaporative reductions are proportional to the decreases in wind speed. Similarly, Wang et al. (2001) found that evaporation was reduced by 14% across a large scale shelterbelt network in

China. Moreover, Hipsey et al. (2004) found that well-designed wind-shelters can reduce evaporation from open water bodies by 20% to 30% as a result of using 8m high tree windbreak. However, wind breakers are effective only for small sized reservoirs and not useful in large reservoirs where their effect is limited to a short distance from the edge of the reservoir. Thus, the central surface water of large reservoirs will still experience the wind effect. Moreover, planted trees can consume large quantity of water due to transpiration. Therefore, wind breakers are usually used only for particular regions undergoing high wind speed (Sinha et al., 2006).

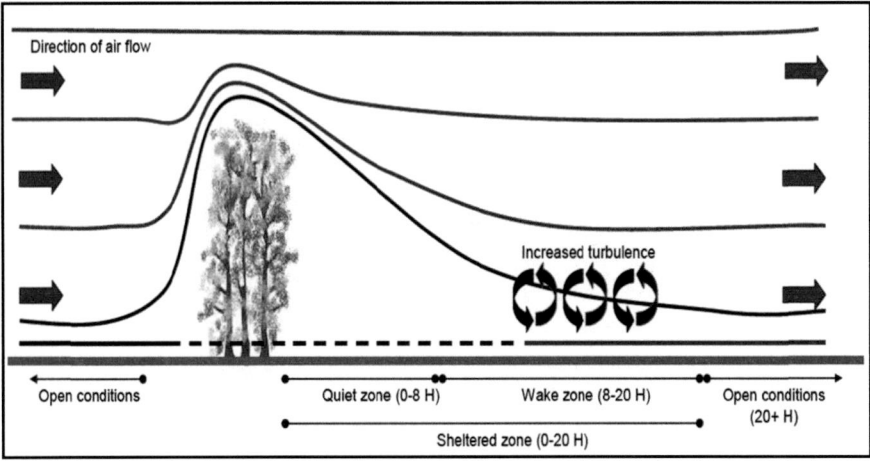

Figure 2.14: Air flow around a windbreak showing the extent of various microclimate zones in terms of multiples of windbreak height (H) (Helfer et al., 2009)

2.3.1.3 Palm fronds

Palm fronds can be used to cover water bodies. Palms are usually available in arid regions that undergo high evaporation rates. Palm fronds are a massive agricultural waste and environmental friendly bio-product and are capable of withstanding extremely hot weather conditions of arid regions (Alam and Al Shaikh, 2013). Al-Hassoun et al. (2011) found that the average reduction in evaporation using a floating cover made up of palm leaves was 55% for the fully covered pool while it was only 26% for the half covered pool. Moreover, the study confirms the effectiveness of the fronds in evaporation reduction, and no harmful effects on water quality as no serious effect on water quality was found. Alam and Al Shaikh (2013) used covers made of palm fronds to reduce the evaporation from the pan as shown in Figure 2.15. They found that using a single layer of cover can reduce up to 47% in evaporation, while using a double layer cover enables less evaporation losses and reduces 58% of the evaporation rate.

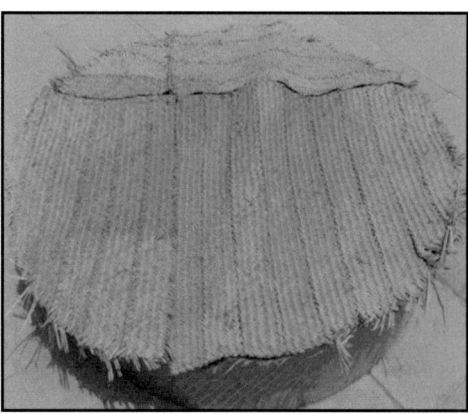

Figure 2.15: Evaporation pan covered with palm fronds (Alam and Al Shaikh, 2013)

2.3.2 Physical shade structures

Physical shade structures such as sheets or covers provide a physical barrier between the water and atmosphere to reduce evaporation losses. These covers reflect energy inputs from the atmosphere and prevent solar radiation, thereby decreasing water temperature. Furthermore, the covers literally trap the air and prevent transfer of water vapor to the outer atmosphere, thus mitigating wind action. Consequently, evaporation losses can be reduced. Numerous shapes and types of covering materials including foam, wax, polystyrene, and bamboo, have been used with varying levels of evaporation reduction and cost effectiveness (Craig et al., 2005, Sinha et al., 2006). Examples of these covers are given below.

2.3.2.1 Suspended covers

Suspended covers are shade structures made of shade cloth suspended above the water surface by a cable structure such as super span type, as shown in Figure 2.16. They can reduce evaporation losses through reducing direct sunlight on the water surface and consequently reducing net radiation and the wind speed across the water as well as increasing the humidity of the air under the cover. The previous studies indicate that this type of covering can reduce evaporation losses on dams from 59% to 76% (Jennison, 2003). Not only the shade cloth is significant in saving the evaporation, but it also causes limited environmental impact. However, it requires high capital outlay and maintenance costs since these structures need to be supported above the water. Hence, it is not applicable to use them for large water storages. The structure has a lifespan of 30 years, and the cloth of 15 years (Craig et al., 2005, Sinha et al., 2006).

Figure 2.16: Suspended covers (super span type) covering the Bemm River in Australia (Tech-Span, 2007)

2.3.2.2 Floating covers and floating objects

Physical floating covers can be found as continuous cover (floating sheets) or modular covers (floating objects). These covers float on the water surface and are made from sun-blocking materials that reflect a proportion of the incoming solar radiation and act as physical barriers to the passage of water vapor, both vertically and horizontally, thereby reducing evaporation (Craig et al., 2005, Jennison, 2003). Unlike suspended covers, the floating covers are supported by the water itself. On large dams, the floating covers must be fixed on the water surface using some form of anchoring mechanism. These covers are manufactured using various materials and have numerous colors and shapes.

Continuous sheeting options such as the E-VapCap cover are designed to float on top of a dam, protecting the water from evaporation. In the event of rain, there are drainage holes throughout the material, allowing water to leak into the storage, as shown in Figure 2.17. Many different materials have been used in the past including polyethylene, wax, foam and polystyrene. These sheets are one of the most effective evaporation reduction techniques and can reduce evaporation by 100%, as well as reduce or prevent algal growth (Craig et al., 2005, Jennison, 2003).

Unlike the continuous floating covers, the modular floating objects are a compound of multiple individual units that enable easier installation and maintenance compared to the continuous covers as shown in Figure 2.18. They often float freely and do not entirely cover the water surface, which allows water to evaporate from the uncovered areas. Thus, the energy input is only somewhat reduced, and wind can still blow away humid air, thereby reducing the evaporation reduction efficiency. Unlike the continuous floating covers, floating objects allow for more dissolved O_2 transfer to occur at the water to air interface due to the open gaps between them. The water-saving efficiency depends on the design and shape of the modules, as well as the material. There are numerous types and shapes of commercial floating objects, they can reduce evaporation from 70% to 90% (Craig et al., 2005, Jennison, 2003).

Figure 2.17: Continuous Floating covers Evapcap (Craig et al., 2005, WPDC, 2010)

Figure 2.18: Aquacaps covers (NYLEX, 2002)

Aquacaps covers are round, dome-shaped floating module manufactured from white polypropylene disks that float semi-submerged on the surface of the water, as shown in Figure 2.18. They are about 1.1 m in diameter with a curved surface on top and together cover up to 80% of a surface area. They are UV treated and have a life of around 20 years. They lower solar radiation that penetrates the water and reduce the level of dissolved oxygen in water. Aquacaps have the potential to reduce evaporation by an average of about 70% annually. Estimated costs of installation of one m^2 are around $22 (Yao. X et al., 2010, NYLEX, 2002).

Aqua-Armour is similar to Aquacap, only with hexagonal shape, so that the units lock together as shown in Figure 2.19. The individual units measure 1.2 m in diameter. Aqua Armour covers are made from food-grade, high density polyethylene with UV stabilized. They stop 95% of algae-inducing sunlight which significantly reduces the risk of algal growth and consequently improves water quality. They have a life span of 20 years. The evaporation reduction efficiency of this cover is about 88%. A square kilometer costs $30 (AQUA, 2006).

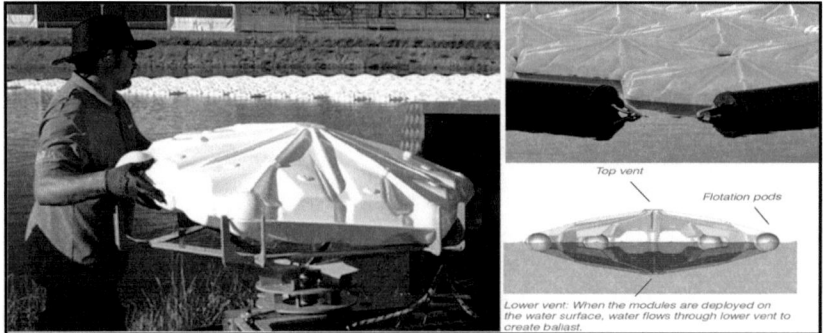

Figure 2.19: Aqua Armour (AQUA, 2006)

Hexprotect tiles are Hexagonal-shape tile covers with 22 cm diameter made of virgin or recycled, high density polyethylene with UV stabilized as shown in Figure 2.20. The cover self ballasts, increasing their weight more than 260%, making it suitable for high wind applications. The installation is simple and can ensure coverage of up to 99%, with an evaporation reduction efficiency of 95%. They reduce the penetration of UV rays and consequently prevent the growth of algae and clogging weeds. They have a life of around 25 years. (AWTT, 2012).

Figure 2.20: Hexprotect tiles (AWTT, 2012)

AgFloats is a system that uses recycled tires as an evaporation barrier, with polystyrene to make the tires float as shown in Figure 2.21. The disadvantage of this system is that the polystyrene is exposed to sunlight and will gradually degrade in spite of that fact that the tires will provide some protection from the UV light. They cost about $10 per square meter, which is much lower than the cost of the other floating objects (Clarke, 2009).

Figure 2.21: AgFloats - recycled tires on a community dam at Blyth, SA- Australia (Clarke, 2009)

ECC floating ball blankets also provide a highly effective cover to water bodies. By placing a sufficient quantity of hollow plastic balls on a water surface, the balls automatically arrange themselves into a close packed formation over 91% of the surface area as shown in Figure 2.22. The ECC ball with its small diameter of 10 cm provides an extremely effective barrier and significantly reduces the mass and heat transfer mechanisms operating between the water and surrounding environment, which allows an evaporation reduction efficiency of 90%. These balls can eliminate algae and weeds by blocking the Sunlight UV rays. It is a quick and simple installation through pouring the balls into the water body and does not require maintenance cost in addition to having a long time span of 25 years (Stuck, 2010).

Figure 2.22: ECC Floating ball (Stuck, 2010)

2.3.3 Chemical layers

Spreading chemicals over the water surface forms a thin, mono-molecular film that is effective in reducing evaporation loss from a water surface. The formed mono-molecular films reflect energy inputs from the atmosphere, and work as barriers between the water body and the atmospheric conditions and hence reduce evaporation. Since, the film allows enough passage of air, aquatic life is not affected. The most useful material is the film developed by using fatty alcohols of different grades. These fatty alcohols used for evaporation control are generally termed chemical water evapo-retardants (WERs) and are available in the form of powders, solutions or emulsions. These chemical water evapo-retardants such as Water$avr have the limitation of high cost of application since it must be applied every 2-4 days. Moreover, the mono-layer breaks OR tends to break at high wind speeds. Therefore OR In addition, the evaporation reduction efficiency is limited between 20% and 40% (Craig et al., 2005, Jennison, 2003). The following chemicals are generally used for water evaporation retardation:

- Cetyl Alcohol (Hexadecanol) $C_{16}H_{33}OH$
- Stearyl Alcohol (Octadecanol) $C_{18}H_{37}OH$
- Ethoxylated Alcohols and Linear Alcohols Linoxyd CS-40
- Acilol TA 1618 (Cetyl Stearyl Alcohol)

2.3.4 Design features

Apart from biological, physical and chemical methods, which are considered short term methods, numerous methods are available to reduce evaporation from water storage areas through built-in design features. Such design features can consist of building deeper storages, reducing the exposed area of surface water and storing water in underground storage. These design features are considered long term methods for reducing the evaporation (Jennison, 2003, Sinha et al., 2006).

2.3.4.1 Deeper storages

The construction of deeper storages allows a larger volume of water to be stored while exposing the same amount of surface area available for evaporation. This reduces the evaporation in proportion to the volume of water stored. Deeper storages enable better water quality. The cost of constructing deeper storages depends on the site conditions and the type of storage that is required, either in-stream or off-stream. This extra cost must be balanced against the volume of water saved. However, this method is more useful in the construction of new storages rather than upgrading existing storages (Jennison, 2003).

2.3.4.2 Reduction of exposed water surface

This method is useful for shallow portions in reservoirs that can be isolated by constructing dykes and water, for example by, diverting or pumping to the deeper pockets so that the exposed surface is effectively reduced. This method is suitable for evaporation control in drought areas. The reduction in exposed water surface can also be achieved with the integrated operation of reservoirs through controlling water levels. Furthermore, reduction is achieved by storage management in reservoirs, so that shallow reservoirs are utilized first and deeper reservoirs later (Sinha et al., 2006).

2.3.4.3 Underground storages of water:

Water can also be stored underground in cavities and aquifers, which do not have higher lateral dispersion losses. This method can save valuable land on the ground, which requires geological investigations in order to identify such underground areas. Although evaporation can be reduced, the water needs to be pumped for use, which requires additional pumping costs and energy consumption. (Sinha et al., 2006).

2.4 Sedimentation in large reservoirs

Dams and their reservoirs play an essential role in people's lives. The purposes of dams manifold: They are not only used for water supply or irrigation, but also for flood control, navigation, water quality, sediment control and hydropower. These multipurpose dams provide domestic and economic benefits to the people in the industrial, developing and rural countries (Asmal et al., 2000, ICOLD, 2009). According to the World Register of Large Dams (ICOLD), dams have been constructed at a rate of 1.2 dams daily worldwide since 1930 (ICOLD, 2009). About 45000 large dams higher than 15 m, with a total reservoir surface of about 500000 km^2 and 7000 km^3 storage have been constructed worldwide by the end of the twentieth century (Asmal et al., 2000, ICOLD, 2009). In general, artificial reservoirs formed due to dams on natural waterways cause problems with regard to sediment inflow and deposition. Sediment consists of solid particles of mineral and organic material transported by water. Estimating the rate of sedimentation is a major problem facing hydropower experts in planning the life time of any dam before the sediment begins to affect the function of the reservoir (Strand and Pemberton, 1982, ICOLD, 2009). Therefore, engineers need to include sufficient sediment storage in the reservoir in order to maintain that all of the reservoir's functions will not be affected during the life span of the project (Tigrek and Aras, 2012, CEQA, 2011). The replacement costs of storage lost to sediment accumulation in the reservoirs are extremely high (Schleiss et al., 2008, Strand and Pemberton, 1982).

2.4.1 The problem of reservoir sedimentation

The current world wide annual loss of storage capacity due to sedimentation is higher than the increase of capacity by the construction of new reservoirs for irrigation, water supply and hydropower. Consequently the sustainable use of reservoirs is not assured in the long run (Schleiss et al., 2008). The mean annual sediment loads of all rivers worldwide is estimated to be between 0.6-0.75 T/1000m^3 of water, while the mean annual sediment loads of all the reservoirs is estimated to be 0.6% of the total annual storage (Xiaoqing, 2003, IRTCES, 2011). These rates vary considerably from river to river and from reservoir to reservoir depending on their respective morphological and hydrological characteristics. For example, in deep and long reservoirs, the sedimentation rate is well below the world mean value (Hargrove, 2008, ICOLD, 2009, Strand and Pemberton, 1982, Strand and Pemberton, 1982). After only three to four decades of operation, sedimentation becomes a serious problem in most reservoirs, even in catchment areas with moderate surface erosion (Schleiss et al., 2008).

The accumulation of sediments in reservoirs can lead to several problems, namely the loss of storage capacity, which in turn, reduces the functional efficiency of the reservoir in the long-run and increases spillway flows and higher flood risk in downstream waterways. It also deposits sediments against the upstream face of dams, which can negatively affect the stability of certain dam structures. In addition, the high sediment load of the water passing through turbines due to the sediment accumulations near power intakes causes erosion and damages to turbines, clogs reservoir intakes and outlet structures, and scours of hydraulic machines. As a result, its efficiency is decreased and maintenance costs are increased. Moreover, the accumulation of sediment in reservoirs causes the loss or impairment of fish, macro-invertebrates, and other aquatic organisms due to the increase of sedimentation loads. It degrades the water quality due to turbidity in the water, which, in turn, limits light penetration and prohibits healthy plant growth on the reservoir bed. In addition, sediments reduce dissolved oxygen levels due to an increased amount of particles in the water and higher water temperatures (Halcrow, 2001, Strand and Pemberton, 1982, Carvalho et al., 2000, IRTCES, 2011, Hargrove, 2008, Boroujeni, 2012)

In 1957, the Sultan Abu Bakar Dam was constructed in Malaysia to store water in the Ringlet reservoir. The original reservoir storage capacity was 6.7 hm^3. 34% of its storage was already filled with sediments after 35 years of operation, leaving the reservoir with a lifespan of 10 years. Sediment levels increased to 45% in 2010, which led to a 62% decrease in efficiency at Jor Power Station (Luis et al., 2013). In 1963, the High Dez dam was constructed in the Zagros Mountains in the southwestern Iran. The original reservoir volume was 3315 hm^3, this volume was reduced to 2600 hm^3 after 40 years of operation due to sedimentation, which greatly reduced the efficiency of the power station (Boroujeni, 2012).

In the Nile basin, reservoirs lose a considerable percentage of their storage capacity due to sedimentation. In 1925, the Sennar dam was constructed on the Blue Nile in Sudan to store 0.93 km^3. The reservoir lost 71% of its original reservoir capacity in about 60 years due to changes concerning operational rules based on irrigation requirements for agriculture upstream and downstream the dam. About 90 years after operations began, the reservoir is no longer used to store a significant amount of water, but to regulate the river flow and to generate hydropower at a limited capacity station of 15 Mw (Ahmed and Ismail, 2008, Awulachew et al., 2009).

Similar problems have also occurred at a larger dam site further south. In 1960, the Koka dam was constructed on the Awash River in the Ethiopian Highlands. The reservoir had an original storage capacity of 1650 hm^3. The dead zone capacity decreased from 180 hm^3 to only 8 hm^3, which means that 96% of the dead storage volume is filled by sediment. Both the reservoir and the dam are threatened by increasing sedimentation caused by environmental degradation with an estimated loss of 32% of capacity in 1999. Similarly, Angereb dam was constructed on Angereb river in the Ethiopian Highlands in 1986 for water supply. The reservoir lost 50% of its storage capacity due to sedimentation in about 20 years, a development that led to serious water shortage problems and adversely affected the people in the surrounding area and their livelihoods.

The general pattern described above can also be observed in the case of the Khashm El-Girba reservoir, which was constructed on the Atbara river in Sudan in 1964. Due to the high sedimentation rates, the reservoir lost about 60% of its capacity, which caused severe water shortages during drought years and resulted in a decline of crop area cultivated Hydropower generation was limited to the flood season, which had the additional negative effect of increased erosion at the turbine intakes and cavitations of the turbines. In 1966, the Roseires dam was constructed on the Blue Nile with initial reservoir storage capacity of 3.0 km^3 raised to 3.3 km^3 in the last century. The high sedimentation rates in the filling seasons caused a loss of 40% of storage capacity and reduced the efficiency of the dam. Again, the increased silt loads ultimately reduced the life span of the entire plant.

2.4.2 Measuring sedimentation rate

Worldwide, there is no precise data on the rates of reservoir sedimentation, but it is commonly accepted that about 1% to 2% of the storage capacity is lost annually (Carvalho et al., 2000, IRTCES, 2011, Flögl, 2010, Hargrove, 2008, Xiaoqing, 2003). The sedimentation rate can be measured by surveying the reservoir. This rate is calculated as follows: a sedimentation estimate is conducted during the planning stage; primary survey is conducted to bottom of the lake before operating the dam; regular surveys are conducted to the bottom of the lake during the operation stage; deposit formations and sedimentation effects are estimated from the changes in surveys. The general procedures for reservoirs surveys have undergone changes due to scientific development and the emergence of new technologies and equipment. The general procedure commonly consists of creating a bathymetric map of the bottom of the lake, which may be compared to an earlier map. The common methods for reservoir survey are reservoir contour and topo-bathymetric lines (Vente et al., 2004, Carvalho et al., 2000, Strand and Pemberton, 1982, Ferrari, 2006).

The contour survey method is used for small reservoirs or for reservoirs that may be shallow or that have already dried up. This method mainly uses the procedures of topographic mapping by aerophotogrametry that is, obtaining photos of the reservoir at several different levels. The method is especially suitable for aerial surveys, when one may schedule flights for different levels of reservoir reduction in a relatively short time span. The cost for this kind of survey is very high, but it is very accurate. On the other hand, the topo-bathymetric survey method, based on the cross-sections survey method, is usually used for medium and large reservoirs. The basic procedures are as follows: planning of sections to be surveyed; determining the survey reduction level, usually the maximum normal level; installing of new reference marks; measuring relative distance and depths using a combination of GPS and echo-sounding to determine the below-water topography; drawing cross-sections; and interpreting, computing, and mapping the bed. The number of sections needed to be planned are based on the size of the reservoir. Modern methodologies allow for a more accurate survey, which may be completed more quickly. The choice of tools depends on the width of the section being surveyed, its depth, the reservoir size, available resources, and other factors. It ranges from simple methods, for example the use of tape measure and ruler, to sophisticated methods, for instance Differential Global Positioning System (DGPS) (Carvalho et al., 2000, Strand and Pemberton, 1982, Ferrari, 2006, Vente et al., 2004).

2.4.3 Procedure against sediment problems

The major reason for losing the water storage capacities in reservoirs is, as explained above, sedimentation. In the past decades, several measures to prevent reservoir sedimentation were developed (Sumi and Hirose T., 2011). Worldwide, there are several techniques in order to deal with the problem of siltation. Some of these techniques deal with inflowing sediment while others try to remove sediment from the bed of the reservoir. The techniques used depend on the reservoir's geological, geographical, and climatic characteristics (Aras, 2009). The following four techniques are used to manage reservoir water storage capacity: first, reducing sediment entering reservoir with the help of watershed management, upstream check dams, and reservoir bypass; second, minimizing siltation in reservoir by density current venting and sediment bypass; third, removing sediment from reservoir by flushing, sluicing, dredging, trucking, and hydrosuction dredging; and, fourth, replacing lost storage by raising dam height, decommissioning the dam, or constructing a new dam (Aras, 2009, Schleiss et al., 2008, Halcrow, 2001, IRTCES, 2011, Tigrek and Aras, 2012).

2.4.3.1 Minimize sediment entering reservoir

Watershed management

The rules and guidelines for the operation of a reservoir can affect sediment deposition. If the water level is high during flood season, then sediment is typically deposited in the upper reaches of the reservoir. However, if reservoir is drawdown during flood season, then sediment is typically deposited in the dead storage zone of the reservoir. Tactical dredging is used for local sediment removal especially at hydropower generation dams to keep the area of the outlets clear and to prevent blockage of the outlets, which may disrupt energy production. As turbines and other mechanical equipments can be damaged by sediment, localized dredging can help to extend the lifespan of the dam and reduce maintenance cost of hydropower plants (Hargrove, 2008, Xiaoqing, 2003, ICOLD, 2009).

Upstream check structures

Check dams are, in terms of size, smaller than the main dam, which is protected by spillway structures. Check dams are used to reduce the sediment flowing from tributaries to the reservoir. They are useful for reservoirs that are used for power generation or water supply. These dams can extend the life of the main dam; however, a sediment management program is needed to operate check dams. In spite of their high cost, they are useful to retain coarse material, which can cause backwater deposits in the main reservoir and lead to several serious problems (Aras, 2009, Halcrow, 2001).

2.4.3.2 Minimize deposition of sediment in reservoir

Density current venting

The technique of density current venting is used to route the sediment-laden flow through the water stored in the reservoir and to get it out through the bottom outlets in the dam. The flow should have enough velocity for fine particles to form turbid flow, and the current needs to reach the dam for density current venting to be successful. The bottom outlet should be open, causing the current to be vented through. The efficiency of density current venting depends on the reservoir's topography, thermal and salinity related stratifications, the conditions of the incoming flow, sediment characteristics, and outlet facilities (Kashiwai, 1998, IRTCES, 2011, Schleiss et al., 2008).

Bypass tunnels

The technique of reservoir bypassing is used to let sediment-laden water pass through a channel, thereby keeping the rest of the water mostly free of sediment. It is composed of a channel beginning in the river before the reservoir. This system is very hard to develop because topographical, hydrological, or economic conditions may impose severe restrictions in terms of design, construction, and operation. Bypass tunnels have many advantages, for example the possibility that they can be constructed at existing dams and that they may prevent a loss of stored reservoir water caused by a lowering of the reservoir water level. In cases of flooding, water is diverted to the bypass system, and thus there is no need for a large-capacity spillway at the main dam. Therefore, bypass tunnels have a relatively small impact on the environment downstream (Kashiwai, 1998, IRTCES, 2011, Aras, 2009).

2.4.3.3 Lost storage replacement

Raising dam height

Dam height is raised to increase reservoir capacity and also to replace storage lost as a result of sedimentation. In spite of the high costs of raising dam height, it is economically feasible, especially in arid regions. In the long term period, it can, however, cause several socio-economic and political problems, for example if people have to be relocated. Raising dam height also increases the water loss due to evaporation and seepage, and dam safety aspects can lead to higher maintenance costs. This approach usually involves raising intake and bottom outlet structures or other major changes to the overall design of the dam, which likewise would require considerable financial investments in order to avoid these costs, approaches such as raising dam height or other techniques need to be considered already during the planning process,, and alternative strategies to increase dead storage, that is reservoir capacity below the intake level, need to be considered well before construction begins. (IRTCES, 2011, Aras, 2009, Halcrow, 2001).

Building a new dam

New dams can be built downstream or upstream of reservoirs or on another river to replace the storage lost at an given dam site. In general, this is a temporary and uneconomical solution (Allen et al., 2005, Stroud, 2012, IRTCES, 2011).

Decommissioning

The process of decommissioning involves the removal of all structures of a given dam project, and it leads to the end of the operational life of this dam. Decommissioning is an economical alternative when the operational costs of the reservoir exceed the benefits gained from it. In addition, decommissioning is also often carried out in order to improve water quality, to protect flora and fauna, or to eliminate a public safety hazard (Aras, 2009, Stroud, 2012, Downs et al., 2009).

2.4.3.4 Remove sediment from reservoir

Flushing

Flushing is a technique for sediment removal that is used to hydraulically scour deposited sediment from reservoir by increasing flow velocity and thereby moving sediment through low level outlets. It is possible to distinguish between two types of flushing: empty (free-flow) flushing or pressure flushing. Empty (free-flow) flushing can be achieved by lowering the water level, emptying reservoir, and routing inflowing water through the reservoir, a procedure that resembles natural fluvial conditions. There are two conditions for empty flushing: flood season and non-flood season. Flood season flushing is more effective since it provides larger discharges to route the sediment. Flushing is generally effective in narrow dams. However, this technique has many limitations because a significant amount of water has to pass through the reservoir for extreme flushing operation and hydropower dams cannot generate energy. Moreover, the water released during this process is characterized by a very high sediment concentration that is much greater than during natural fluvial conditions and that can therefore cause an undesirable environmental impact on the environment downstream. Pressure flushing can be achieved by releasing water through the bottom outlets and keeping the reservoir water level high. This technique can, however, only be used to clear a very limited area in the reservoir. Moreover, free-flow flushing can transport a much greater sediment load than pressure flushing (Amini and Heller, 2014, Wen Shen, 1999, Fruchard and Camenen, 2012, Kashiwai, 1998).

Sluicing

Sediment sluicing is an operational design. In this case, the reservoir is drawn down during flood season and then the inflow carrying sediment is passed directly through the reservoir. As a result, the sediment will not be deposited. In order to work this method, enough water has to be available. As water rises during flooding, the outflowing sediment discharge is always smaller than that of the inflow due to the backwater effect, which causes a decrease in velocity. As water levels are being lowered, the outflowing sediment discharge becomes greater than the inflow because there is no backwater effect and there is erosion in the reservoir. After the flood season, clear water can be stored, and the water levels in the reservoir can be raised for the usage during the next season. Sluicing is preferable to other techniques because the incoming sediment is not deposited in the reservoir. The efficiency of this process is affected by availability of excess runoff, the grain size of sediments, and reservoir morphology. Worldwide, the techniques of sluicing and flushing are frequently combined (Aras, 2009, Halcrow, 2001, Tigrek and Aras, 2012, Yang, 2006).

Dredging

Dredging involves the accumulation of deposited sediment from the reservoir or lake bed and the transportation to another area. Several types of dredging equipment can be used. Dredging systems can be classified as either hydraulic or mechanical. Hydraulic systems lift both sediment deposited in the reservoir and water, and then this slurry is transported from the output point to the point of placement. Mechanical systems use buckets to dig into the reservoir bed and then pick up the sediment. Hydraulic dredging is used more widely than mechanical dredging because the removal cost is lower, production rates are higher, and both fine and large materials can be effectively removed. The need for a dewatering system to process for the slurry and the difficulty to move fine sediments to the point of placement are the two main disadvantages of hydraulic dredging. Dredging methods are suitable for small and medium-sized reservoirs, which do not have enough water for flushing. While dredging is used effectively for the removal of coarse sediment, muddy reservoir deposits with silt and clay, represent a major problem, as they are not easy to remove or use due to their high content of water and organic matter. It is also often difficult to find a proper place for dumping removed sediments. Therefore, the cost of land set aside for the disposal of sediment must be considered when estimating the financial impact of dredging (IADC, 2005, Sciortino, 2010, Helmke, 2010, Halcrow, 2001, Allen and Dunbar, 2005, Allen et al., 2005).

Hydrosuction Removal Systems (HSRS)

Hydrosuction removal systems (HSRS) are siphon and airlift systems, which use the potential energy stored by hydraulic head at the dam and removes sediments through a floating or submerged pipeline to an outlet. The system is composed of barges, pipelines, and valves to control flow. The barges are used to control the flow both upstream and downstream of the pipeline and also to connect the upstream end of the pipe with the suction head. Because it is powered by kinetic energy, there is no need for any equipment to produce energy, which reduces operating costs. In addition, there is no need to find suitable disposal sites, because sediment is discharged to the downstream end of the reservoir, and it can therefore be considered a more sustainable technique than others. HSRS can, however, only be used in relatively short reservoirs, that is, in reservoirs less than 3 km long, and these systems are also dependent on the elevation of the dam and reservoir. This technique is often used due to a lack of water for flushing. The mix of water and sediment released downstream is frequently used for irrigation (IRTCES, 2011, Aras, 2009, Halcrow, 2001).

Trucking (Dry Excavation)

The technique of trucking involves dry excavation of sediment deposited in reservoir, but in contrast to dredging, it requires a drawdown of the reservoir. The excavated sediment is transported to a proper disposal site with standard earthmoving equipment. The method is very expensive, for example more expensive than dredging, because of the costs related to the drawdown, transportation, and disposal, and consequently, this method is not widely used. The lowering of the reservoir during the dry season is required when reduced river flows can be adequately controlled and they do not interfere with the excavation work. Reservoirs used for flood control may be more suitable for sediment management by trucking, for instance at Cogswell Dam and Reservoir in California. The sediment from this reservoir has been excavated with conventional earthmoving equipment and has been used as engineered landfill in the hills adjacent to the reservoir (Allen et al., 2005, Tigrek and Aras, 2012, Kashiwai, 1998, IRTCES, 2011).

2.4.3.5 Evaluation of sedimentation management methods

Not all of above mentioned techniques are sustainable, efficient, or affordable. For example, raising dam height or work on outlets is expensive and does not provide a long-term solution. Allen and Dunbar (2005) showed that the costs of dredging may be twice that of developing a new reservoir. Dredging is mostly used in semi-arid regions while flushing is used mainly in the rainy regions in northern European countries (Aras, 2009). Flood flushing and venting of turbid currents represent the most effective means of reducing deposition in reservoirs, although the successful application of these techniques depends on availability of suitable bottom outlets and an excess of water. The removal of sediment deposits by dredging or excavation is a costly operation, but it may be justified in certain circumstances, for instance in light of the economic value of the water or the impossibility of replacing lost reservoir capacity by other means. In general, the disposal of sediment may also cause difficulties unless it can be used to improve surrounding agricultural land (IADC, 2005, CEQA, 2011, Tigrek and Aras, 2012).

2.5 Summary

This chapter consists of a comprehensive literature review on climate change, methods for estimating the evaporation losses, methods for reducing evaporation losses, sedimentation problems in open reservoirs, and sedimentation management methods.

There has been great evidence that the global climate is hanging due to human-induced emissions of greenhouse gases since 1850. The global temperature has constantly risen and has led to the retreat of some mountain glaciers. Moreover, the average atmospheric water vapor content has increased over land and oceans as well as in the upper troposphere. The global average sea level has risen at an average rate of 1.8 mm annually for the past fifty years. Based on the GCMs models, the IPCC in 2013 expected that the Earth's surface temperature will increase by approximately 0.3°C to 4.8°C by the late twenty first century, and the sea level rise will be 0.42 m to 0.98 m in 2100 relative to 1986-2005

Numerous factors affect the evaporation phenomenon. The main factors are solar radiation, temperature, wind, atmospheric pressure, the nature and shape of surfaces, and the quality of water. Thus, quantifying evaporation of a body of water can be very difficult. Large reservoir evaporation measurements from an inflow-outflow water balance method or using pan measurement methods are time and labor intensive and can be affected by environmental factors, particularly for large reservoirs. Empirical formulas are widely used in estimating the evaporation losses based on direct measurements. Most of these formulas are based on the energy budget approach, bulk aerodynamic approach, and water budget approach. In addition, the infrared bands of satellite images, either Landsat or ASTER images, hold large amounts of data that can be used to estimate evaporation because they have infrared bands.

Over the years, a number of commercial products have also been developed to minimize water losses, especially in Australia, India, China, and USA. The methods of evaporation control can be grouped under two extensive categories, namely short term measures and long term measures with four main classes including biological methods, physical methods, chemical methods, and management and design methods. Biological methods such as aquatic plants, palm fronds and nearby vegetation (wind breaks) can reduce evaporation rates by about from 35 to 60% through reducing any or all of incident solar radiation, surface area or wind speed. Physical shade structures such as sheets or covers provide a physical barrier between the water and the atmosphere to reduce evaporation losses. Suspended covers can reduce the evaporation losses up to 76%, while the floating covers can save up to 100% of the water losses. The floating objects can save from 80 to 95% according to the shape and distributions of these objects. Chemical covers are based on the use of long chain alcohols to form a thin layer on the water's surface to reduce evaporation by an efficiency limited between 20% and 40%. Each of these short term methods has its negative and positive influences on the environment. Therefore, the selection of the appropriate method depends on uses of the open water bodies. Moreover, numerous long term methods are available to reduce evaporation from water storage areas through built-in design features such as deeper storage, thereby reducing the exposed area of surface water and storing water in underground storage.

The mean annual sediment loads of all rivers worldwide is estimated to be between 0.6-0.75 T/1000m^3 of water, while the mean annual sediment loads of all the reservoirs is estimated to be 0.6% of the total annual storage, and consequently the loss of storage capacity due to sedimentation is higher than the increase of capacity by the construction of new reservoirs. These rates vary considerably from river to river and from reservoir to reservoir depending on their respective morphological and hydrological characteristics. The accumulation of sediments in reservoirs causes the loss of storage capacity, the reduction in the functional efficiency of the reservoir, higher flood risk in downstream waterways, erosion and damages to turbines, blockage reservoir intakes and outlet structures, scourge of hydraulic machines, the loss of fish and other aquatic organisms, and degradation of the water quality.

In the past decades, several measures to prevent reservoir sedimentation were developed. Some of these techniques deal with inflowing sediment while others try to remove sediment from the bed of the reservoir, depending on the reservoir's geological, geographical, and climatic characteristics. These techniques are reducing sediment entering reservoir, minimizing siltation in reservoir, removing sediment from reservoir and, replacing lost storage. The selection of the proper method is difficult, because not all of these techniques are sustainable, efficient, or affordable.

Chapter 3

Physiographic, Demographic, and Historical Background

Chapter three provides information on the physiographic, demographic, and historical background of the Nile Basin, Egypt, High Aswan Dam, and its reservoir. It includes a general description of the Nile Basin, its hydrology, its climate, and a history of Nile water agreements. More specifically, this chapter offers an overview of Egypt's geography, geology, geomorphology (especially soil types), demographics, climate, flora and fauna, land uses, and water resources. The chapter also includes a history of the High Aswan Dam and offers an overview of the High Aswan Dam Reservoir's geography, morphology, storage capacity, geology, climate, hydrology, and sedimentation.

3.1 Nile Basin

The land surface area of the earth is covered by 45.3% fresh water consisting of 263 river basins, excluding Antarctica. Worldwide, there are 26 major river basins as shown in Figure 3.1. There are about 61 river basins just in Africa as shown in Figure 3.2, the Nile Basin is one of the major river basins in Africa. It is one of the most complexes of all major river basins because of its size and variety of climates and topographies. Although the Nile River is the longest river in the world, its annual discharge is relatively small as listed in Table 3.1 because half of its flows through countries with no effective rainfall.

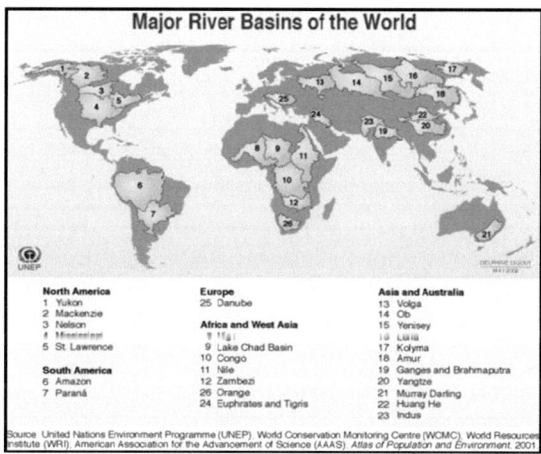

Figure 3.1: Location of the world's major river basins (Digout, 2001)

Physiographic, Demographic, and Historical Background

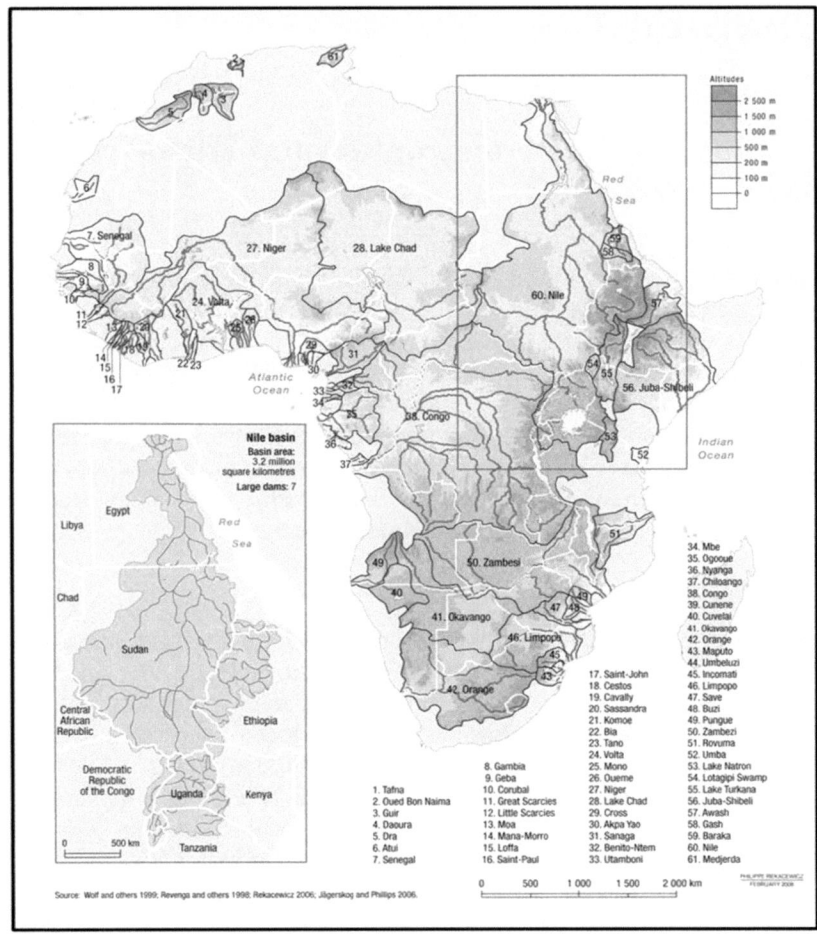

Figure 3.2: The locations of 61 river basins in Africa and the Nile Basin (Rekacewicz et al., 2005)

Table 3.1: Major rivers in the world (Karyabwite, 2000)

River	Length (km)	River Basin (million km^2)	Annual Discharge (KM3)
Nile	6700 - 6850	3.11	84
Amazon	6700	7.05	5518
Congo	4700	3.82	1248
Niger	4100	2.274	177
Mississippi	3705	3.27	562
Danube	2900	0.816	206
Rhine	1320	0.224	70
Zambezi	2700	1,200	223

3.1.1 General description

The Nile Basin is shared by the riparian states of the Nile: Burundi, the Democratic Republic of Congo, Egypt, Ethiopia, Eritrea, Kenya, Rwanda, Tanzania, Uganda, and Sudan, which has recently divided into two countries, Sudan and South Sudan as shown in Figure 3.3. It drains an area of roughly three million km^2 as recorded in Table 3.2. The table shows the distribution of the drainage area of the basin countries and their mean annual rainfall (FAO, 1997). The Nile water originates from 35% of the basin area, while the remainder consists of arid or semi-arid regions where the water supply is minimal and where evaporation and seepage losses are very large. Its length is of about 6700 km, and it is considered the world's longest river (UNESCO, 2008). Its course extends over a wide band of latitudes, from 4°S to 32°N, and crosses different climatic zones as it flows from highland regions in the tropics with high moisture to lowland plains under severe arid conditions as shown in Figure 3.4.

Table 3.2: Nile Basin countries areas and rainfall (FAO, 1997)

Country	Total area of the country (km^2)	country Area in the basin (km²)	As % total area of basin (%)	As % total area of country (%)	Average annual rainfall in the basin area (mm)		
					Min.	Max.	Mean
Burundi	27834	13260	0.4	47.6	895	1570	1110
D.R. Congo	2344860	22143	0.7	0.9	875	1915	1245
Egypt	1001450	326751	10.5	32.6	0	120	150
Eritrea	121890	24921	0.8	20.4	240	665	520
Ethiopia	1100010	365117	11.7	33.2	205	2010	1125
Kenya	580370	46229	1.5	8	505	1790	1260
Rwanda	26340	19876	0.6	75.5	840	1935	1105
Sudan, South Sudan	2505810	1978506	63.6	79	0	1610	500
Tanzania	945090	84200	2.7	8.9	625	1630	1015
Uganda	235880	231366	7.4	98.1	395	2060	1140
Total for Nile Basin	8889534	3112369	100	35	0	2060	615

Within the Nile Basin, there are five major lakes, each with a surface area over 1000 km^2 including Lake Victoria, Lake Albert, Lake Kyoga, Lake Edward, and Lake Tana. The Nile is controlled through six major dams including the High Aswan, Owen, Roseires, Senar, Jabel Aulia, Khashm Elgirba, and Merowe dams, as shown in Figure 3.3. The main tributaries of the river Nile are the White Nile, with origins in the basin of Lake Victoria, and the Blue Nile and the Atbara, which both originate in the Ethiopian Plateau near Lake Tana (Fahmy, 2006, NBI, 2012, NWS 2012).

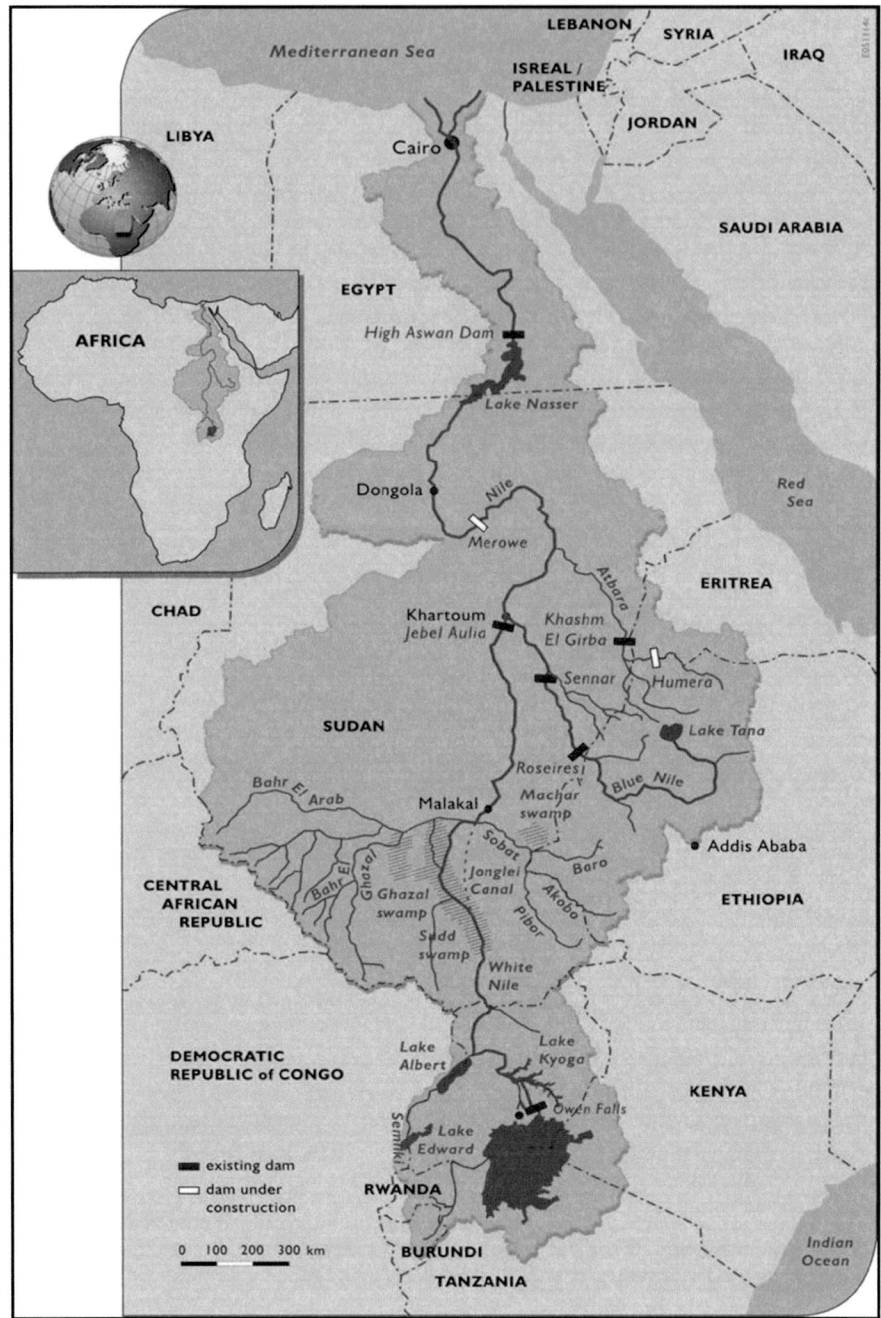

Figure 3.3: Nile Basin catchment area (NBI, 2012)

Physiographic, Demographic, and Historical Background

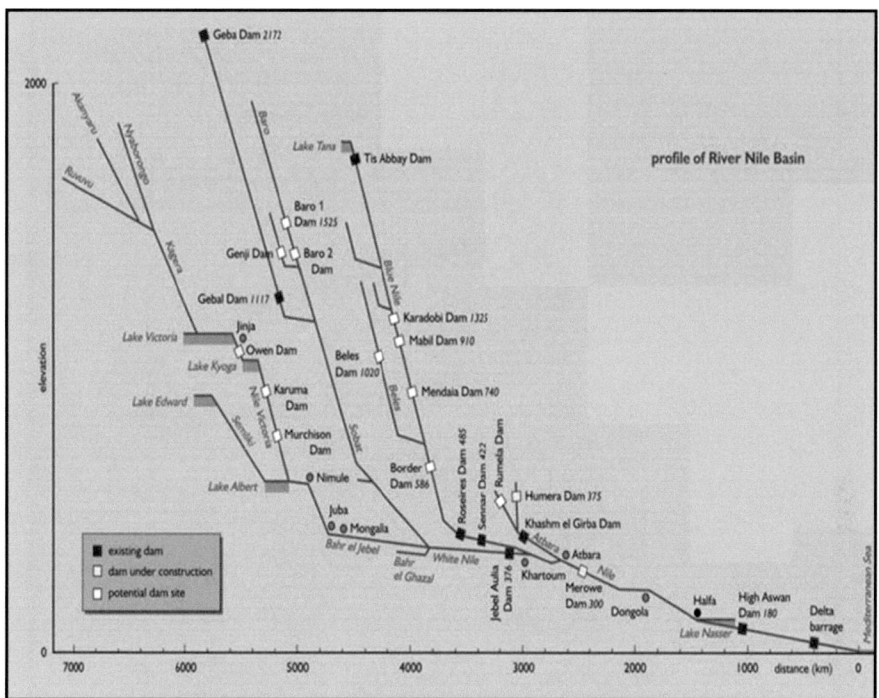

Figure 3.4: Schematic longitudinal profile of the River Nile (NBI, 2012)

3.1.2 Hydrology of the Nile Basin

Fluctuations in the Nile flows are primarily driven by the variation in rainfall over the Ethiopian Highlands. These highlands contribute on average about 83% to the total Nile flow at Aswan, while the Equatorial Lakes' Basin adds only about 17% but are still important as base flow components. Although the Nile is one of the world's major rivers, its flow is limited as it loses a considerable amount of water to evaporation in wetlands as well as in natural and man-made lakes, in addition to its 3000 km course through the arid lands in Sudan and South Sudan. Figure 3.5 shows a scheme of the upper Nile average annual flow up till HAD including the different inputs and losses (Fahmy, 2006, NBI, 2012).

The basin of the present-day Nile can be divided into 5 major regions: the Equatorial Plateau, the Sudd, the White Nile, the Ethiopian Plateau, and the Main Nile. Figures 3.6 to Figure 3.9 show the description of each of these basins while the volume of water with which they supply the main Nile annually is presented in Figure 3.9 and summarized in the following sections.

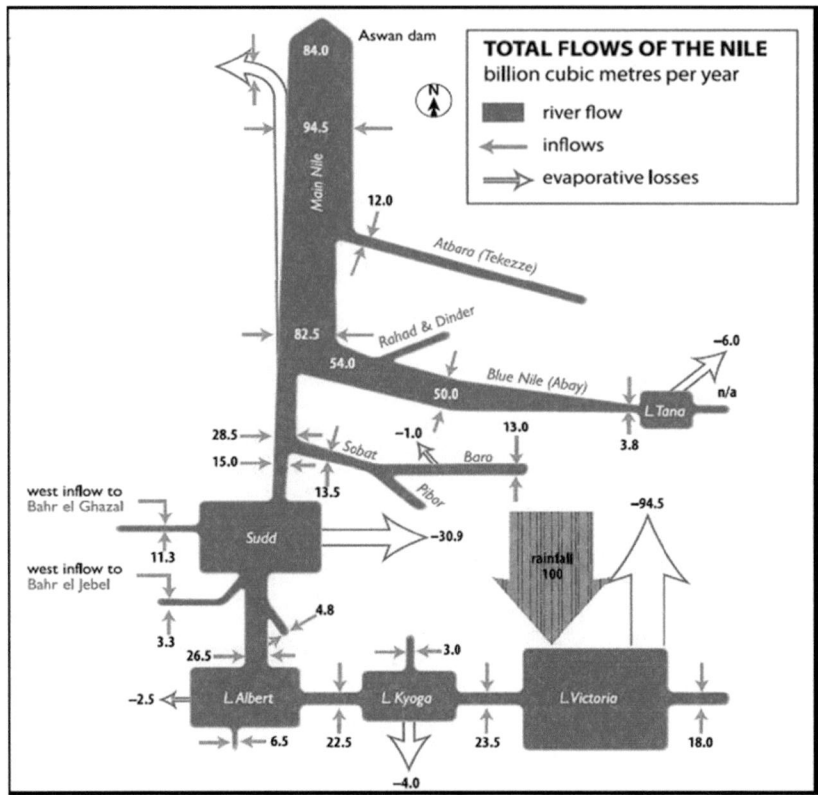

Figure 3.5: Scheme of upper Nile average annual flow (NBI, 2012)

3.1.2.1 The Equatorial Lakes Plateau

The Equatorial lakes plateau with its large catchment area of about 417000 km^2 is located in the centre of the African continent and lies in several countries including Burundi (13260 km^2), Rwanda (19876 km^2), Tanzania (84200 km^2), Kenya (46229 km^2), Uganda (231366 km^2), Democratic Republic of Congo (22143 km^2) (LNFDC, 2008). The basin is composed of several Lakes namely Lake Victoria (67000 km^2), Kyoga (6270 km^2), George (300 km^2), Edward (2200 km^2), and Albert (5300 km^2) (Figure 3.6). The Equatorial Lakes Plateau has a very sensitive water balance to rainfall leading to variability in lake level and river discharge through the year. Numerous floodplains, satellite lakes, and other wetlands are found within the basins of these lakes, leading to high rates of evaporation from the large surface area of these water bodies. The direct evaporation losses from the lake surfaces approximately equal to the direct precipitation onto the lakes. Therefore, the net water flow per unit area is small (Hurst, 1965, Karyabwite, 2000).

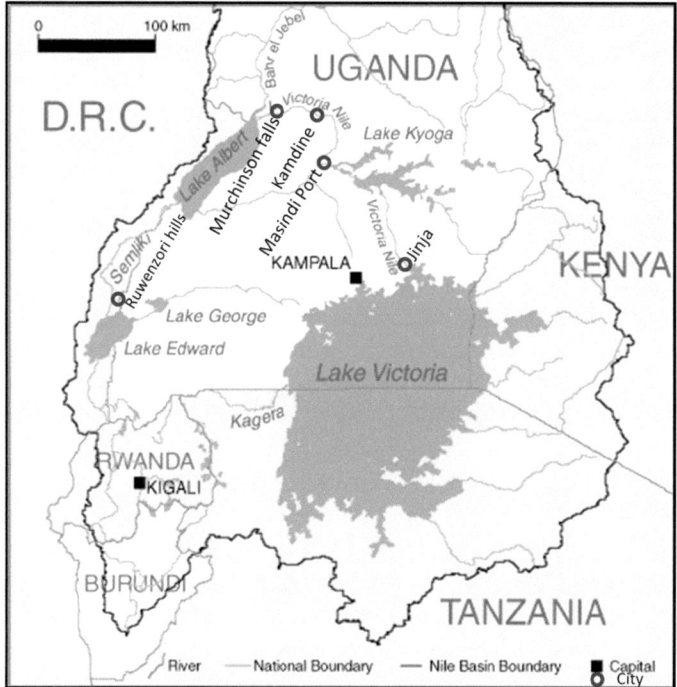

Figure 3.6: Main subbasin of the Equatorial Lakes Plateau catchment area (Karyabwite, 2000)

Lake Victoria

Lake Victoria is the largest lake in Africa and the second largest fresh water lake in the world. The catchment area of the lake is shared by five countries: Burundi, Kenya, Rwanda, Tanzania and Uganda (Figure 3.6). It is a depression with a surface of about 67000 km² at a level of 1134 m AMSL. The average depth of the lake is 40 m, with a maximum depth of 70 m. The mean annual precipitation over the lake catchment area is 1.15 m. The annual evaporation rate from the lake catchment area amounts to 1.12 m. The average annual discharge on the outlet of the lake is 23.5 km³. Three sources contribute to the net supply of Lake Victoria: the outflow of River Kagera, the direct precipitation on the lake surface, and the direct runoff of the catchment's land portion. Upper Victoria Nile is the only outlet from Lake Victoria through the regulated Owen Falls Dam in Jinja, Uganda (Hurst, 1965, Fahmy, 2006, NBI, 2012).

River Kagera

This basin River Kagera is shared by Burundi, Rwanda, Tanzania and Uganda (Figure 3.6). River Kagera is considered the remote and principal source for Lake Victoria, providing nearly a third of the lake's entire inflow. Major part of it is situated between 1200 and 1600 m AMSL, but the country level is at 2500 m AMSL in the west and rises to about 4500 m AMSL to form the peaks of the Mufumbiro Range. The Kagera basin is a complex of rivers and streams of varying order, which are intercepted and interconnected by lakes and swamps. Its discharge is rather low because of the swamps and lakes, which exist in the basin, and the considerable length of streams flowing in it. The mean annual discharge of river Kagera is six km^3 (Hurst, 1965, Fahmy, 2006, NBI, 2012).

Upper Victoria Nile

The basin of the Upper Victoria Nile is located in Uganda (Figure 3.6). The Victoria Nile flows out of Lake Victoria at the northern end over the Ripon Falls, which have been submerged since the construction of the Owen Falls Dam in 1954 into Lake Kyoga. The river is about 130 km long, and the difference in level between its inlet and outlet is about 102 m (Hurst, 1965, Fahmy, 2006, NBI, 2012).

Lake Kyoga

The catchment area of Lake Kyoga is located in Uganda (Figure 3.6). After a series of rapids, the Victoria Nile flows through the swampy Lake Kyoga. It is a shallow depression consisting of a number of arms, many of which are filled with swamp vegetation. The lake has an area of 6270 km^2, including the area of the surrounding swamps. Its depth varies from 3 to 7 m. The annual rainfall on the basin and the lake is 1.3 m. The annual evaporation rate is 1.2 m from the lake, and two meters from the swamps. The average discharge from the lake is 22.5 km^3 annually (Hurst, 1965, Fahmy, 2006, NBI, 2012).

Lower Victoria Nile

The basin of the Lower Victoria Nile (Nile Kyoga) is located in Uganda (Figure 3.6). It leaves Lake Kyoga at Masindi Port. It takes a natural course of a normal slope to a distance of 80 km until it reaches Kamdine. Then, it flows over many falls ending in the Murchinson falls at about 84 km from Kamdine. The difference in water level in the Lower Victoria Nile is almost 410 m. The total drop in water levels between the outlet of Lake Victoria and the inlet of Lake Albert is 514 m (Hurst, 1965, Fahmy, 2006, NBI, 2012).

Lake Albert

The catchment area of Lake Albert is shared by Democratic Republic of Congo and Uganda (Figure 3.6). It has a surface area of 5300 km² corresponding to an elevation of 617 m AMSL. The drainage area is 17000 km². The water depth reaches 50 m at some places in the lake. The annual rainfall on the basin is 1.26 m, while the annual rainfall on the lake surface is 0.81 m. The annual evaporation rate is 1.2 m. The total annual discharge downstream the exit of the lake is 26.5 km^3 (Hurst, 1965, Fahmy, 2006, NBI, 2012).

River Semliki

The basin of River Semliki is shared by Democratic Republic of Congo and Uganda (Figure 3.6). It feeds Lakes Edward and George and connects Lake Edward to Lake Albert, after a flowing distance of about 250 km down the Rift Valley to the west of the Ruwenzori hills. The difference in water level is 295 m. Most of the drop takes place over the rapids, which exist in the upper part of the river course. In the lower part, the river has a width of 150 m in flood reduced to 50 m at a low stage. The average depth of water in these two seasons is 5 m and 3.0 m, respectively. The average annual flow of river Semliki into Lake Albert is 6.5 km^3 (Hurst, 1965, Fahmy, 2006, NBI, 2012).

Lake Edward

Lake Edward is shared by Democratic Republic of Congo and Uganda (Figure 3.6). It is fed by many tributaries from the western slopes of Ruwenzori hills, in addition to some tributaries springing from the south and the east of the lake. The average water level of the lake is 912 m AMSL. Its surface area is 2200 km², while the basin area is 12000 km². The annual evaporation rate is 1.2 m, while the annual rainfall is 1.365 m (Hurst, 1965, Fahmy, 2006, NBI, 2012).

Lake George

Lake George is located in Uganda (Figure 3.6). It is fed by many tributaries springing from the eastern slopes of the Ruwenzori hills and diverts to the south to flow into the north side of the lake, in addition to some tributaries springing from the southern hills and going north to flow into the southern side of the lake. The lake's average water level is 912 m AMSL. The lake surface area is 300 km², and its catchment area covers 800 km². The annual rainfall amounts to 1.365 m while the annual evaporation rate is 1.2 m (Hurst, 1965, Fahmy, 2006, NBI, 2012).

3.1.2.2 Sudd and central Sudan basin

The River Nile is subjected to considerable loss of its discharge by spreading overlarge areas of swamps and lagoons, as shown in Figure 3.7. The two main swamps are located in the Bahr El-Jebel and the Bahr El-Ghazal basins (Fahmy, 2006).

Bahr El-Jebel (Sudd Region)

These swamps located in south of Sudan are called Bahr el Jebel or Sudd Region (Figure 3.7). The area of the Bahr el-Jebel swamps is about 7200 km². The annual evaporation rate from the swamps is about two meters. Thus, the Sudd region causes the loss of 17 km^3 of water on average. This evaporated water is equivalent to 20% of the total mean annual flow of the River Nile at Aswan (Hurst, 1965, Fahmy, 2006, NBI, 2012).

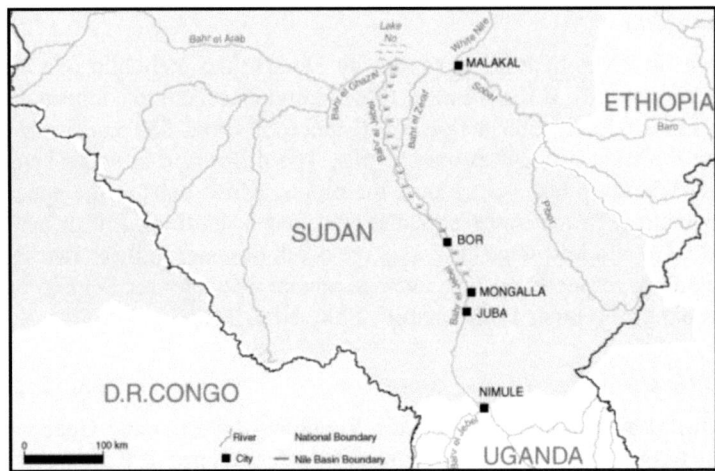

Figure 3.7: Map of the Sudd and Central Sudan Basin (Karyabwite, 2000)

Bahr El-Ghazal (Sudd Region)

The area of the Bahr el-Ghazal basin is estimated to be about 526000 km², and the area of the swamps is estimated to be about 40000 km², as shown in Figure 3.7. There are large areas of swamps on the lower courses of all the tributaries of the Bahr el Ghazal and along the Ghazal itself. Most of the flow carried by the tributaries is lost there. The annual rainfall at the southern part of Bahr el-Ghazal is 1.3 m, and it decreases gradually towards the north until it reaches 0.3 m in the northern part of the basin. The rate of evaporation is about two meters per year. The net flow into the White Nile after the Sudd regions is estimated to be about 15 km³ (Hurst, 1965, Fahmy, 2006, NBI, 2012).

3.1.2.3 White Nile

The stretch of the Nile from Lake No down to its junction with the Blue Nile is known as the White Nile. This river has an extremely flat slope. In the upper 120 km from Lake No to the mouth of the Sobat River, there are several swamps, khors and lagoons. In the remaining 800 km, from Malakal to just upstream Khartoum, the channel of the White Nile is almost free of swamps. The drainage basin of the White Nile extends from the foothills of the lake plateau in the south to the junction of the White and Blue Nile up north and from the foothills of the Abyssinian (Ethiopian) Plateau in the east to the Nile-Congo basins division in the south-west and the Nuba Mountains in the west, as shown in Figure 3.7 (Hurst, 1965, Fahmy, 2006, NBI, 2012).

3.1.2.4 Ethiopian plateau basin

The three major tributaries of the River Nile that originate in the Ethiopian Plateau are the River Sobat, Blue Nile and River Atbara, as shown in Figure 3.8.

Figure 3.8: Map of the Ethiopian Plateau (Karyabwite, 2000)

River Sobat

River Sobat is formed by the junction of its two main tributaries, the Baro and the Pibor. The River Baro, claimed to be the principal feeder of the Sobat, is formed by a number of streams that originate in the Abyssinian Plateau. The Pibor draws the greater part of its supply from the Abyssinian Plateau and the rest from the northern slopes of the Equatorial Plateau and from the Sudan plains. The average annual discharge of River Sobat, at its outfall into the White Nile, is 13.5 km^3. The basin area of River Sobat and its tributaries is estimated at about 187200 km². The rate of annual rainfall on the basin varies between 0.80 and 1.00 m, while it reaches up to 2.00 m on the high land. The average rate of evaporation is 1.42 m annually (Hurst, 1965, Fahmy, 2006, NBI, 2012).

Blue Nile

The Blue Nile, starting from Lake Tana in Ethiopia, receives its water from numerous streams along its course, and by the time it goes to the Roseries Dam, the discharge amounts to 50 km^3. It then receives two more tributaries, the

Dinder three km³ and Rahad one km³ before joining the White Nile at Khartoum, as shown in Figure 3.8 (Hurst, 1965, Fahmy, 2006, NBI, 2012).

River Atbara

River Atbara is the last tributary of the Nile, and it is a strongly seasonal river that enters the Main Nile at about 320 km downstream Khartoum (Figure 3.8). It is 880 km long, and the greater part of its catchment is situated in Ethiopia and Eritrea. River Atbara is fed by two tributaries, namely the Bahr el-Salam and River Steit. The area of the Atbara Basin is 143600 km². The mean annual discharge of River Atbara at its outfall on the main Nile is about 12 km³ (Hurst, 1965, Fahmy, 2006, NBI, 2012).

3.1.2.5 The main Nile

Once the Blue Nile joins the White Nile at Khartoum, the river is known as the Main Nile, as shown in Figure 3.9. Between Khartoum and Aswan, the river crosses six cataracts where the river is not completely navigable (Fahmy, 2006). The first storage facility in the Nile Valley was the old Aswan Dam (OAD). This dam, together with the other storage facilities on the Blue and White Niles, has changed the Nile between Aswan and the Mediterranean Sea into a partially regulated river instead of a naturally flowing one. Full regulation has almost been achieved with the formation of High Aswan Dam Reservoir (HADR) upstream of the High Aswan Dam (HAD) at Aswan in 1965. This huge artificial impoundment of Nile water extends from Aswan to a place a little south of the Dal Cataract. In its natural condition, the length of the river from Aswan to the Delta barrages was 968 km in the low-flow season and 923 km in the flood season. The mean width was approximately 900 m and the mean velocity between 1 to 2 m/s (Fahmy, 2006, NBI, 2012, Karyabwite, 2000).

The discharge of the River Nile at Aswan is subject to wide seasonal variation. The average annual discharge of the Nile at Aswan is 84 km³. This average was estimated as the annual average river inflow during the period from 1900 to1954. The highest recorded discharge was 150 km³ in the water year 1878/1879, and the lowest was 42 km³ in 1913/1914. The water year begins in August with the start of the flood season. About 80% of the total annual discharge is received during the flood season (August to October), and the remainder is spread over the rest of the year, as shown in Figure 3.10. About 86% of this annual discharge originates in Ethiopia and 14% in the Lake Plateau, which is the main source of water to the Nile during the low flow period (MWRI, 2005, NBI, 2012).

Figure 3.9: Map of the Main Nile Area (Karyabwite, 2000)

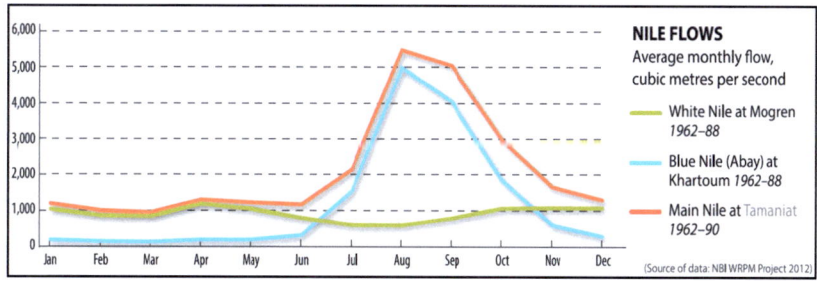

Figure 3.10: Average Monthly Nile Flow (NBI, 2012)

3.1.3 Nile Basin climate

The north–south orientation of the River Nile through 36 degrees of latitude has an extreme influence on the variability in climate. The North countries of Egypt, South Sudan and Sudan are characterized by extreme aridity and extensive desert while in the South and East, strong rainfall results in green vegetation, humid conditions and, in some locations, even tropical rainforests. The temperature varies widely across the Nile Basin, ranging from cold conditions in the desert winter nights, where temperatures fall as low as 4°C and mountainous areas where even freezing conditions are encountered, to searing heat, often exceeding 40°C in the desert regions. In some areas, especially in parts of the Equatorial Lakes Regions, temperatures vary little from month to month, while in the desert regions, they vary widely throughout the year. Most of the basin experiences only one rainy season, typically during the summer months. Only the equatorial zone has two distinct rainy periods. The reliability and volume of precipitation generally declines moving northwards, with the arid regions in Egypt and the northern region of the Sudan receiving insignificant annual rainfall (Camberlin, 2009, Melesse, 2011, NBI, 2012).

The basin receives some 650 mm of rainfall annually corresponding to around 1900 km^3 of water annually. Long-term mean flow at Aswan is only approximately 84 km^3 per year, making the annual runoff coefficient of the basin around 4.5% (Mutua et al., 2005, Nicol and Mamdouh, 2003). This is explained by the fact that a significant portion of the basin is comprised of arid and hyper-arid zones that are large in surface area, yet contribute only negligibly to basin runoff. Additionally, the evaporation losses from major swamp areas cause up to 30% of the basin's rainfall to be lost (NBI, 2012). The distribution of the mean annual rainfall and evaporation rates over the basin vary considerably from upstream to downstream, as shown in Figure 3.11 and Figure 3.12.

The highest rainfall occurs on and around Lake Victoria. High values are also observed along the Nile-Congo and south of Lake Tana. North of the section between Malakal and Roseires, the annual amounts of rainfall reduce drastically to fairly nil north of Dongola till Cairo, where they rise slightly toward the Mediterranean Sea. On the other hand, the evaporation rates increase within the arid zones. The high potential evaporation values in the Nile region ranging from some 3000 mm/year in northern Sudan to 1400 mm/year in the Ethiopian Highlands, and around 1100 mm/year in the hills in Rwanda and Burundi make the basin particularly vulnerable to drought events (NBI, 2012).

The monthly rainfall distributions over the basin are shown in Figure 3.13, Figure 3.14, and Figure 3.15 (Sutcliffe and Parks, 1999). In the upper part of the White Nile basin, the Equatorial Lakes basin located farthest to the south, maximum rainfall occurs in April, with a lower maximum between September and November. Northwards, the months with maximum rainfall are in July-August. In addition, the total amount reduces drastically towards the north (Camberlin, 2009, Melesse, 2011, NBI, 2012).

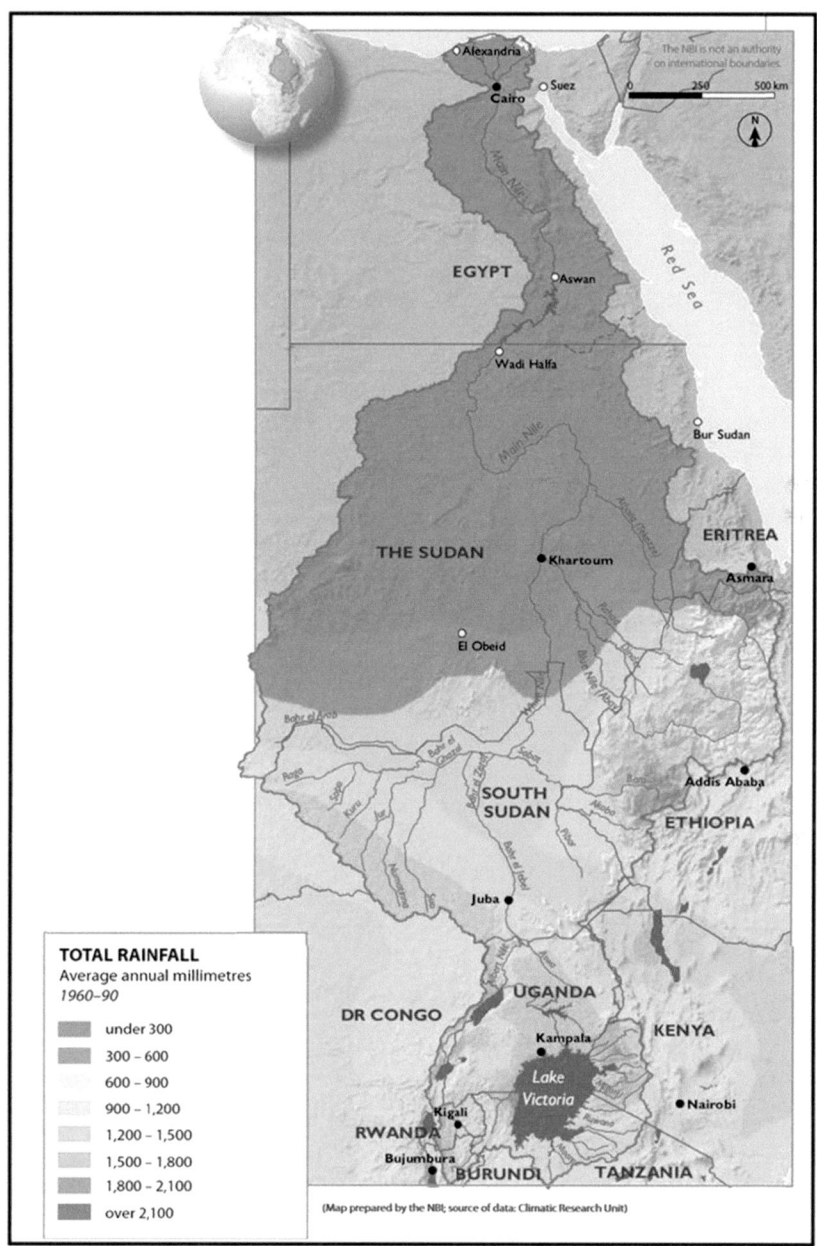

Figure 3.11: The Average Annual Total Rainfall in mm from 1960 to 1990 (NBI, 2012)

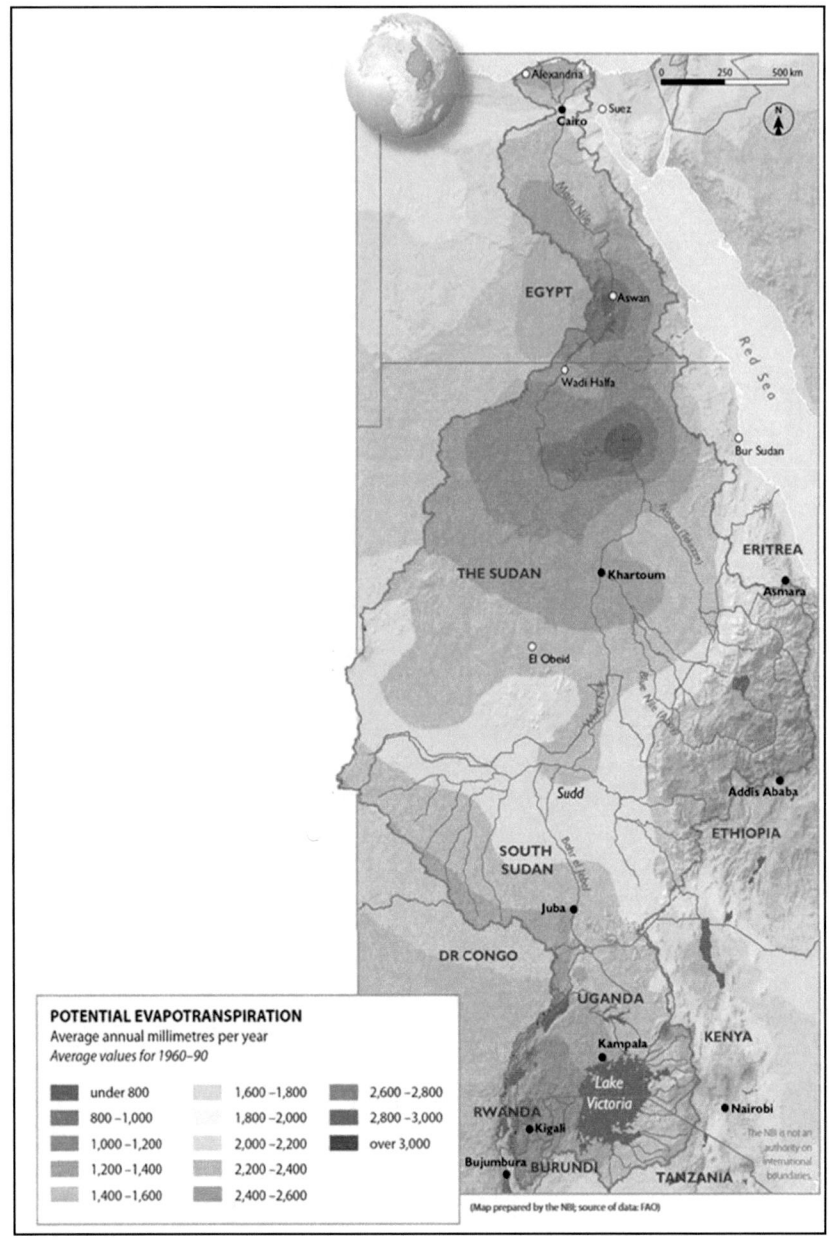

Figure 3.12: The Average Annual Potential Evapotranspiration in mm/ year from 1960 to 1990 (NBI, 2012)

Physiographic, Demographic, and Historical Background   73

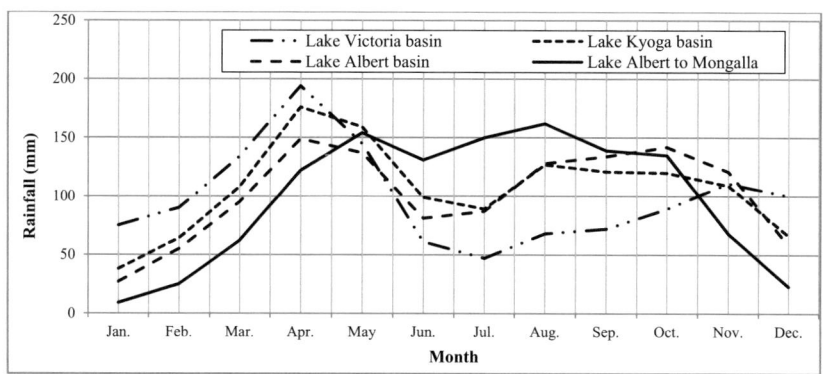

Figure 3.13: Average monthly rainfall over the upper parts of the White Nile basin (Hurst and Black, 1943, Sutcliffe and Parks, 1999)

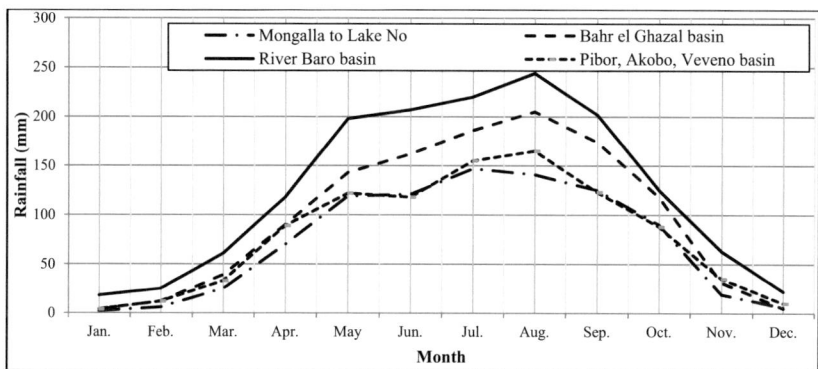

Figure 3. 14: Average monthly rainfall over the middle parts of the White Nile basin (Hurst and Black, 1943, Sutcliffe and Parks, 1999)

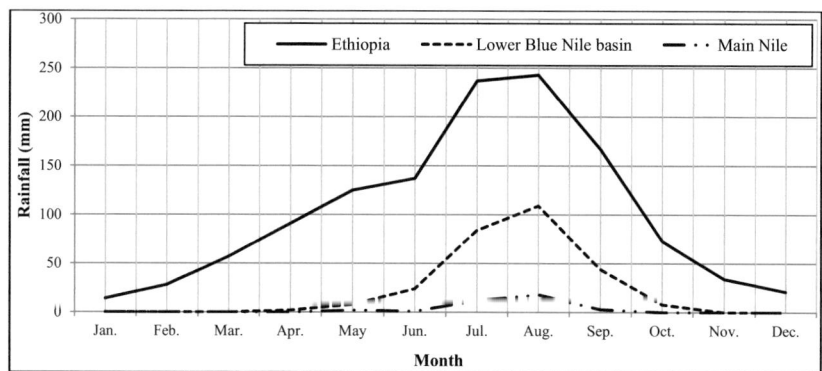

Figure 3.15: Average monthly rainfall over the basins of the Blue Nile and the main Nile (Hurst and Black, 1943, Sutcliffe and Parks, 1999)

3.1.4 Nile water agreement

Egypt, being at the downstream of the Nile basin, has been cooperating with the other countries for many years in agreements. These agreements aim to secure Egypt's share in the Nile Water. The first agreement between Nile Basin countries was signed in 1891 (Maru and Teklehaimanot, 2012). The most important agreements are listed in Appendix A. An important step for this cooperation has been technical cooperation. This started in 1967 in the form of a joint project between Egypt, Sudan, Uganda, Kenya and Tanzania on hydrological and meteorological studies in the basin. The project was joined by other countries at a later stage. In 1992, the ministerial council of the Nile Basin countries initiated the Technical Cooperation Committee for the Promotion of the Development and Environmental Protection of the Nile Basin (TECCONILE) to investigate and study all technical aspects related to the Nile Basin. The TECCONILE continued until 1999 when it was replaced by the Nile Basin Initiative (NBI) as a wider framework with different operating organs aiming at paving the way for a final installation of legal and institutional arrangements. Egypt has been a strong supporter of the Nile Basin Initiative and is planning on continuing to do so (IWLP, 2008, NBI, 2012).

In 1959, Egypt and Sudan signed an agreement about sharing the water of the Nile called the 1959 Nile Water Agreement. The agreement assumes a long-term natural annual flow at Aswan of 84 km^3. According to the 1959 Nile Water Agreement, Egypt is entitled to use 55.5 km^3 and Sudan 18.5 km^3 of this annual yield assuming an evaporation loss in Lake Nasser of 10 km^3 per year. If the average yield increases above these average figures, the increase would be divided equally. Significant decreases would be divided according to the ratio between the shares of Sudan and Egypt. Funding for any project that increases Nile flow (after the High Dam) would be provided evenly, and the resulting additional water would be split evenly. Egypt agreed to the payment of fifteen million Egyptian pounds to Sudan as full compensation for flooding and relocations of Sudanese property, resulting from the storage of water in the HADR to a level of 182 m AMSL. A Permanent Joint Technical Committee (PJTC) to resolve disputes and jointly review claims by any other riparian countries has been established. The Committee has also determined allocations in the event of exceptional low flows (FAO, 1997b, IWLP, 2008).

3.2 Egypt

Egypt is one of the oldest water civilizations, with a millennia-old culture based on the power and fertility of the Nile. It was at the forefront of a series of huge water projects. Today, it holds the key to the future of water sector development in much of eastern and central Africa. The importance of the Nile in Egyptian politics, economics and culture is highlighted by the fact that 98% of the country's population live on just 4% of the land, almost exclusively along the narrow

Nile Valley and in the Delta. An overview of Egypt's geography, geology, morphology, demographics, flora and fauna and water resources is given below.

3.2.1 Geographic situation

Egypt is located in the northeastern fringes of Africa and is bordered by the Mediterranean Sea to the north, by Sudan to the south, by the Red Sea, Palestine and Israel to the east, and by Libya to the west (Figure 3.16). It lies between latitude 22° and 32° North with maximum distances of 1024 km from north to south and of 1240 km from east to west. Its total area is approximately one million square kilometers with a coastline of 3500 km on the Mediterranean and the Red Sea. The land elevations range from 133 m below sea level in the Western Desert to 2629 m above sea level on the Sinai Peninsula. Geographically, Egypt is divided into four main regions: the Nile Valley & Delta, the Sinai Peninsula, the Eastern Desert and the Western Desert. A summary for each region is given below (MALR, 2005).

3.2.2 Geology (Geomorphological Features)

Physiographically, Egypt is divided into six main geographic provinces: the Western Desert, the Nile Valley, the Nile Delta, the Fayum Depression, the Eastern Desert and Red Sea mountains, and the Sinai peninsula, as shown in Figure 3.17 (Panagos et al., 2011).

3.2.2.1 Western Desert

The Western Desert covers an area of approximately 700000 km^2. It lies between the Libyan border on the west and the Nile Valley and Delta on the east, and is bounded from north by the Mediterranean Sea and Sudanese border from south as shown in Figure 3.16 (Nour El-Din, 2013, Meissner and Wycisk, 1993). The Western Desert has the following geomorphologic and demographical characteristics:

- Arid climate with very rare rainfall and seasonally windy weather
- Presence of internal drainage lines
- Sandy wind as the principal geomorphic agent that accentuate its characteristic landforms
- Absence of high relief mountainous areas and prominent wadis except Gabal Uweinat at its southwestern corner
- Presence of a large number of depressions and oases (e.g. Siwa, Qattara, Moghra, Fayum, Bahariya, Farafra, Dakhla, Kharga, Kurkur and Dungul)
- Presence of several sand dune belts and sand sea (Ex. Abu Moharik Sand Dune, Ghorabi Sand Dune, El Hussein Sand Dune and Qazzun Sand Dune).
- People are living in the oases of the depressions cultivating plants depending upon the underground water and springs.

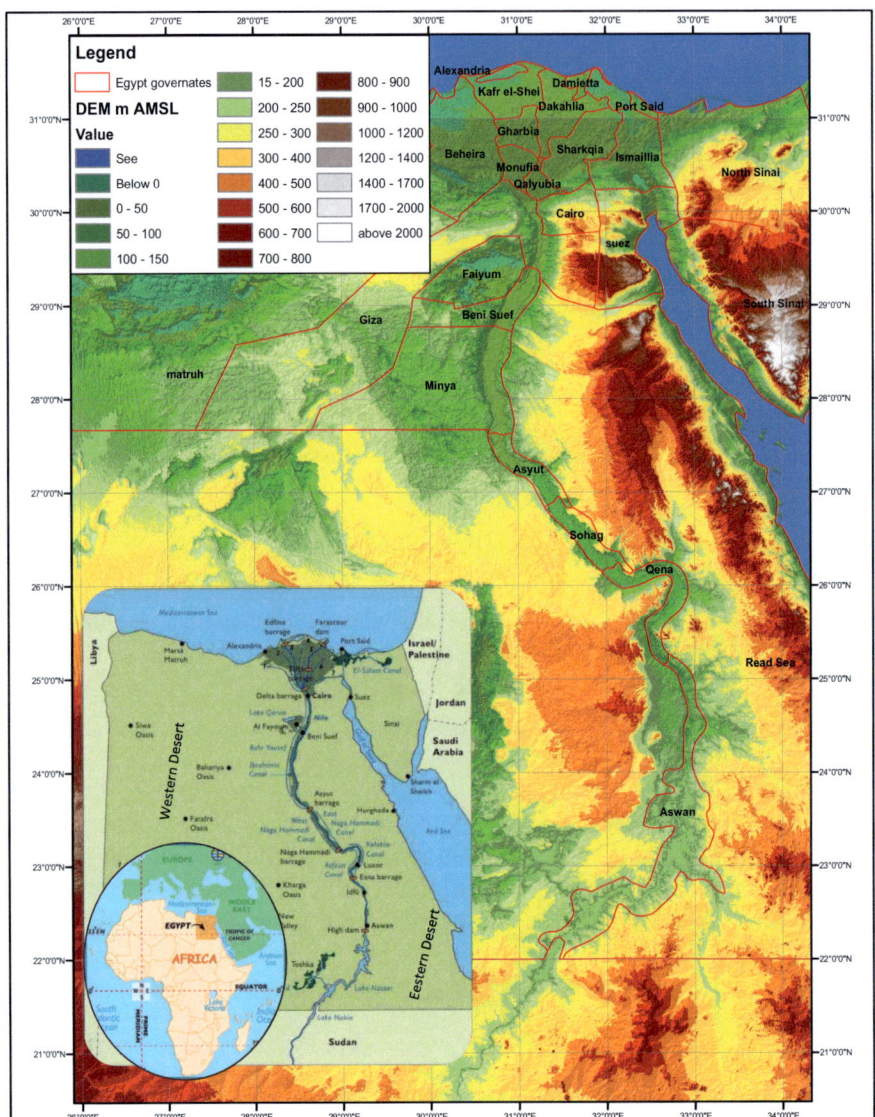

Figure 3.16: Digital Elevation model of Egypt

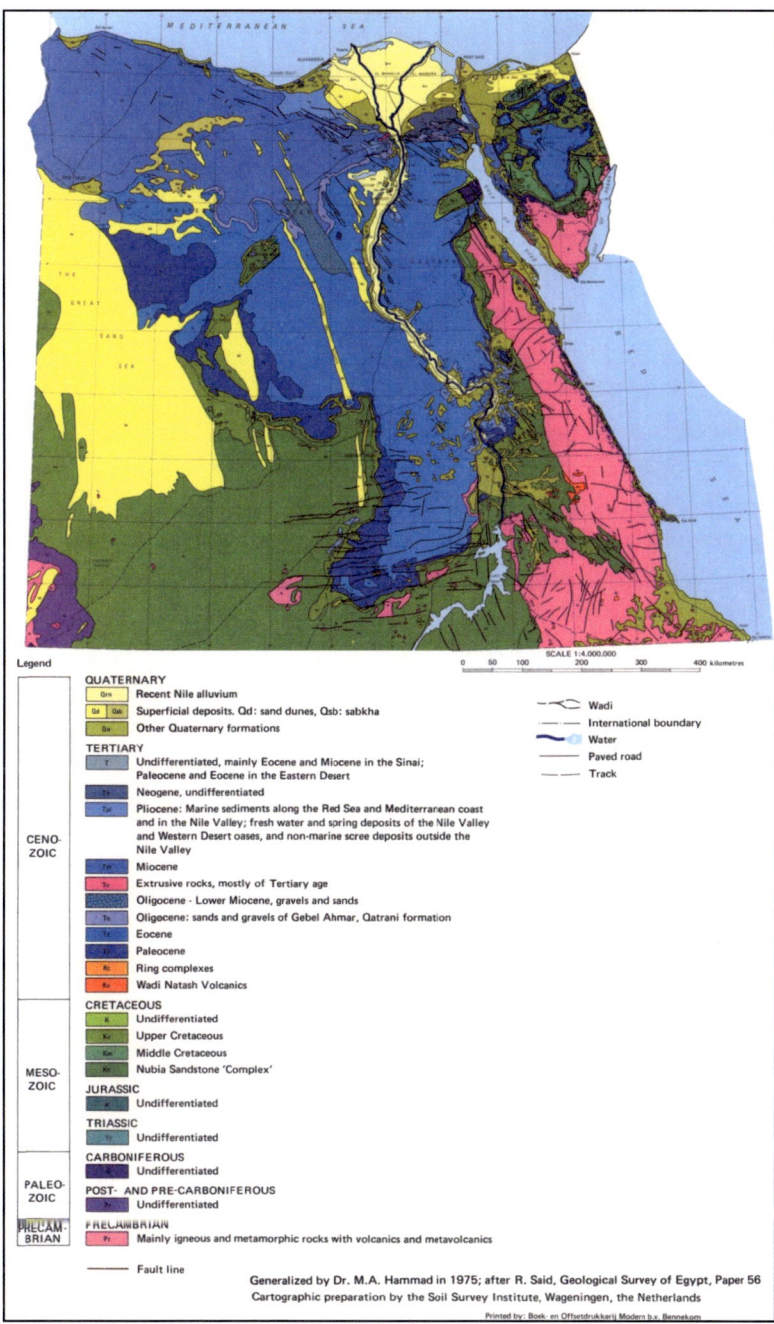

Figure 3.17: Geologic map of Egypt - Generalized by M., A., Hammad in 1975; after R. Said in 1962, Geological Survey of Egypt - Cartographic preparation by the Soil Survey Institute, Wageningen, the Netherlands (Panagos et al., 2011)

3.2.2.2 Western Desert

The Western Desert covers an area of approximately 700000 km^2. It lies between the Libyan border on the west and the Nile Valley and Delta on the east, and is borderedfrom north by the Mediterranean Sea and Sudanese border from south as shown in Figure 3.16 (Nour El-Din, 2013, Meissner and Wycisk, 1993). The Western Desert has the following geomorphologic and demographical characteristics:

- Arid climate with very rare rainfall and seasonally windy weather
- Presence of internal drainage lines
- Absence of high relief mountainous areas and prominent wadis except Gabal Uweinat at its southwestern corner
- Sandy wind as the principal geomorphic agent that accentuates its characteristic landforms
- Presence of a large number of depressions and oases (e.g. Siwa, Qattara, Moghra, Fayum, Bahariya, Farafra, Dakhla, Kharga, Kurkur and Dungul)
- Presence of several sand dune belts and sand sea (Ex. Abu Moharik Sand Dune, Ghorabi Sand Dune, El Hussein Sand Dune and Qazzun Sand Dune).
- People are living in the oases of the depressions cultivating plants depending upon the underground water and springs.

3.2.2.3 Eastern Desert

The Eastern Desert covers an area of approximately 22% of the surface area of the country, namely 147820 km^2 (Abdel Moneim, 2005). The Eastern Desert lies between the Nile Valley and Delta in the west and the Red Sea and Gulf of Suez in the east, as shown in Figure 3.16. Its north and south regions are bordered by the Mediterranean Sea and the Sudanese border, respectively (Nour El-Din, 2013, Abdel Moneim, 2005). The Eastern Desert has the following geomorphologic and demographical characteristics:

- It is a mountainous desert and rocky plateau land with high relief.
- Arid climate with winter rainfall.
- Presence of high relief mountainous areas with external drainage and prominent wadis draining either into Suez Gulf, Red Sea and Nile Valley (Wadi Araba, Wadi Abu Had, Wadi Eltarfa, Wadi Qena, Wadi Assuiti, Wadi Shait, Wadi Kharit and Wadi Elalaky).
- Rainfall, flash floods and wind are the principal geomorphic agents that accentuate its characteristic landforms.
- Absence of depressions and oases.
- People are living in the great wadis depending upon the heavy seasonal rainfall.

3.2.2.4 Nile Valley

The Nile enters Egypt at Adindan Village in Wadi Halfa (at the Egyptian - Sudanese border), and flows northward without receiving any tributaries until draining its load into the Mediterranean via the Rosseta and Damietta branches (Figure 3.16). The Nile has a meandering path with several islands; its valley has different widths and is drained by many large wadis (e.g. Wadi Kalabsha, Wadi Alaqi, Wadi Kharit, Wadi Shait, Wadi Assuti, Wadi Qena, and Wadi Tarfa). The valley is embanked with different rocks that vary from one place to another (Said, 1962). The Nile Valley has the following geomorphologic and demographical characteristics:

- From Adeindan, where the Nile Valley starts, to Aswan, the Nile cuts in the Cretaceous sandstone and shale rocks of the Nubia Group.
- At Kalabsha and Aswan, the Nile cuts through the Precambrian granites covered with thin sandstone beds. Isolated blocks of granites prevents the Nile course from forming cataracts (e.g. Aswan cataract)
- At North Aswan, steep scarps of Nubian sandstone rise, extend North-South, and border the Nile from both sides. These scarps include the economic iron ore of Aswan.
- At Komombo town (North of Aswan), the Nubian sandstone cliffs turn and extend east- westward, where the Nile forms the widest part of its valley known as the Komombo plain. The East-West Komombo plain has a flat surface and is about 20-25 m above the Nile water and is structurally formed like a drain.
- Going downstream from Idfu to Luxor, the Nile Valley is flanked by the Upper Cretaceous rocks capped by the Lower Eocene Carbonates.
- At Qena, the Nile forms its famous bend (the Qena bend) and flows from there to Cairo; it is bordered from both sides by the Eocene carbonates, which form the Mokattam and Giza Pyramid plateaus overlooking Cairo.

3.2.2.5 The Nile Delta

The Nile Delta covers a triangular area of approximately 21000 km^2; its apex is north of Cairo (at El Kanater El Khairia) where the Nile bifurcates into the Rosetta and Damietta branches (Figure 3.16). These branches are the remnants of six pre-existing branches that crossed the Delta at the beginning of the Holocene period and in historic times. The famous old branch is the Pellusi a branch that drained its load into Lake Manzala and in the Mediterranean Sea. Besides Lake Manzala, other lakes such as Lake Burullos, Lake Idku and Lake Marut, exist along the northern margin of the Nile Delta (Said, 1962, MALR, 2003).

3.2.2.6 Sinai Peninsula

The Sinai Peninsula is located at the northeast corner of Egypt. It has a triangular shape covering an area of approximately 61000 km^2 (MALR, 2005). It possesses geomorphological features similar to the Eastern Desert (Said, 1962, Nour El-Din, 2013).

3.2.2.7 The Fayum and Wad Rayan depressions

The Fayum and Rayan depression are connected to the Nile Valley and the Delta geomorphic unit because they are located close to the Nile Valley as shown in Figure 3.16. Further the Fayum depression is connected with the Nile by the water channel Bahr Youssif.

The Fayum depression has a total area of approximately 1700 km^2; Birket Qarun (-45 m, below sea level) occupies its northern part and is boarded by an elongate scarp of Gabal Qatrani from the north trending generally east- west. Lake Qarun occupied much more space in ancient times (pre-historic) proven by the presence of old raised beaches containing archeological evidence for ancient populations (implements) and is known as Lake (Moeris).

The Wadi El Rayan depression lies to the south of the Fayum depression, situated over -60 m below sea level. Now, it is connected with the Fayum depression by a subsurface canal in order to get rid of the drainage water of the cultivated lands of the Fayum depression instead of draining this water into Birket Qarun (Said, 1962, MALR, 2003).

3.2.3 Alluvial Soils of Egypt

According to the "Soil Taxonomy", Egyptian alluvial soils are classified as Vertisols. In general, Vertisols are the most difficult ones of the ten orders described in the Soil Taxonomy to manage. They are defined as "Mineral soils that have 30% or more clay, deep cracks when dry, and either (a natural) gilai microrelief, intersecting slickensides, or wedge shaped structural aggregates tilted at an angle from the horizontal". Different studies have shown that the abundant clay mineral (51% of the clay fraction) is montmorillonite. Other identified components are Kaolinite (31%), Mica (11%), Feldspars (2%) and Quartz (4%). This clay is mainly composed of fine clay fraction, causing large surface area for water absorption, thus creating a high volume of soil water. One of the main soil physical characteristics of these soils is swelling by wetting and shrinking by drying (Said, 1962, MALR, 2003).

The alluvial soils of the delta are stratified and of strong clay, which forms the majority, alternating with strong sandy or peaty horizons. In general, the profile consists of a heavy, very low permeable clay top layer of variable thickness underlain by a better permeable loamy, sandy or peaty layer. Soils in the old-irrigated areas in the Nile valley and delta are described as young alluvium. The depth varies between 5 and 50 meters depending on the location relevant to the main course of the river. Textures of the deep, homogenous profiles are generally light in the south and in fields close to the river and change gradually to heavy in northward areas and those away from the river's course. Salinity is generally low in the south but increase northwards towards the lakes and the Mediterranean Sea, where alkalinity is a characteristic and dominating feature (Said, 1962, MALR, 2003).

The World Reference Base for Soil Resources defines *"soils"* by the vertical combination of soil horizons, properties and/or characteristics occurring within a defined depth and by the vertical organization (sequence) of soil horizons. The main objective of the World Reference Base for Soil Resources is to provide scientific depth and background to the 1988 FAO Revised Legend, incorporating the latest knowledge relating to global soil resources and their interrelationships. According to the World Reference Base for Soil Resources, the main major soil groups of Egypt are: Arenosols (AR), Calcisols (CL) interfered with Gypsisols (GY), Calcisols (CL), Fluvisols (FL), Leptosolos (LP), Regosols (RG), Solonchaks (SC) and Vertisolos (VR) (Table 3.3) and the map in Figure 3.18 shows the distribution of major soil groups of Egypt (FAO, 2005).

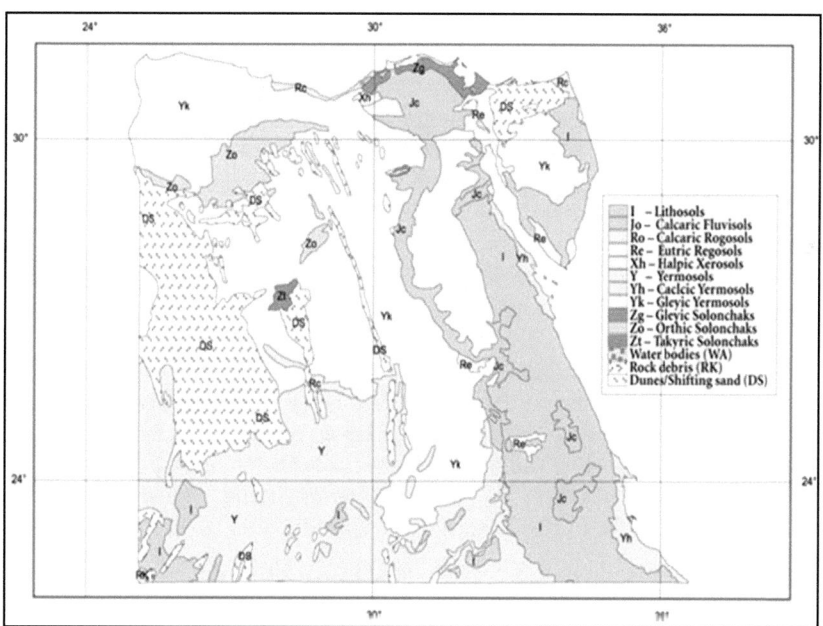

Figure 3.18: Soil map of Egypt (FAO, 2005)

Table 3.3: Egypt, Major Soil Groups and Non Soil Land Covers (%) (FAO, 2005)

Major Soil Groups and Land Covers Units	% Area Percentage
ARENSOLS (AR)	25.80
CALCISOLS (CL), INTERFERED WITH GYPSISOLS (GY)	0.37
CALCISOLS (CL)	9.12
FLUVISOLS (FL)	0.80
LEPTOSOLS (LP)	24.87
WATER BODIES	15.44
OUT SIDE STUDYAREA surveyed	9.59
REGOSOLS (RG)	8.68
SOLONCHAKS (SC)	0.48
VERTISOLS (VR)	4.85

3.2.4 Demographic data

The Population Division in the United Nations Department of Economic and Social Affairs, estimated the Egyptian population to be over 90 million in 2013, and its population is predicted to grow to approximately 123 million by 2050, and to 198 million by 2100, with an annual growth rate of about 1.8% (FAO, 2014, UN, 2011). In Egypt, the population density is high along the Nile, stretching from Aswan to the Mediterranean Sea. Virtually, all Egyptians (99.7%) live in the Nile basin (FAO, 2011). About 20% of the population is concentrated in the Greater Cairo area, 6% in the coastal governorates, 40% in the Delta governorates, 33% in the Upper Egypt governorates and 1% is distributed among the remaining areas of the country. The population density varies from 15000 to 0.4 person/km^2, in Cairo and the New Valley Governorate respectively (MALR, 2005). The growth of Egypt's population is among the most pressing challenges for the country. The population has grown from approximately 25 million in 1950 to 73 million in 2004 and keeps on growing with a growth rate between 1.88% (high variant) and 1.54% (low variant). An average growth rate of 1.66% is expected. Figure 3.19 illustrates the resulting population projections for 2017 based on these scenarios. For the middle scenario, a population of 83.1 million in 2017 was predicted based on observed trends until 1999. Current trends suggest that the middle scenario is no longer valid as the population has already reached over 90 million in the beginning of 2013, according to the Central Agency for Public Mobilization and Statistics (CAPMAS, 2013).

Most of Egypt's population lives in the small strip along the Nile and in the Delta, leading to an extremely high population density. This population density in Egypt has many consequences. From a water management point of view, these consequences include the increasing demand for good drinking water and the need of irrigation water to produce food.

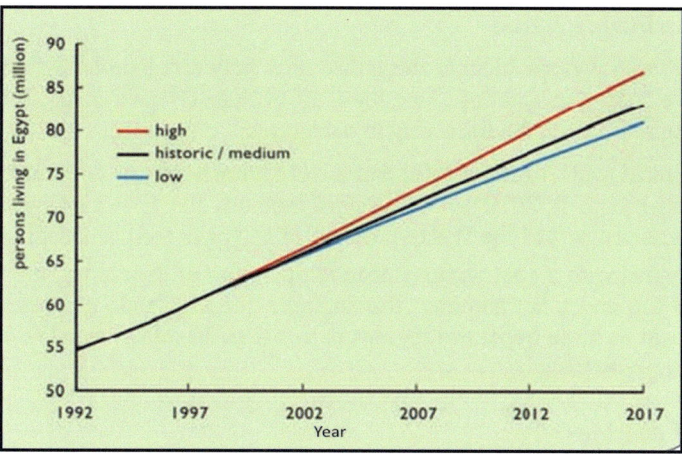

Figure 3.19: Expected population growth in Egypt (MWRI, 2005).

3.2.5 Climate

As shown in Figure 3.20, Egypt is located in the hyper-arid regions of North Africa and West Asia, ranging from the Sahara to the Arabian Desert, with annual rainfall rates in most parts of less than 50 mm (EEAA, 2001). According to the aridity index P/ETP (P = precipitation and E_{TP} = potential evapotranspiration, calculated by Penman's formula), the arid regions are classified as hyper arid (P/E_{TP} < 0.03) and arid (P/E_{TP} = 0.03 – 0.20) (MALR, 2005, Robert, 2006). These classes are, in turn, subdivided according to the mean temperature of the coldest month and that of the hottest month of the year. Consideration is also given to the time of rainy periods relative to the temperature regime (Maliva and Missimer, 2012). On these bases, four climatic regions (two hyper arid and two arid regions) in Egypt are distinguished (Hegazi et al, 2005).

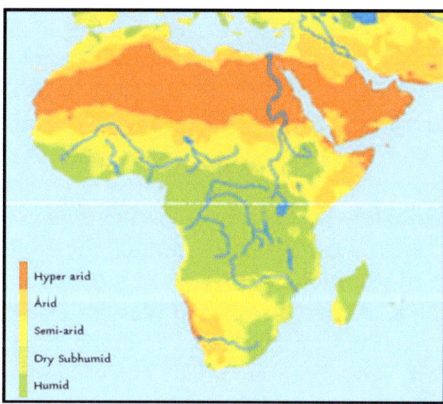

Figure 3.20: The hyper-arid regions of North Africa (Matsuura, 2003)

a) Hyperarid regions:

The hyperarid regions include the entire area between latitude 22° and 30°N, except the coastal mountains along the Gulf of Suez (Hegazi et al, 2005). These are distinguished into the following to categories:

- Hyperarid with a mild winter and a hot summer (mean temperature of the hottest month is 20°C to 30°C); this includes the Eastern Desert and the northeastern part of the Western Desert of Egypt as well as Gebel Uweinat.
- Hyperarid with a cool winter (mean temperature of the coldest month is 0°C to 10°C), and a hot summer; this includes the highlands of southern Sinai. The rain in these hyperarid regions is less than 30 mm yr^{-1} and is occasional and unpredictable.

b) Arid Regions:

The arid regions include the northern section with winter rainfall; they extend along the Mediterranean coast and the Gulf of Suez (Hegazi et al, 2005). This section is distinguished into two regions, namely:

- The coastal belt region under the maritime influence of the Mediterranean, with a shorter dry period (attenuated),
- The more inland region with a longer dry period (accentuated) and an annual rainfall from 20 to 100 mm. Both regions are characterized by a mild winter and a hot summer.

The main features of the basic climatic elements in these regions can be summarized as follows:

3.2.5.1 Temperature

Figure 3.21 shows the distribution of the average annual mean of daily temperatures along Egypt in °C. Generally, summers are hot (mean of the very hottest month ranges between 20°C and 30°C) or very hot (mean of the hottest month is more than 30°C). Winters are either warm (mean of the coldest month is 20°C to 30°C) or mild (mean minimum of the coldest month is 10°C to 20°C) except on the highlands where winters are cool with a mean minimum of the coldest month between 0°C and 10°C. The temperatures along the Red Sea coast vary between a mean minimum of the coldest month of approximately 10°C towards the North and approximately 20°C towards the South, and a mean maximum of the hottest month of approximately 33°C towards the North and 40°C towards the South. The range of variation becomes even greater further inland (from approximately 4°C to 38°C in the oases of the Western Desert). In continental locations, temperature extremes of less than 4°C in the coldest month (e.g. oases of the Western Desert) have been recorded. The coldest month is usually between December and February, and the hottest month is usually between June and August in hyperarid and arid provinces, respectively (Hegazi et al, 2005, EEAA, 2001).

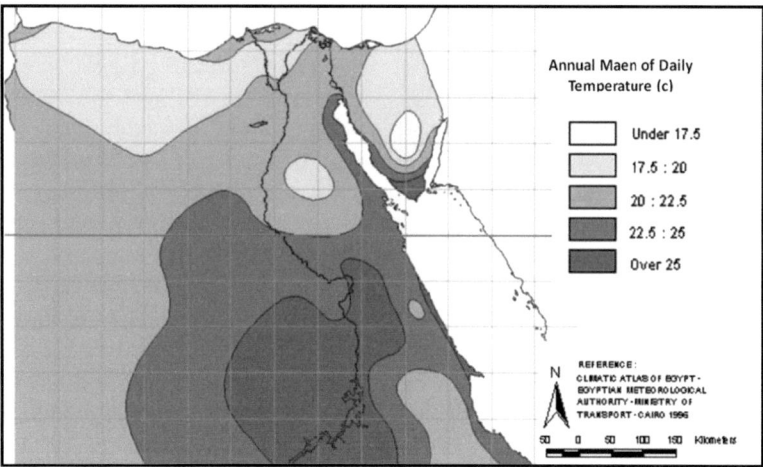

Figure 3.21: Average annual mean of daily temperatures (°C) (EEAA, 2001)

3.2.5.2 Relative humidity

Figure 3.22 shows the distribution of the average annual mean of daily relative humidity in Egypt in %. The relative humidity is affected mainly by the relative proximity to the Mediterranean and the Red Sea. The lowest records are those of inland locations in arid and hyperarid regions, and the highest ones are those of locations closer to the Mediterranean coast and in the Nile Delta within the arid regions (e.g. mean minimum of 38% in Aswan and mean maximum of 79% in North Sinia). The lowest records of relative humidity are generally measured in late spring, whereas the highest records are measured in late autumn and early winter (Hegazi et al, 2005, Khalil1 et al., 2011).

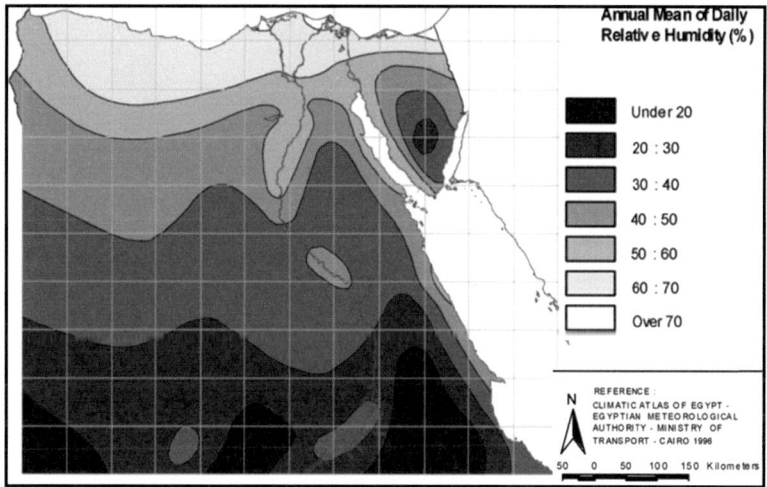

Figure 3.22: Annual mean of daily relative humidity (%) (Hegazi et al, 2005, EEAA, 2001)

3.2.5.3 Rainfall

Figure 3.23 shows the mean annual precipitation in mm along Egypt. In general, three rainfall belts may be distinguished in the deserts of Egypt. The first is the Mediterranean coastal belt, the second is middle Egypt with latitude 30°N as its southern boundary, and the third is upper Egypt. The first and second belt experience winter rainfall (Mediterranean regime). In this region, the rainy season extends from November to April, although it is mainly concentrated in December and January. These belts correspond roughly to the attenuated and accentuated arid regions of northern Egypt, where the average annual rainfall ranges from 100 to 150 mm in the attenuated arid province (FAO, 2005) and from 20 to 100 mm in the accentuated arid region. They extend rather south along the Gulf of Suez to latitude 26°N due to the orographic influence of the Red Sea coastal mountains. The third belt is almost rainless; it corresponds roughly to the hyperarid regions. Rain in this belt is not an annually recurring incident; 10 mm may occur once every ten years. The rainfall increases gradually to the North until it reaches approximately 20 mm at the borders with the arid regions (at Giza) (Hegazi et al, 2005, EEAA, 2001).

One of the major features of rainfall in arid and semi-arid regions other than being rare is its great temporal variability. The average deviation of annual precipitation from the mean, expressed as percentage of the mean, is greatest in the hyperarid regions (e. g. Siwa 83%). In the arid region, the percentage variability is 65% at Giza, which is close to the hyperarid regions (Hegazi et al, 2005).

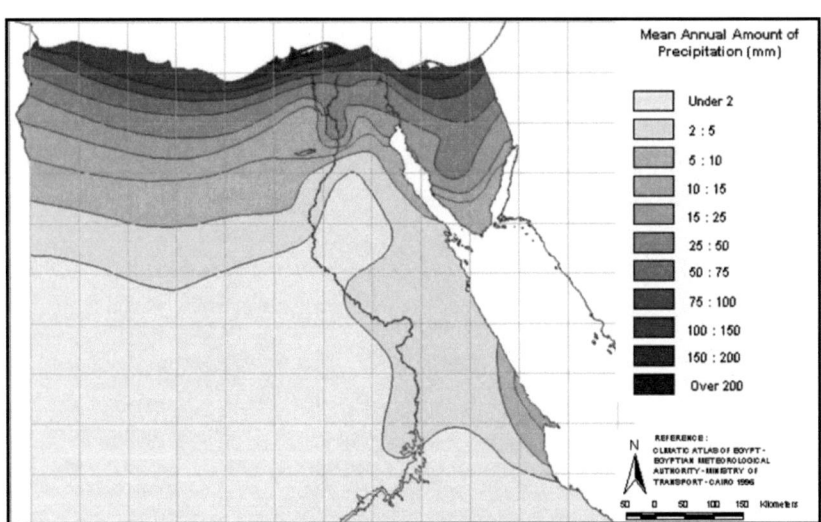

Figure 3.23: Mean annual precipitation (mm) (EEAA, 2001)

3.2.5.4 Wind

Figure 3.24 shows the predicted wind systems of Egypt determined by mesoscale modeling. In winter, the Sahara high-pressure system dominates the circulation, and the north wind northerlies bring cool dry air from the North African continental source region, although occasionally, the Arabian high brings warmer air to the eastern parts of the Sudan (MALR, 2005). Both of these types are occasionally interrupted by East - West depressions along the Mediterranean and replaced by cold dry air from the Eurasian landmass. In spring and autumn, the Arabian high is more dominant in the East, and the effect of the Mediterranean depression is rarely felt, as air from both the North African and the Arabian sources is considerably warmer than in winter. In summer, the Saharan high is again dominant, bringing hot dry air. Occasionally, very hot dust-laden winds blow (Khamaseen), which have numerous environmental effects on the climate, soil formation, groundwater quality and crop growth. They may create problems including substantial degrees of deflation and erosion.

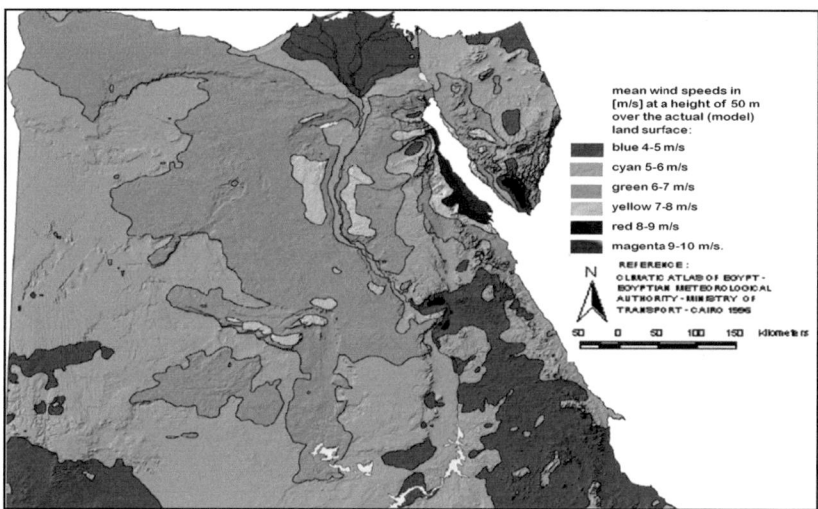

Figure 3.24: The predicted wind speeds of Egypt determined by mesoscale modeling (Mortensen et al., 2006)

The Egyptian Meteorological Authority and WMO have issued a Wind Atlas for Egypt. This includes a simulated wind resource map at the resolution of the model simulations (Mortensen et al., 2006) shown in Figure 3.24. The project results are available in the Wind Atlas for Egypt Measurements and Modeling 1991-2005. The Wind Atlas for Egypt confirms the existence of a widespread and particularly high wind resource along the Gulf of Suez. The Wind Atlas further indicates that the wind energy resource in the large regions of the Western and Eastern Desert – in particular west and east of the Nile valley between 27°N and 29°N, but also north and west of the city of Kharga – have mean wind speeds ranging between 7 and 8 m/s and power densities ranging between 300 and 400 W/m^2 (Mortensen et al., 2006, Hegazi et al, 2005).

3.2.5.5 Evaporation rates

Current evaporation rates in Egypt range between 7 mm/day in Upper Egypt to approximately 4 mm/day along the Northern Mediterranean coast. Table 3.4 illustrates monthly average annual potential evapotranspiration in the eight main agro-climatic regions of Egypt as determined by the Water Management Research Institute of the National Water Research Center (WMRI-NWRC) in 2002. Figure 3.25 illustrates the average monthly potential evapotranspiration rates for these agro-climatic regions in mm/day.

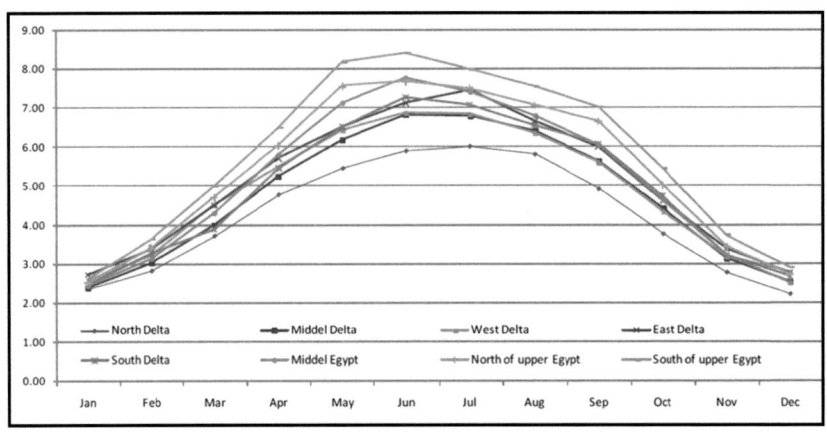

Figure 3.25: Potential monthly average daily evapotranspiration rates for the different agro-climatic regions in mm (MWRI – NWRC, 2002)

Table 3.4: Monthly average annual potential evapotranspiration rates in Egypt in the different agro-climatic regions in mm/day. (MWRI – NWRC, 2002)

Agro-Climatic Region	Mean Annual E_{To} (mm/year)
South Upper Egypt	1722
North Upper Egypt	1610
Middle Egypt	1531
South Delta	1485
East Delta	1522
West Delta	1457
Middle Delta	1417
North Delta	1266

3.2.5.6 Climate change's effect on Egypt

It is recognized that increased temperatures and changes in wind and humidity will affect the evapotranspiration. As reported by Eid (2007), a temperature rise of 1°C may increase the evapotranspiration rates by about 4% to 5%, while a rise of 3°C may increase the evapotranspiration rates by about 15%. This indicates that if the Egyptian agricultural sector is consuming 41 km^3, an increase of 1°C would lead to an additional amount of approximately 2.0 km^3 in evapotranspiration to maintain the same level of productivity. Besides, it was reported that a 10% increase in the annual evapotranspiration rate can result in a 6% decline of groundwater recharge (Eid et al., 2007).

3.2.6 Flora and Fauna

Until 2010, 30 protectorates and conservation areas in Egypt distributed all over the country have been declared, which cover more than 15% of the total area of Egypt, as shown in Figure 3.26. These are concentrated mainly in the Eastern desert and Sinai. The most important ones are: Ras Mohamed National Park, Zaranik protectorate, Ahrish protectorate, El Omayed protectorate, Elba National Park, Red Sea Islands (22 islands), Al Abraq protectorate, Al Daniep protectorate, Elba protectorate, Saluga, Ghazal protectorate, St. Katherine protectorate, Ashtum El Gamil protectorate. Lake Qarun protectorate, Wadi El Rayan protectorate, Wadi Alalaqi protectorate, Wadi El Asuity protectorate, Hassana Dome protectorate, Petrified forest protectorate, Sannur Cave protectorate, Nabaq protectorate Abu Galum protectorate, Taba protectorate, Lake Burullus protectorate, Nile Islands protectorate (144 islands), Wadi Digla protectorate, Siwa protectorate, White desert protectorate, Wadi El Gemal protectorate, and Hamata protectorate (EEAA, 2005). More information in summary about the flora and fauna in Egypt is given below.

3.2.6.1 Flora

Egyptian flora comprises about 2100 species and over 150 endemic species belonging to 121 families, unevenly distributed over the different agro ecological zones of the country. The highest species density is in the Mediterranean coastal zone, the Nile Valley, Gebel Elba and the mountains of Sinai (Figure 3.26). Egypt's flora is characterized by the large number of genera in proportion to that of the species, amounting to about three species per genus. This is a very low figure compared to the average global proportion, which amounts to about 14. The generic index, i.e. the number of genera per 100 species, is relatively high, which points to the marginal conditions in Egypt with respect to many genera and also indicates the lack of accumulation and differentiation centers in Egypt (MALR, 2005, EEAA, 2005).

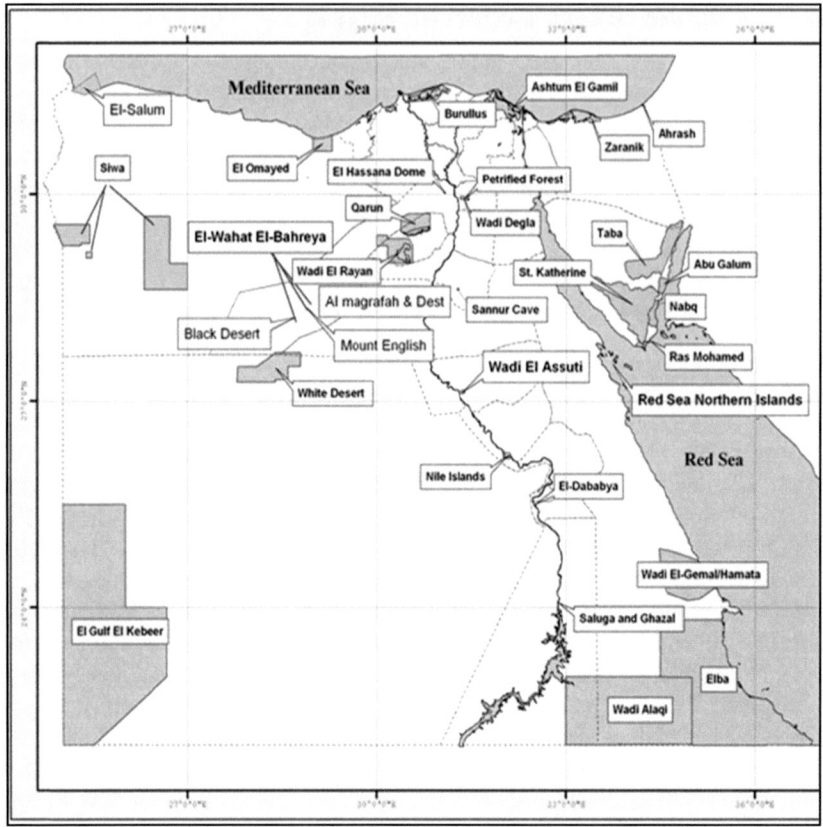

Figure 3.26: Map of Nature Protectorates (EEAA, 2005)

3.2.6.2 Fauna

The presence of various terrestrial and aquatic ecosystems in Egypt, e.g. the arid and semi-arid areas, the Mediterranean Sea, the Red Sea, the River Nile, lakes, ponds and oases, create a suitable environment for more than 175000 species of living animals, from invertebrates to vertebrates and from aquatic to terrestrial animals. For example, the herpetofauna includes 98 species of reptiles and 7 species of amphibians, including 51 species of lizards, 37 species of snakes, 8 species of turtles and one species of crocodilian. Amphibians include 4 species of toads and 2 species of frogs and one species of tree frogs. Until now, about 105 species of mammals have been detected, and about 515 species of birds, out of which 200 species are resident types and the others are migratory and about 200 species are considered as extinct. In addition, there are about 85 species of fresh water fish, of which 22 species are common, but 49 species are rare, and 14 species are considered as extinct (MALR, 2005, EEAA, 2005).

The continuous damage to the Egyptian fauna habitats has negative impacts on the national wild life. About 700 species out of 18000 species of the Egyptian fauna (EEAA, 2005) have become endangered over the last few years. Therefore, around 30 protectorates and conservation areas have been declared to protect the different flora and fauna in Egypt, as shown in Figure 3.26.

3.2.7 Land uses

More than 95% of Egypt's land is desert. About 80% of the remaining 5% is used for agriculture. Due to the arid climate, agriculture relies on irrigation. Agricultural land use is thus concentrated in the Nile River valley and delta, the oases of the New Valley supplied with ground water, and the Mediterranean coastal plains partly rain fed and partly irrigated with ground water or Nile River water. The total land area of Egypt is approximately 998000 km^2 (250 million feddan), about 3% of which have been cultivated. 0.4% is used for public utilities (roads, etc.) and buildings, 1.6% is considered as permanent cropland (covered with trees and shrubs) and woodlands, and the remaining 95% is classified as other land such as desert and semi-desert areas, salt flats, hills, and sand dunes. Land use change takes place all the time as a result of virtually every form of human activities, whether it is urban cover, industrial development, exploration and exploitation of non-renewable resources, development of transportation networks, or the setting aside of land for recreational or conversation purposes (MALR, 2003).

In Egypt, the availability of land has never been a problem. The main constraint has always been water to make the land productive. At the beginning of the 19th century, Egypt's cultivated area was estimated at two million Feddans, of which only approximately 250000 Feddans could be cultivated in the summer. By 1900, the cultivated area had increased to approximately five million Feddans (Biswas and Tortajada, 2004). Egypt is counted among the world's poor countries in cropland base. The presently cultivated area constitutes only about 3% of the total area of Egypt (245 million Feddans). Statistics from the Ministry of Agriculture and Land Reclamation (MALR) indicate an increase in the agricultural land: from 5.67 million Feddans in 1950 to approximately 6.62 million Feddans in 1982. According to an estimation of the MWRI, Egypt's cropland area consists of approximately 7.3 million Feddans, based on the annual quantities of irrigation water, in addition to an area of 0.2 million Feddans outside the Nile basin, in the Egyptian territory in 1991. Most of these croplands are classified under the second and third grades (45.5% and 38.5% respectively), representing good and fairly good lands. First grade lands constitute only about 9.2% of the total area of cultivated lands while low quality lands (fourth grade) constitute about 9.6% (MALR, 2003).

In 1996, arable land per capita in Egypt was 0.13 Feddans, which is the lowest in any African country. In 1970, shortly after the HAD became operational, the Egyptian population was 33.2 million, and the arable land was 6.0 million Feddans, which meant that the per capita arable land available was 0.18 Feddans. Thus, between 1970 and 1996, even though arable land increased from 6 million Feddans to 7.8 million Feddans, an overall increase of 30%, the Egyptian population increased from 33.2 million to 59.3 million, by 78.6%. This meant that in spite of the contribution of the Dam, arable land per capita has actually declined by 27.7% because the population increased at a much higher rate than the arable land. Thus, without the HAD, the agricultural and food situation in Egypt would have been significantly worse than what it is today (Biswas and Tortajada, 2004).

3.2.8 Water resources

Water resources in Egypt are limited to the Nile River, rainfall and flash floods, deep groundwater in the deserts and Sinai, and potential desalination of sea and brackish water. Each resource has its limitation to be used as irrigation or drinking water as well as for other possible purposes, whether these limitations are related to quantity, quality, space, time, or exploitation costs. On the other hand, Egypt's water demand is increasing rapidly due to population growth, technological developments, and climate change effects.

Egypt faces major challenges with respect to water management. From a quantity point of view, the country has to cope with a growing population while the water supply can hardly be increased. This leads to decreasing water availability per capita as indicated in Figure 3.27. The horizontal line in the figure at 1000 m^3/capita is often used to indicate water scarcity. The Egyptian government follows an integrated water resources management (IWRM) approach in developing and managing its water resources (MWRI, 2005, NWRP, 2005, MWRI, 2010). A summary of Egypt's water supply, demand and the resulting water balance is given below.

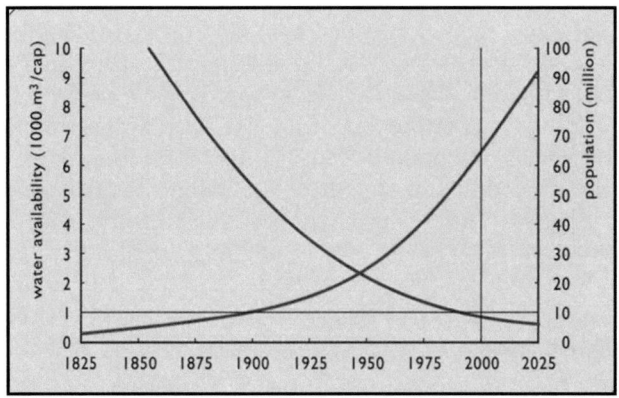

Figure 3.27: Population growth and water availability (NWRP, 2005)

3.2.8.1 The supply system

The water supply system in Egypt is divided into two main groups, mainly conventional and non-conventional water resources that overcome the gap between water supply and water demand as explained below. The conventional water resources include the Nile River, rainfall and flash floods, and deep groundwater in the deserts and Sinai, while the non-conventional water resources include reuse of drainage water, reuse of sewage water, and potential desalination of sea and brackish water. Figure 3.28 shows the quantities of Egypt's water resources (MWRI-PS, 2001, MWRI, 2010).

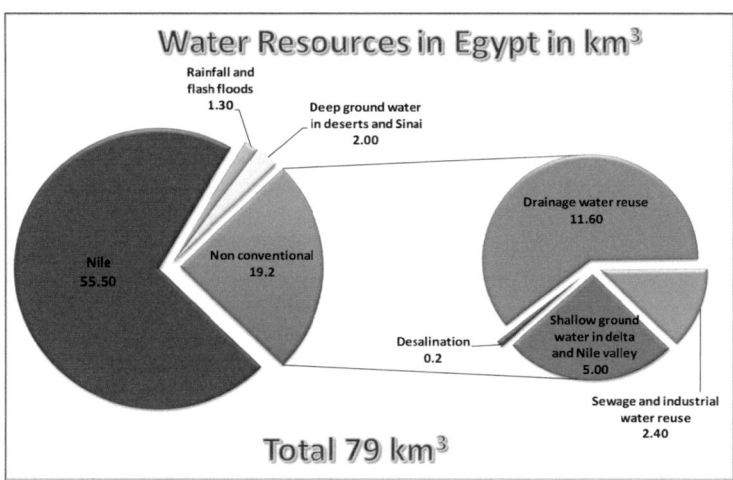

Figure 3.28: Water Resources in Egypt in 2010 (MWRI, 2010)

Nile River

Egypt relies on the available water storage of the HADR to sustain its annual share of water, which is fixed at 55.5 km^3 by the 1959 agreement with Sudan. This share accounts for over 96% of its fresh renewable water resources that come from outside its international borders. The average annual yield of the river is estimated at 84 km^3 at Aswan. This yield is subject to wide seasonal variation (MWRI-PS, 2001, MWRI, 2010).

Rainfall

Rainfall in Egypt is very scarce except in a narrow band along the northern coastal areas, where an insignificant rain-fed agriculture is practiced. Rainfall occurs in winter in the form of scattered showers along the Mediterranean shoreline. The total amount of rainfall is around one km^3 annually. Flash floods occurring due to short-period, heavy storms are considered more a source of environmental damage than a water resource, especially in the Red Sea area and Southern Sinai (MWRI-PS, 2001, MWRI, 2010).

Deep Groundwater in deserts and Sinai

Groundwater is an important source of fresh water in Egypt, both within the Nile system and in the desert. Groundwater exists in the fossil deep aquifers in the Western Desert region and Sinai with the current total estimated rate at only two km^3 per year. On the other hand, most of the available groundwater in the desert is non-renewable and associated with high development costs (MWRI-PS, 2001, MWRI, 2010).

Shallow groundwater in Nile Valley and Delta

One of the main non-conventional water resources in Egypt is the shallow groundwater in Nile valley and delta. Groundwater in the Nile aquifer cannot be considered an additional source of water as it gets its water from percolation losses from irrigated lands and seepage losses from irrigation canals. Therefore, its yield must not be added to the country's water resources, but rather must be considered as a reservoir in the Nile River system with approximately five km^3 annually of rechargeable live storage (MWRI-PS, 2001).

Drainage water reuse

One of the major non-conventional water resources is the reuse of drainage water in the Nile Delta. It has been adopted as an official policy since the late seventies. The policy calls for recycling agricultural drainage water by pumping it from main and branch drains and mixing it with fresh water in main and branch canals. It is estimated to yield 11.6 km^3, including the unofficial reuse in areas with irrigation shortage problems (MWRI-PS, 2001, MWRI, 2010).

Sewage and industrial water reuse

One of the minor, non-conventional water resources is the reuse of treated domestic and industrial wastewater. It is used for irrigation with or without blending it with fresh water. The increasing demands for domestic water due to population growth and improvement in living standards and the growing use of water in the industrial sector due to the expansion in the Egyptian industry will increase the total amount of wastewater available for reuse. For example, in 1996, 0.7 km^3 of treated sewage water were used in irrigating approximately 2500 Feddans northeast of Cairo, while 2.4 km^3 of treated sewage and industrial water were used in 2010 (MWRI-PS, 2001, MWRI, 2010).

Desalination of Sea Water

The desalination of seawater in Egypt has been given low priority as a source of water because the cost of treating seawater is high compared to other sources. However, it is sometimes feasible to treat seawater to provide domestic water especially in remote areas where the cost of constructing pipelines to transfer Nile water is relatively high. Egypt has approximately 2400 km of shorelines on both the Red Sea and the Mediterranean Sea. Therefore, desalination is used as a sustainable water resource for domestic uses in many locations such as tourism villages and resorts where the economic value of the unit of water is high enough to cover the costs of desalination. The desalinated water resources were estimated to be 0.2 km^3 in 2010 (MWRI, 2010, NWRP, 2005).

3.2.8.2 Different water uses

Egypt's water demand increases rapidly due to the increase in population and the improvement of living standards as well as the government policy to reclaim new lands and encourage industrialization. Demands for water can be categorized into three main categories as shown in Figure 3.29.

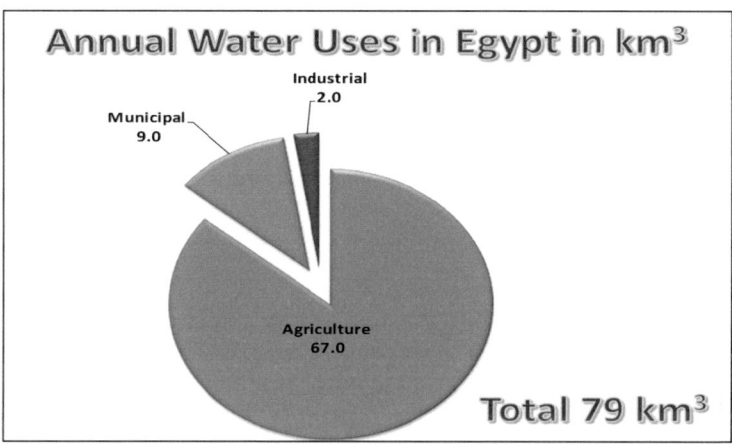

Figure 3.29: Water uses in Egypt in 2010 (MWRI, 2010)

Agriculture

The agricultural sector is the largest user and consumer of water in Egypt, with its share exceeding 85% of the total demand for water. The agricultural land base consists of old land in the Nile Valley and the Delta, rain fed areas, several oases, and lands reclaimed from the desert. The total area of irrigated land in the year 2010 was approximately 8.6 million Feddans and is expected to increase to 11 million Feddans (4.6 million hectares) by 2017 due to horizontal expansion and the implementation of the two mega projects of the El-Salam Canal at North Sinai and Toshka at south valley (NWRP, 2005). Consequently, the agricultural demand for water is expected to increase from 67 km^3 in 2010 to approximately 72 km^3, due to the increase of irrigation efficiency by extending the irrigation improvement projects to cover most of the old lands and applying modern irrigation techniques, e.g. sprinkler and drip irrigation, in the new reclaimed lands (MWRI, 2010).

Municipalities

Compared to the agricultural water demand, the municipal water demand is small, but given the health aspects involved, this supply will receive priority over all other users. The government policy with respect to drinking water plans to provide full coverage to both urban and rural areas including further improving the quality of the services. The estimated value was approximately nine km^3 in 2010 and is expected to increase to reach 16 km^3 annually by 2050 (MWRI, 2010).

Industry

Industry is a growing sector in the national economy of Egypt. There is no accurate estimate for the current industrial water requirement. However, in 1990, the general authority for industry conducted a survey that covered 90% major factories of the public sector to estimate industrial needs and requirements. The industrial requirements reached approximately two km^3 in 2010 exclusive the water used in the cooling system (MWRI, 2010).

Navigation

The main inland waterway is the Nile and few other main canals. Inland waterways are used by traditional sailing boats to transport building materials, by river barges and hotel boats. The main navigation activity consists of Nile touristic cruises between Aswan and Luxor and the transportation of commodities between Upper and Lower Egypt. Approximately 0.26 km^3 was released in 1996 to facilitate the navigation on the Nile. Due to change in the water management, no more water is required for navigation purposes (MWRI-PS, 2001).

3.3 The High Aswan Dam (HAD)

The annual flow of the Nile varies considerably from year to year and from one season to another. This varying supply can make it difficult to meet the irrigation requirements, threatening Egypt with water shortage, as happened in the year 1913/1914, when the annual supply did not exceed 42 km^3. On the other hand, the supply may be so high, exposing Egypt to destructive inundations, as happened in the year 1878/1879, when the supply was 151 km^3. The regulation of the varying natural flow of the Nile River to reasonable limits has been the major problem faced by the Egyptians for centuries.

The solution of this problem has been achieved in consecutive steps. Barrages have been constructed to bring the water to command levels. Later on, reservoirs of limited capacities were built for annual storage, particularly to store parts of the late flood waters to supplement the deficient supply in the following season. These reservoirs used to offer partial protection against dangers of high floods, but did not offer complete control of the Nile waters. The Nile's average annual discharge amounts to 84 km^3 (1901/1902 – 1953/1954) out of which 52 km^3

were used for irrigation purposes in Egypt and Sudan. This means that in a normal year, the wasted amount that has been discharged into the Mediterranean Sea was as high as 32 km^3, a little over a third of the total average yearly supply (UNESCO, 2008, NBI, 2012).

The total annual "River Supply" varies from 60 to 150 km^3, with an exception in the year 1913/1914, when the supply was only 42 km^3. The "natural River" during the summer months, from the beginning of February to the end of June, varies from 5 to 26 km^3. No sound policy of development can be based on such a variable supply. Consequently, the idea of over-year storage has been introduced. Extensive research and vast studies have been carried out since 1920. In 1946, the MWRI issued volume VII of the Nile Basin, in which the idea of over-year storage was fully studied and established on a scientific basis (MWRI, 2000).

Thus, in order to safeguard against individual low years and a succession of low years as well, over-year storage was found to be the best solution. It insures the storage of all water surpluses in excess of requirements in years of high floods to be used in years of low floods. For this purpose, a large reservoir is necessary since a considerable capacity to cope with deposited silt and an additional capacity for flood protection has to be added to the over-year storage and annual working range (MWRI, 2000).

3.3.1 Preliminary investigations about building the HAD

On October 19, 1952, a preliminary report of the HAD scheme was prepared by the Nile Control General Inspectorate. The object of the report was to assess the validity of two propositions made by two engineers, Adrian Daninos and Luigi Galioli. The report called for further technical research as well as for studies of the economic aspects of the project (Biswas and Tortajada, 2004).

A committee consisting of members from the Armed Forces General Headquarters, the MWRI, university professors and American Technical AID was assigned to select appropriate sites between Aswan and Halfa for the prospective dam. The committee selected two sites with two alternatives for each. The first site was located between 5.0 and 6.0 km while the second site was located between 14.0 and 14.6 km south of the Old Aswan Dam. Both sites were to be the subject of comprehensive research and study. Investigations were carried out by German engineers of Hochtief, Dortmunder Union with the assistance of Egyptian engineers of the MWRI. A committee headed by the Minister of Public Works and professionals from the National Production Council supervised the research and investigation (Biswas and Tortajada, 2004).

By the end of February 1953, the chairman of the International Bank for Reconstruction and Development (IBRD), accompanied by a group of bank specialists, paid a visit to Aswan with the purpose of inspecting the investigations completed by that time. They were impressed by the idea of the project and its potential for the economic progress in Egypt (Biswas and Tortajada, 2004). By mid-1959, the final design of the project was completed after introducing certain modifications to the original design. The site was then prepared to start the execution, and the first stone was laid on January 9, 1960, after 7 years of investigations, research, experiments and studies (Biswas and Tortajada, 2004).

3.3.2 Location

The site of the HAD is located 6.5 km south of the old Aswan Dam (OAD). This site was considered most suitable and appropriate due to the relative narrowness of the Nile's course, its closeness to the sources of raw materials necessary for construction, the huge storage reservoir, the area's topography, which would decrease the volume of excavation work needed, as well as the New Port Sudan being closer to Aswan city. The HAD and OAD are actually part of one system. The HAD has the main function in this system by storing the water in High Aswan Dam Reservoir (HADR) and by producing hydropower. The OAD is used to re-regulate the HAD releases on a daily and weekly basis. It also generates some electricity. Figure 3.30 shows the locations of HAD, and OAD and the city of Aswan.

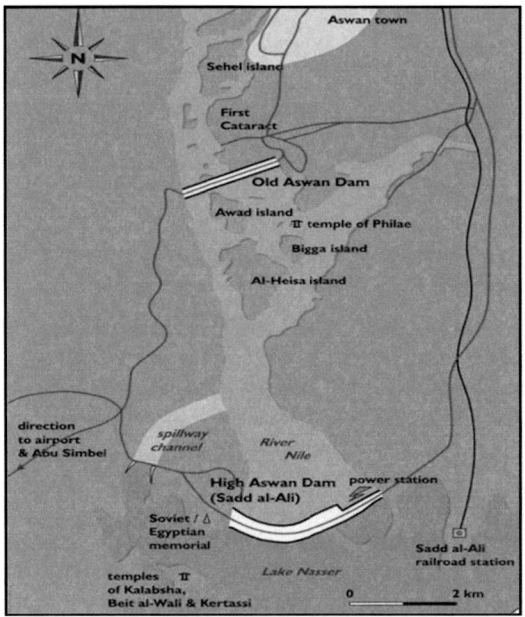

Figure 3.30: Location of the HAD and OAD (NWRP, 2005)

3.3.3 The design of the HAD

The design criteria of the dam were developed according to the Egyptian Government's requirements, results obtained from earlier investigations and research accumulated from 1959 to 1962, and experimental work carried out on the site. Figure 3.31 shows the cross section of the HAD. The design requirements were:

- A rockfill dam of a height of 111 m, crest elevation 196 m AMSL and top width of 40 m.
- The dam is provided with at least two systems of anti-seepage protection. The first is a dam core and grout curtain that ensures damping of about 70% of the head. The second is the core blanket, the sand prism in the profile, the prism of graded stone sluiced with sand and the clay part of the fore-apron.
- On the western bank of the Nile, there is an additional blind spillway with an elevation of water flow 180 m calculated for a discharge capacity of 2300 m^3 s^{-1} at a storage elevation of 182 m AMSL.
- The smallest safety factor for the upstream slope is 1.8 as a whole and 1.5 for the headwater part located below at 130 m AMSL. For the downstream slope, it is 1.7 as a whole and 1.5 for the head water part located below the berm 114 m AMSL.

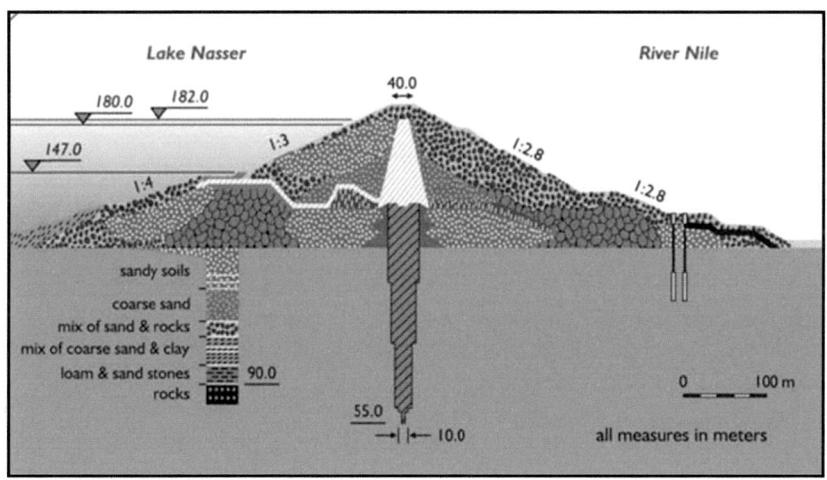

Figure 3.31: Cross-section of the High Aswan Dam, generalized by NWRP (2005) after Abul-Atta (1978)

Seepage studies and tests of the HAD were performed by the method of electro-hydro dynamical analogy on two-dimensional models. According to the test data, at normal operation, all of the anti-seepage structural elements and drainage, the filtration regime in the body, and the foundation of the dam are absolutely safe. About 23% of the seepage head is damped in the borders of prism of low permeable material, 59% of the head is damped at grout curtain (the actual head loss is 96%). The head gradient inside the sandy and sandy-rock prisms of the dam on the downstream side does not exceed 1:10. Each of the anti-seepage protection elements lines separately. In case of failure of any element, the other layers will possess sufficient stability to protect the dam from destruction. At the complete loss of the grout curtain effectiveness, the head loss in the borders of upstream prisms of low-permeable material increases to 55% and 40% on the downstream prism. In a dam without a prism of low-permeable material, without grout relief wells and without a downstream zone of rockfill sluiced with sand, the filtration conditions were also satisfactory. The drop of seepage head at the curtain equals to 76%.

The rockfill dam has a length of 3830 m of which 530 m are within the river channel and the rest in the shape of two wings along both sides of the river. The length of the right wing is 2520 m while the left wing is 780 m. The dam width at the bottom of the river bed is 980 m and 40 m at the crest. The height of the dam above the river bed is 111 m. The bulk volume of materials used in building the HAD reached 43 hm^3, about 17 times the size of the great Giza Pyramid "Cheops". The body of the dam is constructed of granite blocks, sand and clay, in the midst of which is a clay core to prevent seepage of water. The core is connected in the upstream part with a horizontal blanket of clay for the same purpose.

Since the Nile bed, on which the dam is built, consists of loose sedimentary deposits, the dam is provided with a vertical injected curtain extending 170 m under the main core until it reaches the solid impermeable geological formation. The injected curtain has been built of special materials consisting of Aswan clay and other chemical materials in order to prevent the seepage of water. The width of the injected curtain is 40 m under the main core and decreases until it reaches five m at the point where it meets with the solid formation.

The core is penetrated by three galleries, constructed with reinforced concrete. During construction, the galleries were used for completing the vertical injected curtain; now, they are being used for inspection and maintenance purposes. Various measuring devices have been installed in these galleries to measure vertical and horizontal movements, pore pressure in clay and seepage. The dam has, below the end of its toe, a row of vertical relief wells to drain the water that might seep through the dam, while another designed row will be executed if necessary (Abul-Atta, 1978).

The HAD (emergency) spillway is a safety canal on the western side of the dam, making a direct link between HADR and the pond between the HAD and the Old Aswan Dam (Figure 3.32). The spillway is blocked by a regulator under the road that runs over the dam (Figure 3.33). The sill of the weir lies at a level of 178 m AMSL. Although this level has already been reached in the past, the spillway has never been used. In case of extra water, it has been released downstream the HAD; this has been done through the turbine tunnels in the dam to utilize the spillage in generating electricity.

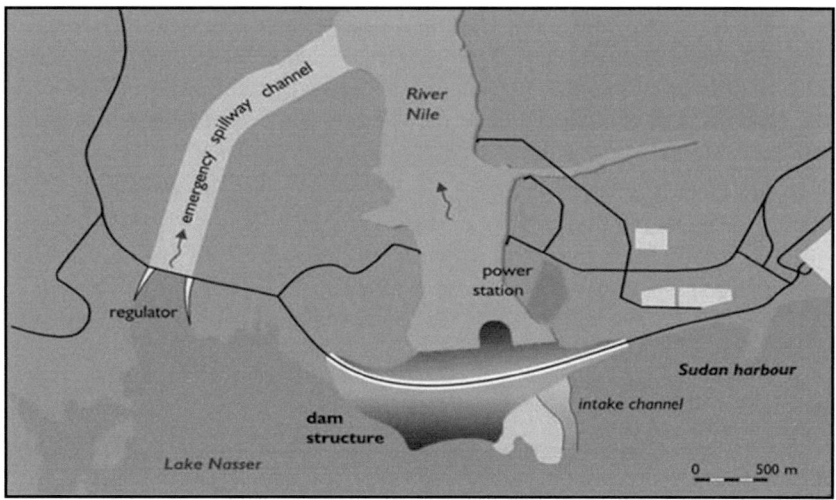

Figure 3.32: Map of HAD facilities (NWRP, 2005)

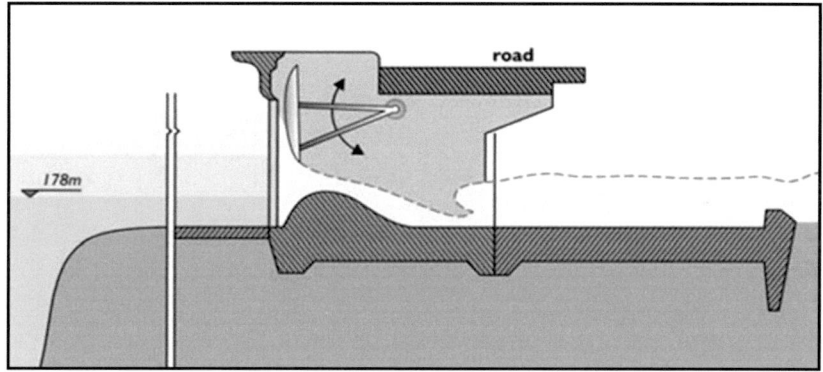

Figure 3.33: Cross section of the emergency spillway (NWRP, 2005)

Six spillway tunnels have been constructed to link the upstream and downstream canals. The average length of each tunnel is 282 m with 15 m circular cross-sections of internal diameter lined with reinforced concrete of a minimum thickness of one meter. Each of the six spillway tunnels has two inlets, a down inlet and an upper inlet. The down inlet starts from the upstream canal bed level. It was temporarily used during construction and then blocked with reinforced concrete before the completion of the dam and the filling of the reservoir with water to use only the upper inlets permanently. The gates of the down inlets moved to the upper inlet after the blocking. The upstream of each inlet is divided into two openings of five m width, provided with a separate gate lifted by an electric crane of 450 tons capacity. Each tunnel is divided vertically into two branches before its connection with the electric power station. These branches are divided again by a horizontal wall into two water passages, one of them supplies water to generate units and the other is controlled by sector gates for passing the surplus water needed during the period of peak water requirements. The six tunnels were designed to release a discharge of 11000 m^3 s^{-1}, which stands for approximately one km^3/day (Abul-Atta, 1978).

The hydro-electric power station was built at the outlets of the tunnels. It contains twelve generating units of the Francis type, each with a capacity of 175000 KW. The total generating capacity is 2.1 million KW, producing ten billion KWH annually. A transformer station was installed to raise the electric power from 15.75 KV to 500 KV for transmission to Cairo and 132 KV for local distribution in the area. The electric power generated at Aswan is transferred to Cairo by two high voltage transmission lines of 500 KV (MWRI, 2010; INTEA, 2010).

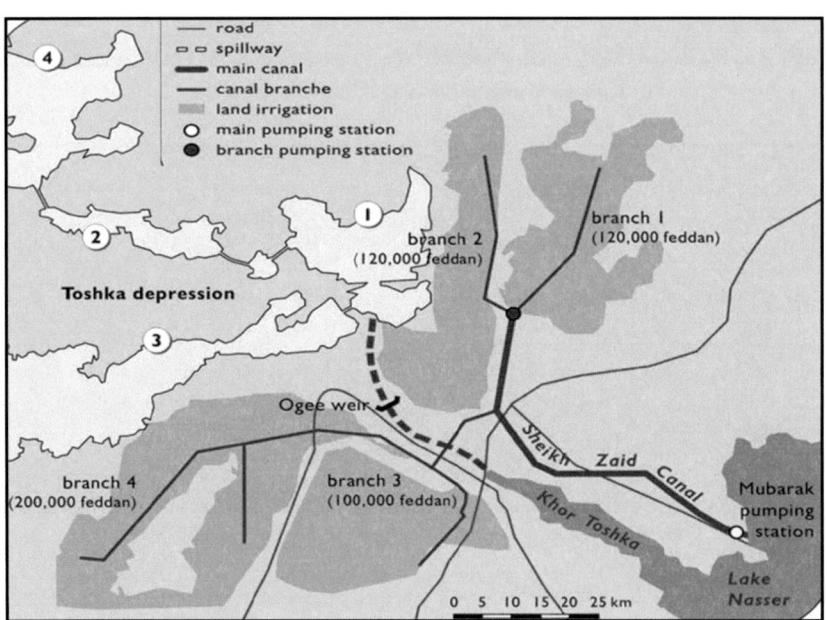

Figure 3.34: Toshka spillway and depression (including the Toshka South Valley project) (NWRP, 2005)

The Toshka spillway plays an important role among the overall spilling facilities of the HADR. The Toshka spillway is an emergency spillway in addition to the spillways available at the HAD. Since 1998, the spillway was used for four consecutive years. The Toshka spillway has three components, namely Khor Toshka, Toshka Channel, and Toshka Depression (Figure 3.34). Khor Toshka and Toshka Channel convey the water to the Toshka Depression when the reservoir levels exceed 178 m AMSL. The Toshka Depression is located approximately 250 km south of the HAD. It consists of four relatively deep cut single basins, separated from each other by natural sills as illustrated in Figure 3.35. The storage volume at a level of 170 m AMSL is approximately 124 km³, with an area of over 6000 km². If filling continues, the basins will become connected and will form one big lake. Ultimately, the depression will be full and start overflowing from basin 4 towards the New Valley (see Figure 3.16) or from basin 3 into the desert. Dams (e.g. at the end of basin 4 - see Figure 3.35) and the dredging of the sills are considered to control the filling of the basins and prevent damage in the New Valley area (NWS, 2012, NWRP, 2005).

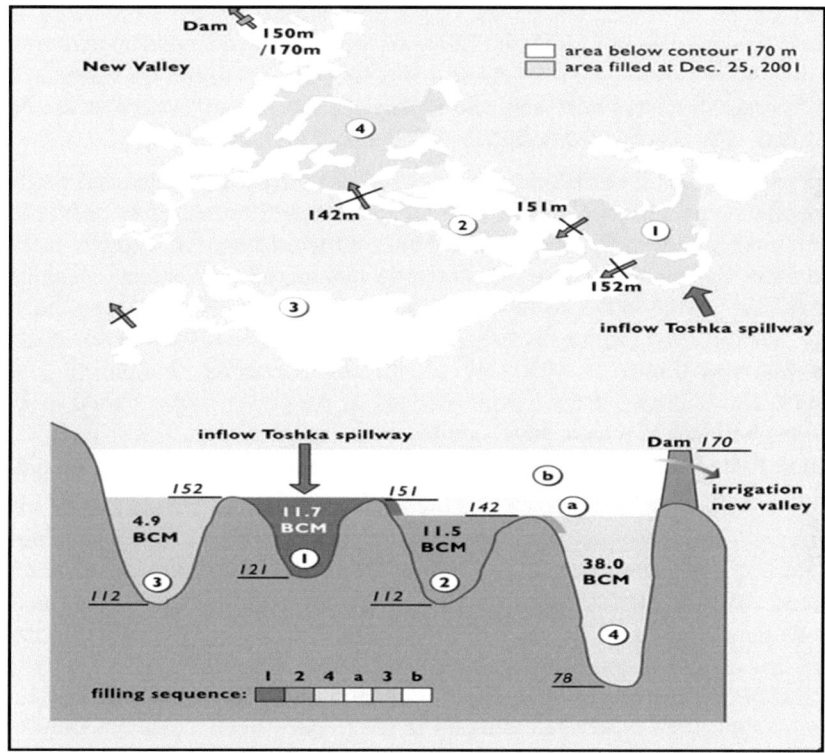

Figure 3.35: Overview over the Toshka depression and filling sequence in times of flooding (NWRP, 2005)

3.3.4 Construction

In the first stage of construction, a portion of the dam was built across the channel of the Nile up to an elevation of 130 m as well as water passing structures on the right bank in order to close the river and store 4.0 km³ of water in the reservoir during the flood of 1964 in addition to the ordinary storage of 5.0 km³ in front of the old Aswan Dam (from 1960 to 1964).

In the second stage, the dam was filled up to the design elevation, completing all the civil and building works on the hydroelectric scheme, and 12 units of the power house were put into operation (from 1967 to 1970). A summary of technical data of the dam's design is shown in Table 3.5 (Abul-Atta, 1978). Table 3.6 shows the quantities of work divided between the two stages.

Access roads and dwelling settlements were the first facilities constructed during this period. The railway from Shellal station to the construction site and 40 km long motor roads on both banks were completed. The work on the main structures started in 1960 with soil and rock excavation from the canals and the power station pit. In 1961, construction of the tunnels and the intake structures started. At the beginning of 1963, the filling of the dam and concrete operations had started. In May 1964, the river channel was already closed, and the filling of water passing structures with concrete took place for passing waters of the Nile. The dam crest reached the design level in 1968 (Abul-Atta, 1978).

The passing of water through the structures was rather complicated and was performed with careful preparation. First, in 1964, the entire run-off passed through the temporary lower openings of the intake structures and the tunnels. In 1965 and 1966, the concrete and water discharge was turned to the upper openings of the intake. Owing to the accumulation of water in the reservoir during the very high flood in 1964 (with a discharge of 12000 m^3 s^{-1}), the discharge downstream the dam was limited to 10500 m^3 s^{-1}, thereby preventing flooding of Lower Egypt. The building of the turbine stations of the power house started in 1965 and the building of the electrical equipment in the power house with an installation in 1966. The power station has been operated partially since October 1967.

HAD construction had offered employment opportunities for thousands of laborers. Work for the project and its training centers created a new generation of engineers, technicians and skilled workers who learned specific techniques and gained valuable experience. This labor force received attractive opportunities inside the country and abroad after HAD completion. Highly qualified firms in civil and electrical engineering were established, benefitting from teams who gained great experience during the construction of HAD (Abul-Atta, 1978). Table 3.7 shows the number of workers in the project while Table 3.8 shows the distribution of these workers.

Overall, the construction of the HAD has contributed to the economic growth of the Aswan Governorate through large-scale activities in different fields. The necessary infrastructure has been provided in the form of education, health care and transportation. As a result of the rapid development in tourism, industry, fishing, and services, the Aswan Governorate has not suffered from unemployment and offers many employment opportunities for professionals from other parts of the country. Before the construction of the HAD, the Aswan Governorate was subject to a decrease in population through immigration. Now, the Aswan area has changed to an attractive area for work and production instead of being an area that saw little economic influx (NWS, 2012).

Table 3.5: Summary of Technical Data of the Project (Abul-Atta, 1978)

1. Hydraulic data about the Nile:	
Annual maximum discharge at Aswan	150.0 km^3
Annual minimum discharge at Aswan	42.0 km^3
Annual average discharge at Aswan	84.0 km^3
2. The Dam:	
Length of the dam at crest	3830 m
Length of the dam within the river bed	530 m
Width at the base	980 m
Width at the top	40 m
Main level of the river bed	85 m. AMSL
Level of the dam's top	196 m. AMSL
Height of the dam over the river bed	111 m
Depth of vertical curtain under river bed	170 m
Width of vertical curtain at the top	40 m
3. The Diversion Canal:	
Maximum discharge of water	11000 m^{3}s^{-1}
Length of the upstream canal	1150 m
Length of the tunnels and power station	315 m
Length of the downstream canal	485 m
Total length of the diversion canal	1950 m
4. The Tunnels:	
Number of the main tunnels	6
Internal diameter of the main tunnels	15 m
Thickness of concrete lining	1 m
The average length of the tunnel	282 m
The designed discharge of tunnel	1 km^3/day

Table 3.6: The quantities of work divided between the two stages (NWS, 2012)

Operation	Volume %	
	First stage	Second stage
Excavation of soil and rock	85	15
Construction of various embankments	37	63
Usage of concrete	49	51
Erection of hydro mechanical equipment and steel structure	18	82

Table 3.7: Number of workers on the project (NWS, 2012)

Department / Company	Average number of workers at the beginning of the year								
	1961	1962	1963	1964	1965	1966	1967	1968	1969
Construction Authority	70	100	150	238	306	592	571	521	500
Hydro-mechanization Dept.	178	203	545	842	690	292	2	-	-
Tunnel Dept.	118	402	2946	1976	1413	723	-	-	-
Injection Dept.	68	150	449	758	974	1887	1639	1278	1157
Mechanical Erection Dept.	67	132	962	1251	1352	1374	1501	1382	1152
Spets-Hydro-Power-Erection Dept.	-	-	-	-	-	-	228	267	262
Chief Electric and Hydro-Energy-Erection Dept.	330	450	977	788	742	715	744	755	639
Mechanical and Transport Dept.	800	1200	2624	2568	2950	2845	3025	2891	2659
Misr Concrete Company	700	1000	1247	8565	5257	5620	6124	3816	3753
Arab Contractors Company	2316	6807	9621	9516	9686	9084	8420	7779	7743
Total	4647	10444	19221	26502	23370	23132	22258	18689	17865

Table 3.8: The Distribution of Workers (NWS, 2012)

Description	Number of workers			
	July 1961	January 1962	July 1962	January 1963
I. Major Workers	1748	2897	5366	5951
At construction of canals	896	1420	1850	2400
At construction of auxiliary and housing facilities	104	367	1866	1050
At operation of the auxiliary facilities	492	1110	1650	2001
II. Auxiliary Workers and Apprentices	568	837	1436	2086
Total	2316	3750	6802	8037

3.3.5 Function and operation

The HAD follows the main governmental policy and objectives on water resources management in Egypt. The national objective is to support the socio-economic development of Egypt on the basis of sustainable resource use (surface water and groundwater) while protecting and restoring the natural environment. Behind the HAD, there is a reservoir (HADR) with an active storage volume of nearly 160 km^3. This means that the lake can contain almost twice the annual yield of the River Nile. The purpose of this massive reservoir is to ensure over-year storage instead of annual storage. With over-year storage, it is possible to have the same reservoir yield every year because differences in inflow can be buffered by active storage mechanism. The release is determined depending on the variability of the inflow and the level of allowed uncertainty.

Figure 3.36 shows the average natural inflow to the HADR in relation to the water demand in Egypt. The lowest lake level usually occurs when the two lines cross. The figure illustrates that the reservoir is able to fully control the Nile flows after building the HAD. The high flows during August and September are completely eliminated downstream the HAD, and maximum discharges are now limited to 270 hm^3/day, i.e. less than one third of the earlier peak values before building the HAD (NWS, 2012).

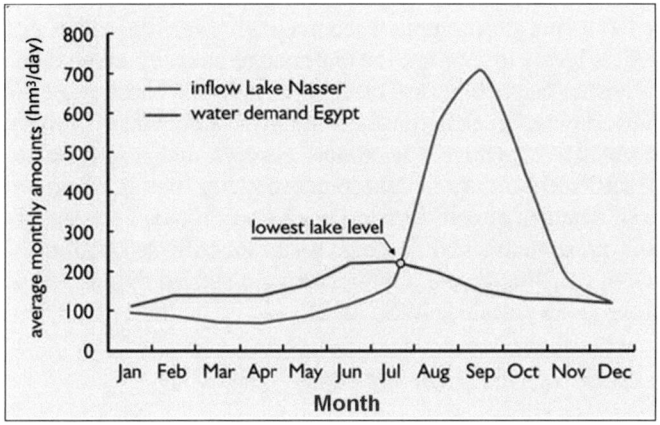

Figure 3.36: Inflow to the HADR and water demand in Egypt (NWRP, 2005)

Although the operation of the HAD does not use any explicit rule protocol, the following formal rules are applied: (Abul-Atta, 1978)

- By 1st of August, the reservoir level should not be higher than 175 m AMSL to allow sufficient room to receive the coming flood. The 1st of August is defined as the start of the hydrological year in Egypt.
- The release from the dam is determined by agricultural demand, which is fixed at an annual amount of 55.5 km^3. A maximum amount of 230 hm^3/day is released in May, whereas in January, when maintenance work is carried out in the irrigation system, a minimum amount of 80 hm^3/day is discharged to satisfy minimum flow requirements.
- The reservoir level should not exceed 182 m AMSL, with an upper limit of 183 m AMSL in emergency situations.
- The reservoir level must be above 150 m AMSL, to enable the operation of the turbines.
- If the reservoir level exceeds a height of 178 m AMSL, water from the reservoir is spilled to the Toshka depression.
- In case of low reservoir levels, releases are reduced, and a sliding scale is applied to Egyptian and Sudanese water quotas (only applied once since 1968).

During the years of exceptionally high levels in the HADR, the application of these rules forced Egypt to release more water than its share to guarantee that the level on August 1st would be indeed equal to or lower than the 175 m.

3.3.6 Safety and monitoring:

It was considered in the design that the behavior of each element of the HAD would be monitored during construction and operation as well. Different types of instruments are installed for monitoring the following phenomena: seepage discharge; piezometeric heads; surface displacements; internal displacements; strain; and total stresses and pore pressures. After the earthquake on the 14th of November 1981, five strong motion seismographs were erected at different locations as well as levels to monitor the earthquake intensity at the dam site. Regular periodic visual inspections are carried out as well. These inspections include the dam powerhouse, tunnels, intake structures, emergency spillway, and inlet and outlet canals. At seepage locations, surveys and notes are taken on the abutments, settlement cracks in the asphalt roadway over the dam crest, and any other signs of changes or deterioration in site conditions. Samples of water from the reservoir, piezometers and drainage wells are collected periodically. Chemical analyses of pH, silicon and alumina ions are carried out as well to assess the stability of the grout curtain (NWS, 2012).

3.4 The High Aswan Dam Reservoir (HADR)

The HAD forms a large man-made reservoir which is called High Aswan Dam Reservoir (HADR). HADR is one of the biggest artificial lakes in Africa and the world. Worldwide, it is the third largest artificial reservoir in terms of storage capacity after Lake Kariba in Zimbabwe and Bratsk Reservoir in Russia (ICOLD, 2009, Chao, 2008, Zwieten et al., 2011, McCartney et al., 2012). It is the second largest artificial reservoir in terms of the surface area, worldwide, after Lake Volta in Ghana (ICOLD, 2009, Chao, 2008, Zwieten et al., 2011, Spalding-Fecher et al., 2014). In Africa, the HADR is the second largest artificial reservoir in terms of storage capacity and surface area (Bernacsek, 1984). Detailed description of the HADR is given below.

3.4.1 Geography

The HADR is located upstream the HAD, 17 km south of the old Aswan dam. It is approximately 500 km long and 35 km wide at its widest point, near the Tropic of Cancer. It covers a surface area of 6514 km^2 at altitude 182 m AMSL and has a storage capacity of 162 km^3 for water. The reservoir is partitioned into Lake Nasser in Egypt (approximately 350 km long) and Lake Nubia (150 km long) to the south in Sudan. It is confined between latitudes 23°58'N at the High Dam and 20°27'N at the Dal Cataract in Sudan and between longitudes 30°07'E and 33°15'E, as shown in Figure 1.1 (Abu-Zeid and El-Shibini, 1997).

3.4.2 HADR morphology and storage capacity

The HADR Morphology is recognized with the presence of numerous side extensions of the reservoir (embayments), known locally as khors, which are considered the key feature of the reservoir. There are over 100 khors, 48 of which are located on the eastern side of the reservoir. The largest khors are Kalabsha, El-Alaky, and Toshka West as shown in Figure 3.37.

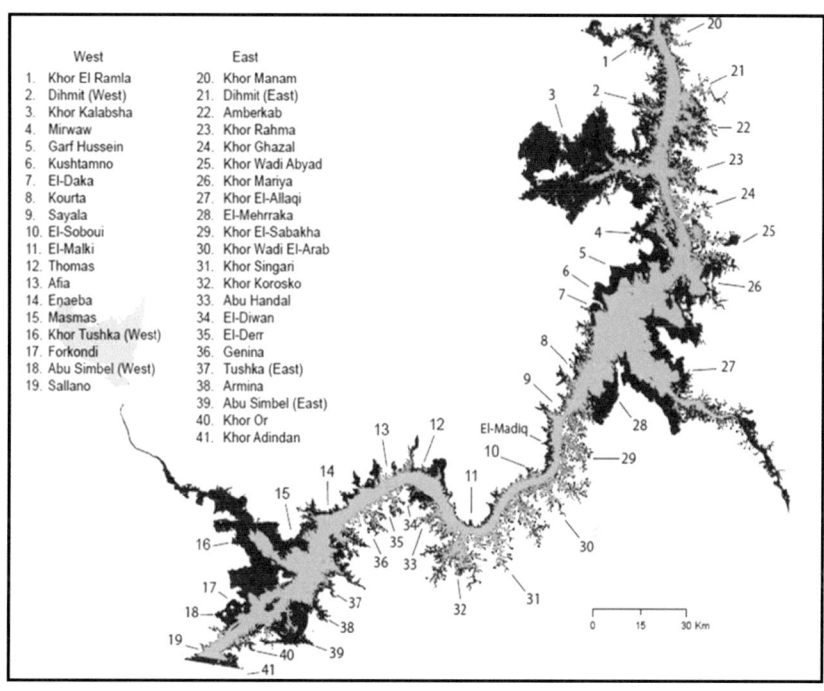

Figure 3.37: Khors' names and locations on the HADR (Zwieten et al., 2011)

The existence of the khors greatly increases the length of the shore, which is estimated to be 12000 km at altitude 182 m AMSL. These khors cover approximately 3000 km^2 or 50% of the whole lake area (Said, 1993). Thus, these shallow khors are considered one of the major causes for high evaporation losses. The total amount of water that can be stored within the reservoir at a level of 183 m AMSL is 169 km^3. This is about twice the average annual yield of the Nile at Aswan, which is estimated at approximately 84 km^3. The reservoir surface area and capacity are shown as functions of reservoir level in Table 3.9 (HDA-MWRI, 2009). There are basically four storage zones as shown in Figure 3.38. A summary of these zones and its subdivisions are listed in Table 3.10.

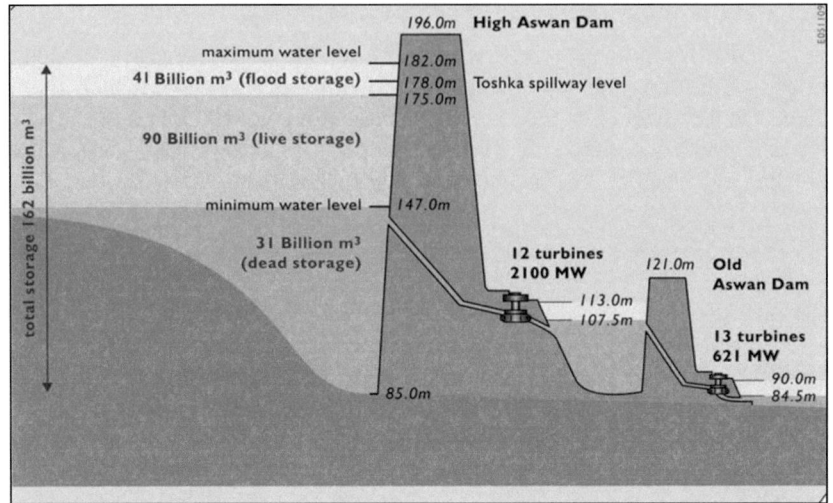

Figure 3.38: Schematic overview of the storage zones in the HADR system (NWRP, 2005)

Dead storage zone (inactive zone)

The inactive zone is the storage below the level of 147 m AMSL; from this zone, no release will take place. This zone is reserved to be filled up with the sediment that is to be expected in the first 500 years of the HADR operation (Abul-Atta, 1978). Sedimentation is currently also taking place in areas outside this inactive zone, forming a small delta at the upstream end of the lake (NWS, 2012).

Live storage zone

The zone between 147 m and 175 m AMSL is the active storage. The zone is sub-divided into two zones:

- **Conservation zone:** The zone between 160 m and 175 m AMSL is the active zone from which the releases are made in accordance with the schedule, required mainly by the demand for irrigation.
- **Buffer zone:** Between the level of 147 m and 160 m AMSL, the storage volume is kept as a buffer zone. When the level of the HADR is in this zone, releases will be reduced according to heuristic rules, instituted by the Permanent Joint Technical Committee, as stated in 1959 Agreement between Egypt and Sudan (IWLP, 2008).

Flood storage zone

The zone between 175 m and 182 m AMSL is the flood storage. The zone is sub-divided into two zones:

- **Flood control zone**: This zone is reserved for the annual flood and should be emptied by the start of the water year (August 1st). On the first of August, the level should be below 175 m AMSL. If the level is above 178 m AMSL, spilling starts happening.
- **Surcharge storage zone:** The maximum allowable level at the HAD is 182 m AMSL. Above that level, flooding in Sudan starts to occur for which Egypt will have to pay compensation. The maximum upper limit is 183 m AMSL.

Table 3.9: Original Surface area and stored volume as functions of HADR level (HDA-MWRI, 2009, NWS, 2012, Abul-Atta, 1978)

Level (m) AMSL	Area (km^2)	Volume (km^3)
120	450	5.2
130	749	11.3
140	1242	21.2
145	1589	28.3
147	1685	31.6
150	1962	37.2
155	2414	48.1
160	2950	61.5
165	3581	77.9
170	4308	97.6
175	5168	121.3
178	5714	137.5
180	6118	149.5
182	6514	162.3
183	6726	168.9
184	6962	175.7

Table 3.10: Summary of the storage zones within the HADR (HDA-MWRI, 2009)

Zones	Subzones	Levels (m) AMSL	Zone storage (km^3)	Cumulative storage (km^3)
Flood zone	Surcharge storage zone	178-182	24.8	162.3
	Flood storage zone	175-178	16.2	137.5
Live zone	Conservation zone	160-175	59.8	121.3
	Buffer zone	147-160	29.9	61.5
Dead zone	Dead storage zone	Below 147 m	31.6	31.6

3.4.3 Geology of the HADR area

Figure 3.39 shows the basic geological map of the HADR area (Said, 1993). The major exposed rock units around the reservoir are Nubia Sandstones. Sand sheets and sand dunes are also relevant for the area. The Sinn El Kaddab carbonate plateau lies at the western side of the HADR. Granites and metamorphic hills are found on the reservoir's eastern side. The Nubia Sandstone plateau in the north eastern side of the reservoir displays an average elevation of 450 m AMSL, while on the other, western side of this water body, isolated hills of Nubia Sandstone are scattered over a wide area with an average altitude of 370 m AMSL. The sediments around Kalabsha Khor consist of fluvial deposits with cross bedded sand stone and have an average altitude of 250 m AMSL. The granite and meta-volcanic hills, in the northern part of the eastern side of the reservoir, have elevations ranging from 390 m to 590 m AMSL with eroded steep slopes. Oligocene basalt extrusions in the area are found in a North West direction around Abu Simbel and Toshka, in structurally controlled tectonically disturbed zones, within Cretaceous Nubia Sandstone exposures. Sand dune encroachments into the reservoir are frequently distributed over major sectors along the northern reservoir boundaries (Said, 1993).

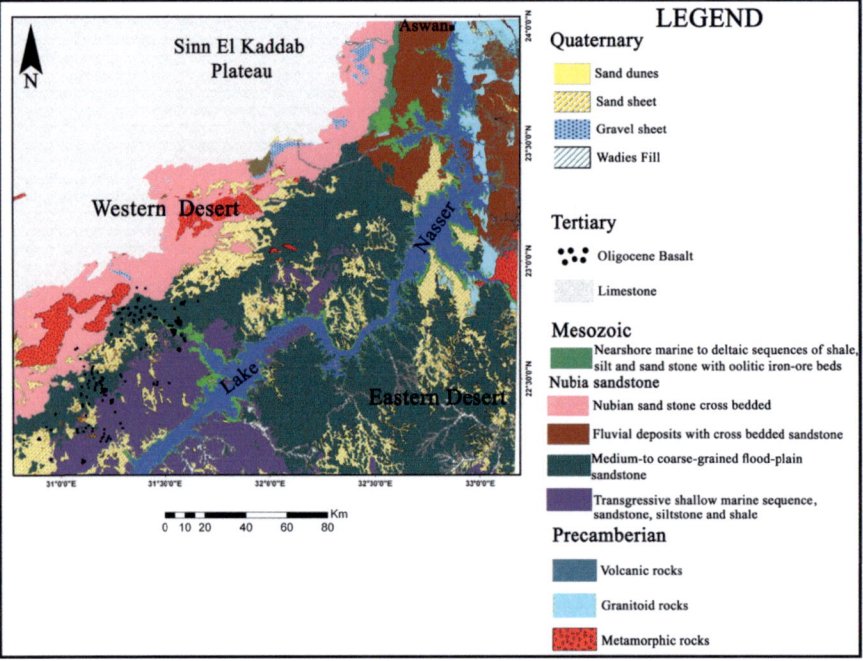

Figure 3.39: Simplified geological map of the HADR area - (Said, 1993) modified by Elba 2014

From a geomorphologic point of view, the area around Aswan is almost flat, with a topography varying from 150–350 m AMSL. It is dominated by faults and the presence of several alkali granites and a syenite ring complex. Tectonic features, dominated by East West and North South fault systems, as well as a regional uplift characterize the northern part of the HADR. Geological data of these features indicate that a right-lateral strike-slip movement is the dominant mechanism along the Kalabsha Fault zone. Igneous intrusions and associated contact metamorphism may be found in a few localities on the western side of the HADR (Khalil et al., 2011).

From a hydrogeological point of view, the Nubian formation in the area consists of three horizontal units of total maximum thickness of 400 m. The lowermost unit, overlying the granitic basement, is an aquifer of fluvial sandstones and has a large-scale (several kilometers) horizontal permeability of $0.32 - 0.43$ μm^2. The middle unit is an Aquiclude unit (A hydrogeologic unit which, although porous and capable of storing water) that extends unbroken under the lake and leaks at times for several years. The uppermost unit is the phreatic aquifer, which is composed of 25% – 30% of porosity sandstones, interspersed with clay stone lenses, and has a large-scale permeability of $1.0 - 1.5$ μm^2 (Said, 1993).

3.4.4 Climate

The HADR reservoir is situated in a desert area where the climate is extremely arid. The area is in the transition zone between the tropical climate with summer rain and the Mediterranean climate with winter rain (Springuel, 2005). Rainfall is extremely scarce during winter season, but sometimes occurs in very small amounts when the area is affected by active cold fronts of the Mediterranean depressions. Rainfall records during of the past 50 years, from 1961 to 2010, are grouped in six categories as listed in Table 3.11 (NWS, 2012, Nour El-Din, 2013).

Table 3.11: Rainfall records from 1961 to 2010 at the HADR (NWS, 2012)

No. of years	Rainfall (mm)
22	Zero
11	Trace
8	Less than 1.0
5	1.0 to 7.1
3	7.2 to 8.7
1	27.3

The area around the HADR receives virtually no rainfall, except for occasional thunder storms in the winter, roughly once every ten years. The relative humidity is highest (40 – 41%) in December and January and lowest in May and June (13% – 15%). The wind speed does not vary greatly throughout the year, as the mean ranges from approximately 15 to 19 km/h. Its direction is mostly NW-NE. Evaporation is very high, approximately 3000 mm/year (NWS, 2012, Nour El-Din, 2013).

There are six floating meteorological stations on the lake and one on the bank. One of the stations is a telemetric station in Abu-Simbel city (280 km upstream HAD), referred to as Toshka Station, and the others are over water (floating) meteorological stations operated by the High Dam Authority (HDA). The names of these stations are Aswan Raft Station (2 km upstream HAD), Kalabsha station (40 km upstream HAD), El-Alaky Station (75 km upstream HAD), Amada Temple station (185 km upstream HAD), and Abu-Simbel Station (280km upstream HAD). Figure 3.40 shows all meteorological stations on the HADR (NWS, 2012).

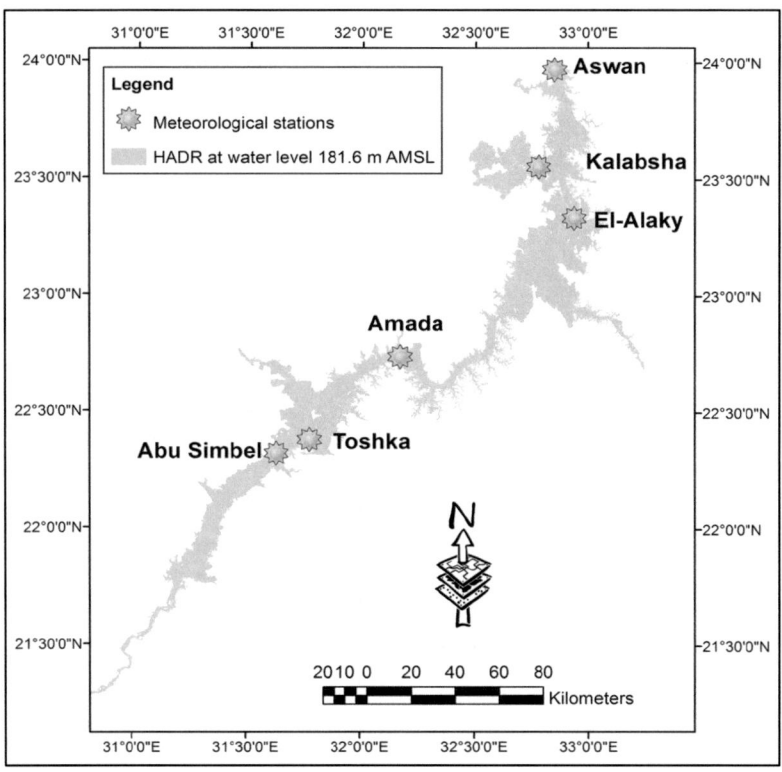

Figure 3.40: Distribution of Meteorological Network Stations on the HADR

Data of the temperature of lake water, air temperature, wind speed and direction, relative humidity, and barometric (atmospheric pressure) are available for all stations except Toshka station, which does not record atmospheric pressure but Net radiation, collected only at Toshka station. Data from Abu Simbel and Toshka stations are hourly data while the available data from other stations are daily data. Net radiation is available at Toshka meteorological station only for about three years. In addition to these stations, the HADR's size and irregular geometry required that detailed thermal surveys be conducted to accurately measure the heat content of the lake. Lake water temperature data were collected at various depths at different stations throughout the HADR, providing data necessary to calculate average lake temperature using the energy budget method. The analysis of raw data is presented through plotting the raw data with respect to time and determines the trend line under confidence limits of 95% (NWS, 2012). Description of these meteorological stations is available in Appendix B.

3.4.5 Hydrology of the HADR

The reservoir water levels change according to inflow and outflow rates. The highest water level was recorded in November 1999 at 181.6 m AMSL, while the lowest was in July 1988 at 150.6 m AMSL (Figure 3.41). Figure 3.42 shows the annual release and the HADR water levels through from the 1966 to 2006. The actual annual release often deviates from 55.5 km^3. Less than 55.5 km^3 was released in all five years between 1987 and 1991. More than 55.5 km^3 was released in between 1977 and 1984 as well as in all the years between 1998 and 2006. Between 1998 and 2006, the high releases were associated with very high inflows and, consequently, resulted in high reservoir levels (NWS, 2012). The reservoir water velocity decreases as it approaches the HAD. At the entrance of the HADR, the flow velocity is approximately 0.5 m/s. This velocity is gradually reduced within a few kilometers to 0.1 - 0.2 m/s (MWRI, 2001).

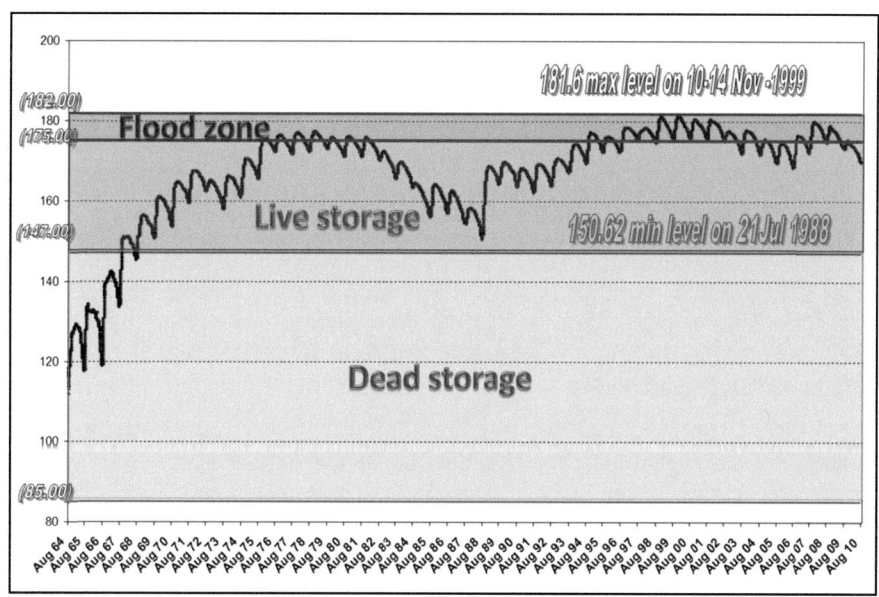

Figure 3.41: Daily Water Levels in HADR from August 1964 to January 2010- (NWS, 2012)

As mentioned before, the Permanent Joint Technical Commission (PJTC) for the Nile Waters has suggested a sliding scale for lake operation during dry periods. According to this sliding scale, releases from the lake should be reduced if the lake storage is less than 60 km^3. The storage of 60 km^3 corresponds with a lake level between 159 and 160 m AMSL. Figure 3.42 shows that this sliding scale has not been strictly followed. For example, between 1985 and 1987, the lake levels were below 159 m AMSL. This means that the annual release should have been less than 55.5 km^3. However, during these years, at least 55.5 km^3 was released. From 1990 to 1992, the releases were lower than 55.5 km^3 although this was not necessary according to the sliding scale (NWS, 2012).

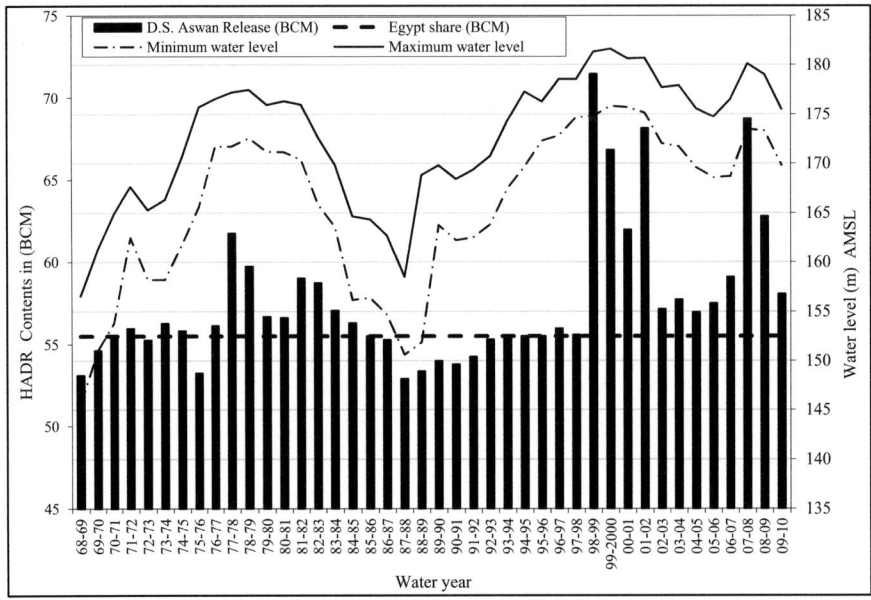

Figure 3.42: History of annual release of the HAD (NWS, 2012)

According to the operation rules, the level on August 1st should not be higher than 175 m AMSL in order to store the coming flood. However, from 1998 to 2007, the lake was operated well in the flood storage zone (Figure 3.43). The lake level on the first of August was near 176 m AMSL instead of the agreed 175 m AMSL (Figure 3.44).

According to decree 203/2002, two km around the lake is the buffer zone, to protect the water quality of the HADR. The decree states that it is forbidden to undertake actions or activities or procedures which would lead to the destruction, damage, or degradation of the natural environment within this buffer zone (Zaghloul et al., 2011).

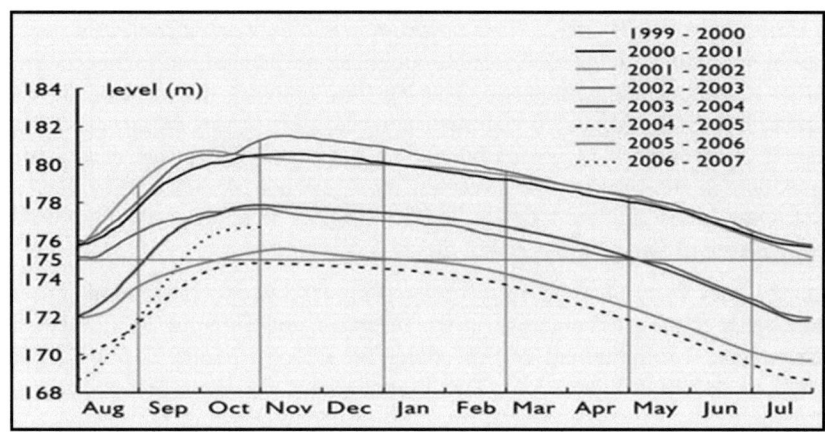

Figure 3.43: Actual Levels of HADR between August 1999 and October 2007 (NWS, 2012)

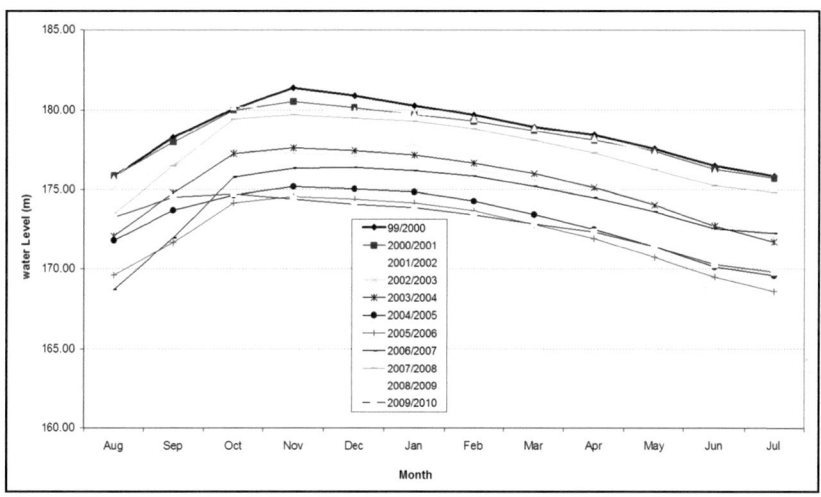

Figure 3.44: Minimum Water Levels of HADR between August 1999 and October 2010 generated by Elba 2014 (NWS, 2012)

3.4.6 Sedimentation in the HADR

The Nile's average annual inflow of 84 km^3 in the HADR contains approximately 134 million tons of suspended matter, mostly silt. It is heavily loaded with inorganic clay, silt, sand, as well as other organic debris (detritus). 98% of the annual sediment load is moved during the flood season (Abul-Atta, 1978). Almost all the clay, sand, and other detritus brought by the River Nile is deposited in Lake Nubia, most of these being deposited south of Halfa, where a new delta is in formation, and only the fine silt enters the HADR. 97% of the sediment load entering the HADR is from the Ethiopian Highlands, with 72% from the Blue Nile and 25% from the Atbara. Only 3% of the sediment load comes from the White Nile (NWS, 2012). This breakdown and concentration levels vary according to season. The peak discharge and peak suspended sediment concentration do not occur simultaneously. The lag time between the peak of the water discharge and the suspended sediment concentration varies from year to year, and but on average it is approximately 10 days (El-Moattassem et al., 2010). Deposition in the reservoir is governed by a number of factors: the most important one is the sudden decrease of flow velocity as soon as the river reaches the open area of Lake Nubia (NWS, 2012).

Almost Every year, a bathymetric survey of the HADR is conducted using echo-sounding at fixed cross-sections in the reservoir and GPS for positioning. For each survey, a longitudinal section along the deepest points in the HADR is plotted, as shown in Figure 3.45. This Figure shows the development of the sedimentation process in the HADR over the past five decades, and shows that a new delta has emerged in the entrance reach of Lake Nubia (NWS, 2012, Negm et al., 2010). The maximum sediment thickness, or the area of most intense deposition, recorded in Lake Nubia until 1998 was at the Kaganarty section, which is located approximately 394 km upstream from the HAD. The layer thickness was over 60 m in 2008 (HDA-MWRI, 2009). Ahmed and Ismail (2008) stated that "It has been proved that the characteristics of rocky-narrow channel of the southern part of Lake Nubia work as a place for a new delta formation on the sides of the lake". The sediment not only fills the dead storage, but also the life storage. The High Dam Authority (HDA) have repeatedly claimed that the life storage will be recovered when reservoir inflows are high. However, the results of the survey indicated that high inflows in recent years simply have extended the delta and reduced the live storage of the reservoir (HDA-MWRI, 2009).

Several studies were conducted to simulate the scouring and silting processes in the HADR. Some of these focused on collecting and analyzing field data to study reservoir characteristics, while others aimed to determine the relationships between flow and sediment load. Recent studies developed mathematical models describing the motion of both water and sediment flow and simulating the water surface and bed profile in the longitudinal direction (Negm et al., 2010). Hurst et al. (1965) calculated the mean composition of the suspended loads carried by the flood based on the sediment concentrations measured during the period from 1929 to 1955. They showed that coarse sand was not deposited; however, the sediment contained 40% silt, 30% sand and 30% clay.

Shalash (1980) also studied sediment transport along HADR and its lifetime. He used the suspended sediment concentration measured during the period from 1968 to1979. He found out that the mean annual sediment rate of inflow was 130 million tons, while the mean annual rate of outflow was only six million tons, and he concluded that 124 million tons were deposited annually. He estimated the deposited sediment to be 1570 million tons in 15 years, and based on this calculation, he suggested that the lifetime of the dead zone of the reservoir to be about 362 years, which is less than the intended lifetime of 500 years. In a follow-up study, Shalash (1982) updated the results of his earlier paper and estimated the total annual inflow to be as high as 142 million tons, the average rate of outflow as high as 6 million tons with a net sedimentation within the Lake of 136 million tons. Using an average sediment density of 1.56 g cm^{-3} corrected for compaction (dry weight density of 2.6 g cm^{-3} and a porosity of 40%), the amount of annually retained sediment of 136 million tons of suspended sediment corresponds to an accumulated volume of 87 hm^3/year.

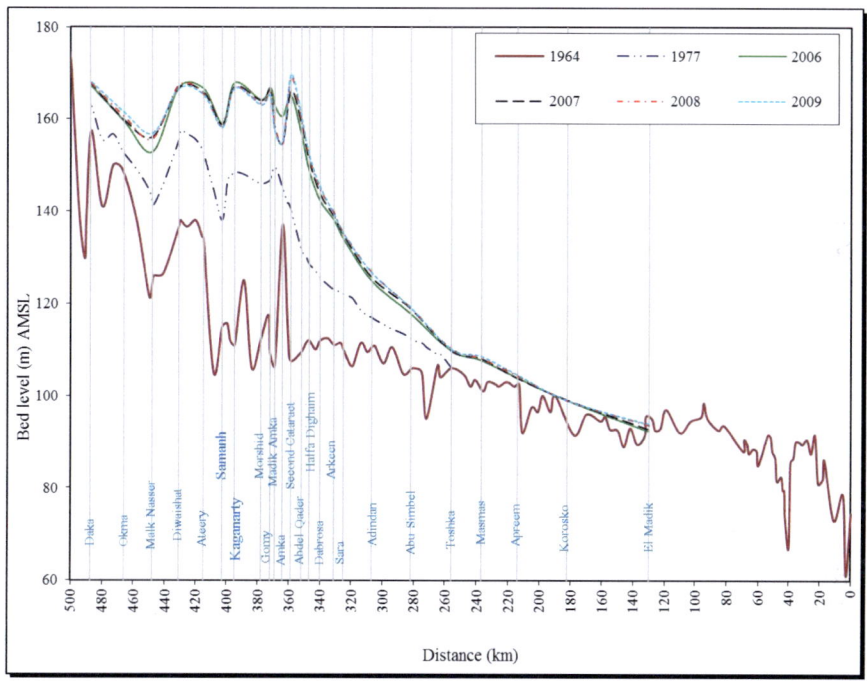

Figure 3.45: Longitudinal section along the deepest points in the HADR (HDA-MWRI, 2009)

Abdel-Aziz (1997) developed a one-dimensional numerical model to simulate and predict the bed profile in HADR in the longitudinal direction based on the principal equations of water volume conservation, water momentum conservation, and general sediment transport equations. The model showed 2.650 km^3 of deposits accumulated inside the reservoir for the period from 1964 to 1988, a result that is very similar to that of field studies, which indicated that 2.760 km^3 were deposited. He expected that the dead zone will be filled in 311 years and the live zone will be filled in 1202 years.

Ahmed and Ismail (2008) released the results of a report entitled "the sediment in the Nile River system," which was funded by the International Sediment Initiative, a UNESCO International Hydrological Program. Based on data available for the HADR, they found out that suspended sediment dominates the total sediment in transport. About 90% of the total sediment loads are suspended in the water, which means that less than 10% is left for the bed load. They estimated the mean annual suspended sediment inflow over the period from 1965/66 to 1977/78 to be as high as 103 million tons, and they assumed all the suspended sediment is deposited in the reservoir plus 10% for the bed load (10 million tons). Consequently, the expected lifespan of the HAD is over 1000 years. They also stated that the dead storage will be filled in about 200 years; however, most of the sediment, if not all, is deposited in the live storage, which slightly reduces the function of the dam.

Negm et al. (2010) used a two-dimensional numerical model (CCHE-2D) to study the scouring and silting processes in HADR. The study focused on Lake Nubia from km 500 to km 350 upstream of the HAD. The model was applied to predict the changes to cross sections along Lake Nubia in 2010, 2020, 2030, and 2040. The results showed that the lifespan of the dead zone is only 254 years, while the lifespan of the life zone is 985 years. El-Moattassem et al. (2013) also developed a theoretical model, but they investigated the effect of upstream structures on the development of delta deposits in the HADR with the help of surface water modeling system (SMS). The future natural flow (annual and monthly average), the sediment load, and water levels in the HADR were calculated and used in the SMS model. They found out that the construction of Merowe Dam in Sudan will change the discharge and sediment regime downstream and affect the rate of deposition in HADR. The rate of deposition in HADR will also decrease by about 16% every year and, as a result, the lifespan will increase by the same percentage, which would have a very positive effect on both Sudan and Egypt.

The life time span of the HADR has been investigated frequently before and after operations began at the HAD. In 1964, Russian engineers estimated the life time span of the dead zone to be 500 years, the German company Hochtief, in 1970, expected the dead zone to be filled in 750 years, and in the same year the American Building Authority calculated the life time span of the dead zone to be 1000 years (Abul-Atta, 1978, Moussa, 2013). Abul-Atta (1978) estimated the life time span of the dead zone to be 440 years. Recent studies, however, expected that the dead zone will be filled in 200 to 254 years, as stated above, that is about 50% of the intended life time span of the original design.

3.5 Summary

The Nile Basin is one of the major river basins in Africa, and one of the most complex major river basins in the world because of its size and variety of climates and topographies. Fluctuations in the Nile flows are primarily driven by the variation in rainfall over the Ethiopian Highlands, which on average contribute about 83% to the total Nile flow at Aswan, while the Equatorial Lakes' Basin adds only about 17% as base flow components. The basin receives some 650 mm of rainfall annually, corresponding to around 1900 km^3 of water annually. Approximately 84 km^3 per year on average arrives at Aswan with an annual runoff coefficient of the basin of around 4.5%. Climate change, especially in the countries where the River Nile originates, will affect the river flows reaching Egypt in different ways. Furthermore, Egypt, as a downstream user, will face significant problems due to the reliance of the Upper Nile basin countries on the abstraction of water from the Nile. Since 1891, Egypt, being the country in the most downstream location in the basin, has been cooperating with the other countries for many years through agreements to secure Egypt's share in the Nile Water. The 1959 Nile Water Agreement is the most important agreement between Egypt and Sudan about sharing the Nile's water. The treaty entitled Egypt to use 55.5 km^3 and Sudan 18.5 km^3 of this annual yield. The increase is divided equally, while the decrease in natural flow is divided according to the ratio between the shares of Sudan and Egypt. Funding for any project that increases Nile flow would be provided evenly, and the resulting additional water would be split evenly too.

Egypt is located in the hyper-arid regions of North Africa and West Asia. It has arid climate with very rare rainfall and seasonally windy weather. 98% of the country's population lives on just 4% of the land, along the narrow Nile Valley and in the Delta. The availability of land has never been a problem; the main constraint has always been water to make the land productive. The total land area is approximately one million km^2, about 3% of which has been cultivated. 0.4% is used for public utilities and buildings, 1.6% is considered as permanent cropland (covered with trees shrubs) and woodlands, and the remaining 95% is classified as other land such as desert and semi-desert areas, salt flats, hills, and sand dunes. Egypt has been facing major challenges with respect to water management. From a quantity point of view, the country has to cope with a growing population while the water supply can hardly be increased. In 2010, 79 km^3 water resources were used including 55.5 km^3 from the Nile River, 1.3 km^3 from rainfall and flash floods, and 2 km^3 from deep groundwater in the deserts and Sinai, 11.6 km^3 from reuse of drainage water, 2.4 km^3 reuse of sewage and industrial water, and 0.2 km^3 from desalination of sea and brackish water. 85% of the total water demand was consumed by the agricultural sector, 12% by the municipal sector, and 3% by the industrial sector.

The High Aswan DAM (HAD) construction started on January 9, 1960, offering employment opportunities for thousands of laborers, contributing to the economic growth of the Aswan Governorate through large-scale activities in different fields, and providing the necessary infrastructure including education, health care and transportation. The HAD operational rules followed the main governmental policy and objectives on water resources management in Egypt to sup port the socio-economic development on the basis of sustainable resource use, protecting and restoring the natural environment.

The HAD construction created one of the biggest artificial lakes in Africa, the HADR. It is approximately 500 km long and 35 km wide at its widest point, near the Tropic of Cancer, in a desert area with extremely arid climate. It covers a surface area of 6514 km² at 182 m AMSL water level and has a storage capacity of some 150–165 km³ of water. The reservoir's morphology reveals the presence of numerous side extensions of the reservoir (embayments), known locally as khors. There are over 100 khors, 48 of which are located on the eastern side of the reservoir. The largest khors are Kalabsha, El-Alaky, and Toshka in the west. The existence of the khors greatly increases the length of the shore, which is estimated at 12000 km at altitude 182 m AMSL. These shallow embayments cover about 50% of the entire lake area and are considered one of the major causes for high evaporation losses, which are estimated to be approximately 3000 mm annually. The major exposed rock units around the reservoir are Nubia Sandstones, sand sheets and sand dunes. Granites and metamorphic hills are found on the reservoir's eastern side. The Nubia Sandstone plateau is located on the north eastern side of the reservoir. The surrounding sedimentary around Kalabsha Khor consists of fluvial deposits with cross bedded sand stone. The highest water level was recorded in November 1999 at 181.6 m AMSL, while the lowest was recorded in July 1988 at 150.6 m AMSL.

The Nile's average annual inflow contains approximately 134 million tons of suspended matter, loaded with inorganic clay, silt, sand, and other organic debris. 98% of the annual sediment load is moved during the flood season from the Ethiopian Highlands. Most of these sediments deposit in Lake Nubia south of Halfa, where a new delta is in formation. Deposition in the reservoir is governed by a number of factors: the most important one is the sudden decrease of flow velocity as soon as the river reaches the open area of Lake Nubia.

Chapter 4

Material and Methods

This chapter demonstrates the materials and methods used in this study. The data used is described including the data for meteorology, hydrology, seepage, sedimentation and topography from old topographic maps, bathymetric survey, satellite images and aerial photos. The methodology is then explained in detail. The required software is provided, the databases established for HADR's characteristics are presented including; meteorological, hydrological, seepage and sedimentation databases, and the geo-databases for contour lines and spot heights are described. Finally, the methods for establishing the digital elevation model and the HADR's hydrological characteristics models are explained.

4.1 Materials

In order to achieve the research goals, several inputs were required to model the characteristics of the High Aswan Dam Reservoir (HADR). A scheme of this research workflow including the used inputs and the produced outputs, is shown in Figure 4.1. The HADR was modeled using hydrological, topographical and meteorological geo-databases. The data and methods used are explained below.

Hydrological, topographical and meteorological geo-databases are essential in constructing mathematical models for the HADR. In general, Hydrological databases include several items such as lake water level, lake flow, seepage losses, and absorption losses. Topographical geo-databases include data on water surface area and lakebed level. In order to create these databases, the gauge water level, satellite images covering the lake, Lake Bathometric survey, piezometers data, and absorption data are needed. The meteorological database has been established using several meteorological parameters such as temperature, relative humidity and evaporation rates.

After the High Aswan Dam (HAD) was constructed, it has been difficult to carry out continuous conventional hydrometeorological measurements along the lake because of the obscurity of accessing some remote areas. The mutual cooperation between the Permanent Joint Technical Commission (PJTC) for Nile Waters, the Aswan Reservoir and the High Dam Authority (ARHDA), and the Egyptian Meteorological Authority (EMA), resulted in the first plan for installing hydro-meteorological network stations along the lake, which has been started in 1973. A historical review about the lake area network stations and the methods used for estimating the different parameters are given below.

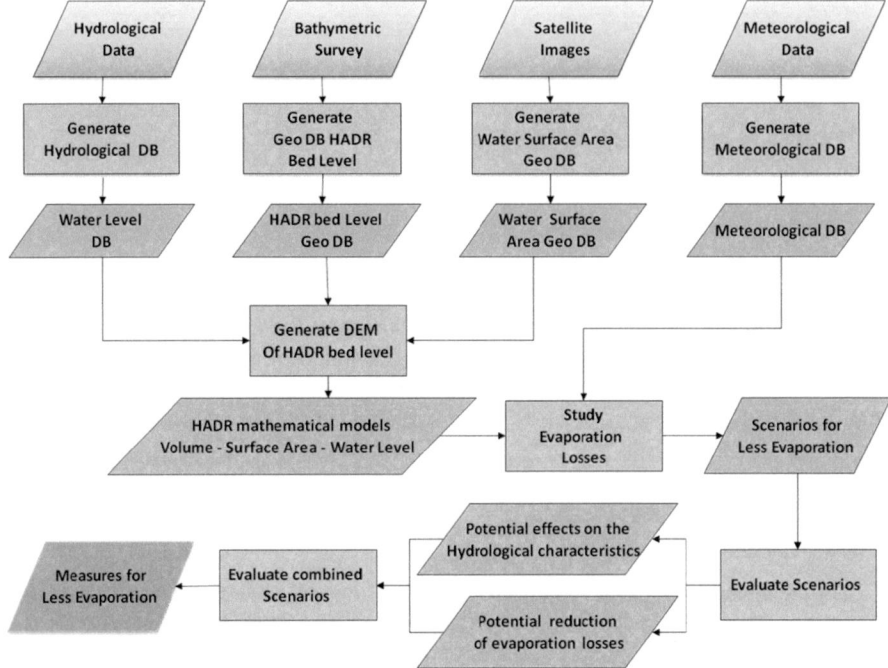

Figure 4.1: Workflow of the study inputs and outputs

4.1.1 Meteorological data

The EMA operates a network of meteorological stations consisting of about 100 stations covering the entire area of Egypt. Of these stations, six are located on the surface area of the HADR upstream of the HAD. These stations are operated either by the weather authority or the ARHDA. The locations of these stations with respect to the Lake boundaries are illustrated in Figure 3.40. The actual data of these stations are used to compute the amount of water lost by evaporation from the HADR (Hassan et al., 2007). A description of the data measured by each station and the estimated parameters are explained below.

4.1.1.1 Meteorological stations

Some measurements were made at one hour intervals and averaged to daily values. The hourly measurements were collected from six meteorological stations. The data available from each station are shown in Table 4.1. Data for each station is available for different time periods; however data for all stations were available for the period from 2004 to 2008. Consequently, only this period is plotted in this table. The complete data sets are presented in green color. Some data sets were not completely filled where it is common to find gaps several days between measurements. These months are presented in red color. Months where no data were recorded are drawn in white color.

Table 4.1: Data available from each meteorological station along the HADR

Station name	Coordinates (Decimal Degree) N	Coordinates (Decimal Degree) E	Distant from HAD km	year	Month 1	2	3	4	5	6	7	8	9	10	11	12
Aswan	23.972	32.850	2	2004												
				2005												
				2006												
				2007												
				2008												
Kalabsha	23.556	32.779	40	2004												
				2005												
				2006												
				2007												
				2008												
El-Alaky	23.336	32.936	75	2004												
				2005												
				2006												
				2007												
				2008												
Amada	22.740	32.169	185	2004												
				2005												
				2006												
				2007												
				2008												
Toshka	22.383	31.771	240	2004												
				2005												
				2006												
				2007												
				2008												
Abu Simbel	22.324	31.625	280	2004												
				2005												
				2006												
				2007												
				2008												

☐ No data avaiable   ■ Part of data available   ▨ All data avaiable

Table 4.2 shows summary of the measured parameters and their conditions and measuring units at the different meteorological stations at the HADR. The available raw data from the six stations are hourly wind speed, wind direction, air temperature, barometric pressure, and relative humidity measured at two and four meters. Water temperature is also measured at water surface level and various water depths as the water depths vary from one station to another. Aswan, Kalabsha, and Abu Simbel meteorological stations measure the average net radiation at two meters as well, and use, consistently, the heat budget method for estimating the evaporation losses, while the other stations use the Harbeck method for estimating them. The data before 2000 are available as hardcopy reports that include daily averages of the reported data. Since 2000, the data are available in electronic form and include hourly data. In this study, the average of each day has been calculated to estimate the daily mean data using the hourly data. The average of each month has been calculated using the estimated daily mean data to compute the monthly mean data.

Table 4.2: Summary of the measured parameters by the meteorological stations on the HADR

Parameters	Conditions at	Unit	Aswan	Kalabsha	El-Alaky	Amada	Toshka	Abu Simbel
Barometric Pressure		MB	✓	✓	✓	✓	✓	✓
Air Temperatures	2 meters	^{0}C	✓	✓	✓	✓	✓	✓
	4 meters	^{0}C	✓	✓	✓	✓	✓	✓
Water Temperature	0 meters	^{0}C	✓	✓	✓	✓	✓	✓
	different depths	^{0}C	✓	✓	✓	✓	✓	✓
Relative Humidity	2 meters	%	✓	✓	✓	✓	✓	✓
	4 meters	%	✓	✓	✓	✓	✓	✓
Wind Speed	2 meters	m/s	✓	✓	✓	✓	✓	✓
	4 meters	m/s	✓	✓	✓	X	✓	✓
Wind direction	2 meters	degrees	✓	✓	✓	✓	✓	✓
	4 meters	degrees	✓	✓	✓	X	✓	✓
Net Radiation	2 meters	(W/m^2)	✓	✓	X	X	X	✓
Daily Evaporation		mm/day	✓	✓	✓	✓	✓	✓

✓ Measured
X Not Measured

a) Aswan Raft station:

At Aswan raft station (Figure 3.40), the available series of data has been recorded daily from 01//06/1974 to 31/12/1977 available as 9209 records, and from 01/01/1986 to 22/01/1991 available as 12250 records. From 01//01/2000 to 31/12/2008, the available series of data has been hourly recorded. In this study, the hourly data have been converted into 45888 daily average records. Thus the total available daily data series amounts to 67347 records. Samples of hourly and monthly reports for this station are given in Appendix C.

b) North El-Alaky automatic recording buoy station

At the North El-Alaky Buoy Station (Figure 3.40), the available raw data of the El-Alaky station are hourly based data series collected between 01/01/2000 to 16/11/2007. In this study, the hourly records were converted into 33989 daily records.

c) Toshka station

At Toshka station (Figure 3.40), the available series of data has been recorded hourly from 10/10/2004 to 14/07/2006. In this study, the hourly records were converted into 4642 daily records.

d) Khor-Kalabsha station

At Khor-Kalabsha station (Figure 3.40), the available series of data has been recorded hourly for the time span from 01/06/2004 to 31/12/2008. In this study, the hourly records were converted into 29329 daily records.

e) Amada _Wadi El-Arab automatic recording buoy station.

At Amada Buoy station (Figure 3.40), the available series of data has been recorded hourly for the period from 01/08/2004 to 31/12/2008. In this study, the hourly records were converted into 15830 daily records.

f) Abu Simbel automatic recording buoy station.

At Abu Simbel Buoy station (Figure 3.40), the available series of data has been recorded hourly for the time span from 01/01/2000 to 31/12/2007. In this study, the hourly records were converted to 32171 daily records.

4.1.2 Hydrological data

All hydrological data used in this study was gathered by the ARHDA and provided by the Nile Water Sector (NWS). This data include inflows to the HADR and water levels. The reservoir has only one inflow source, the Nile River, at its upstream end. The inflow data were specified as daily average values from one gauging station, Dongola station, about 782 km upstream of the HAD. The water levels are measured at the upstream of the HAD and at Wadi Halfa gauge, about 351 km away from the HAD. Historical samples of the readings of the gauges are given in Appendix D, which includes the following information:

- the HADR daily average inflows in Million cubic meters (hm^3), outflows spilled into the Toshka depression in hm^3, outflows released downstream of the HAD in hm^3, and average daily water level in meters above mean sea water level (AMSL) at the upstream of the HAD between August 1964 and July 2010, at HAD
- the monthly mean gauge readings in meters AMSL between January 1968 and July 2007, at Wadi Halfa station
- the monthly average discharge measured at Dongola between January 1962 and December 2007.

4.1.3 Seepage data

In 1964 the ARHDA planned a special program, executed by a Yugoslavian Company, to carry out core drilling, rock permeability investigations on site for shallow and deep boreholes on both banks of the reservoir. The purpose of this study was to collect data aiming at more realistic estimation of seepage losses. Twenty-nine shallow boreholes and five deep boreholes hitting the basement rocks were drilled. Twenty three shallow boreholes and all the deep boreholes were completed as piezometers and equipped with automatic water level recorders at four cross sections across the Nile at Garf Husein, Afia (at the western bank only), Toshka and Adindan (100, 200, 240, and 300 km respectively, upstream of the HAD) as shown in Figure 4.2. The integrated length of the boreholes reaches 4617 m of which 1747 m were tested for permeability (HDA-MWRI, 2009, NWS, 2012).

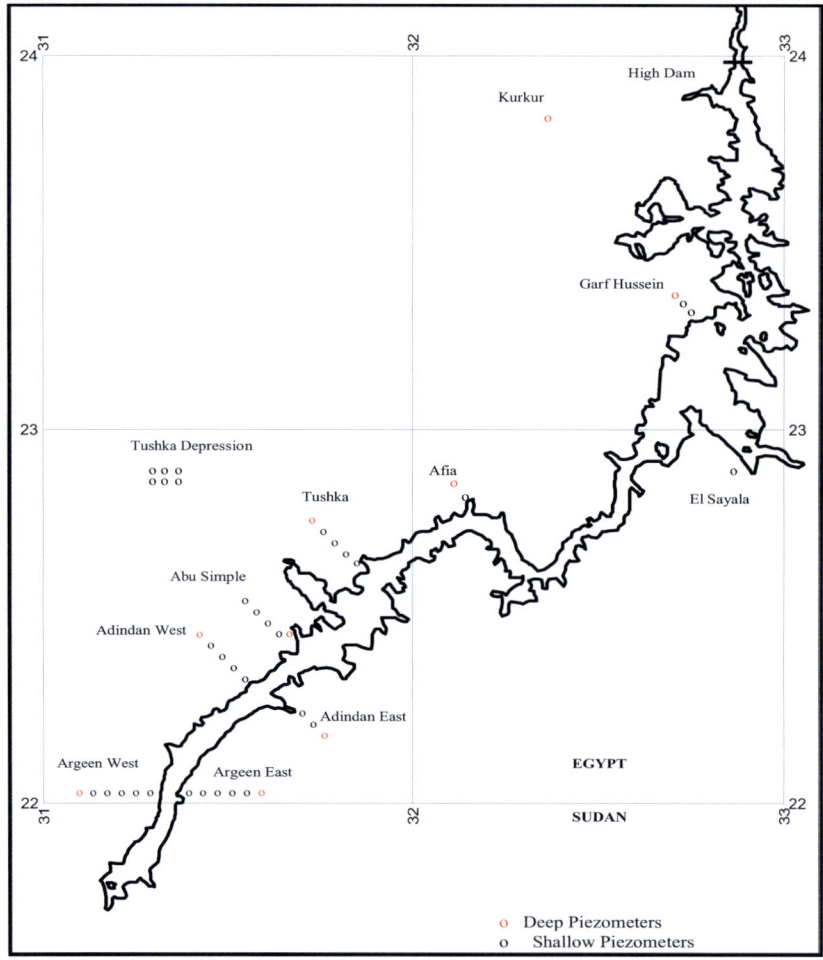

Figure 4.2: Distribution of piezometers around the HADR (HDA-MWRI, 2009)

These piezometers are used to study and monitor seepage from the lake through the banks. They are divided into two groups according to their depth: shallow piezometers ten meters below the lake bottom, and deep piezometers reaching the basement rocks. Based on field investigations, it was found that the overall thickness of the Nubian sandstones in the synclinal basin is about 400 m, and underlying by granite basement. The overall porosity of the Nubian sandstones is in the range of 25%. The average horizontal and vertical permeability coefficient amounts to 3×10^{-4} cm/sec, and 5×10^{-5} cm/sec, respectively. The water levels in the shallow piezometers are mainly affected by the fluctuations of the HADR water levels (HDA-MWRI, 2009, NWS, 2012). The geological formation of the lake banks composed of unhomogenous layers (confined aquifer), as shown by the geological description of the Adindan cross section (Figure 4.3) (HDA-MWRI, 2009).

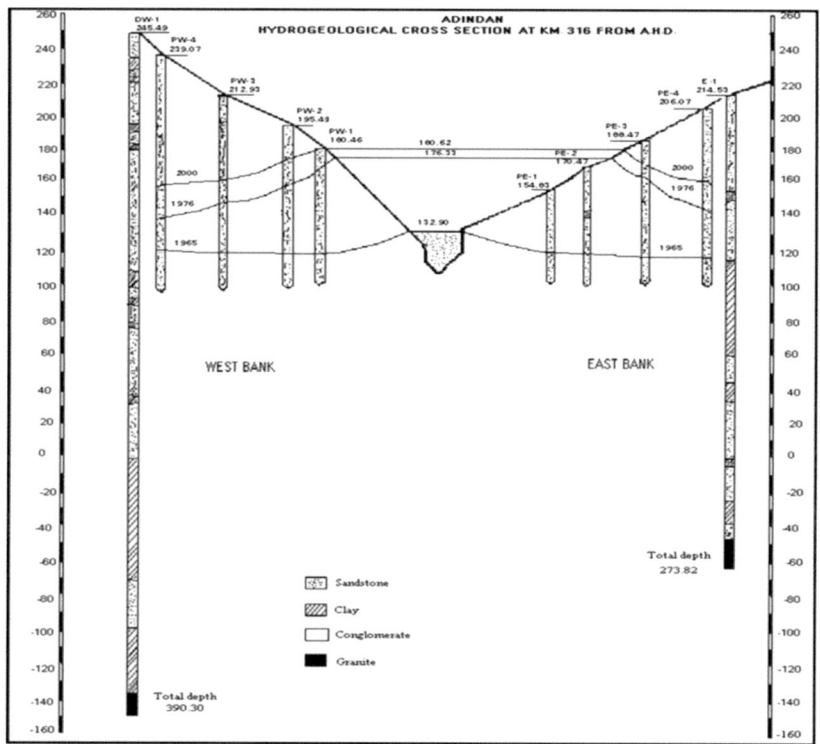

Figure 4.3: Cross Section at the Adindan Area (HDA-MWRI, 2009)

4.1.4 Topographic data

To calculate other hydrological characteristics of Lake Nasser such as lake volume and lake surface area, a Digital Elevation Model (DEM) was required. Consequently, topographic detailed data such as contour lines and spot height for estimation of the lakebed were required. The latest topographic maps dating back to the 1960s with a scale of 1:100,000 show the lake boundary at elevation 160 m AMSL. As the HADR morphology has changed due to erosion and sediment events since the operation of the HAD in 1969, these old topographic maps were used only for the elevations above 182 m AMSL outside the lake margin. The lake bathymetric survey was used to estimate the altitudes of the lakebed. For the contour lines of the lakebed, Landsat images were used in delineating the lake water surface at different water levels. Details of the used data are explained below.

4.1.4.1 Old topographic maps

The latest available topographic maps with a scale of 1:100,000 for Lake Nasser and the surrounding area were produced in the 1960s prior to the construction of the HAD. New up-to-date topographic maps are not available so far. Moreover, these old maps did not include the lakebed or lake elevations under 160 m AMSL. The lake morphology was changed due to sediment and erosion events since the HAD construction. Therefore, these maps were only used to delineate the contour lines for the regions outside the lake edges at its maximum water level of 181.6 m AMSL. The contour interval is ten meters. Figure 4.4 shows the Landsat images of the HADR area with an overlay of the topographic maps' boundaries, and the contour lines. These maps were provided as digital layers from the NWS.

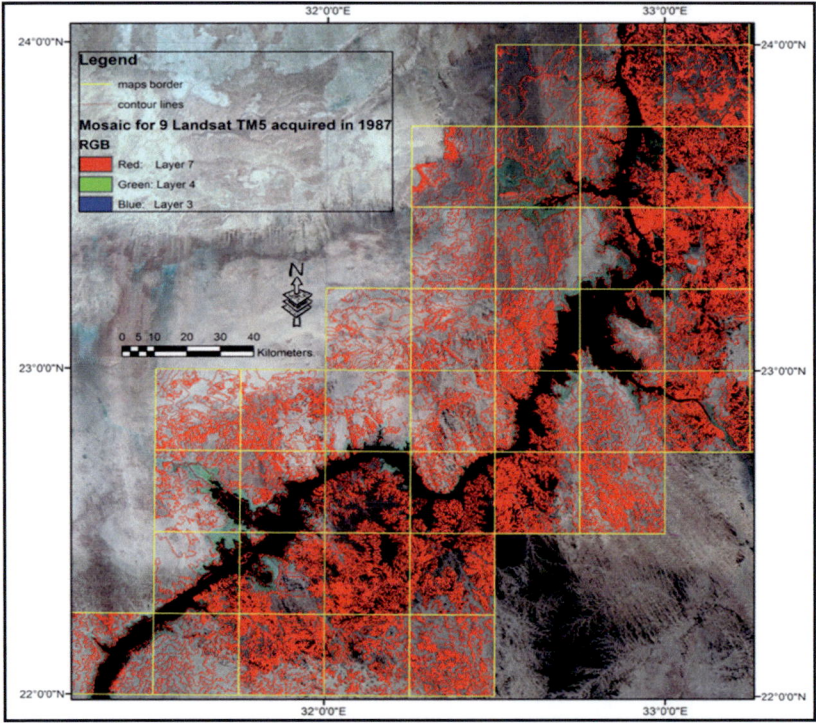

Figure 4.4: The contour lines available from the 1:100000 topographic maps and the boundary of each map overlying a mosaic of Landsat images of the HADR with false colors bands 7,4,2.

4.1.4.2 Bathymetric survey

Since 1973, the MWRI had conducted the bathymetric surveys for only a few known cross sections due to the difficulty of investigating the reservoir sedimentation progress, using the traditional survey method because of the reservoir's morphology. However, since 1999, the MWRI has used a hydro acoustics system with a Differential Global Positioning System (DGPS) and Echo Sounder to collect depth measurement data and locations as an alternate method of mapping the reservoir bottoms. The new technology provides bathymetrical data in a suitable format that can be used to create digital maps. The MWRI used the Digital Echo Sounder to measure depths and an OMNISTAR 6300A receiver incorporating a Leica Global Positioning System (GPS) to obtain real-time differential GPS data with an accuracy of 0.5 to two meters. In addition, The Coastal Oceanographic HYPACK MAX software was used for survey and control. The longitude and latitude with its depth was recorded for each sample point. Usually, the survey data have been collected along known cross sections of the reservoir area as shown in Figure 4.5.

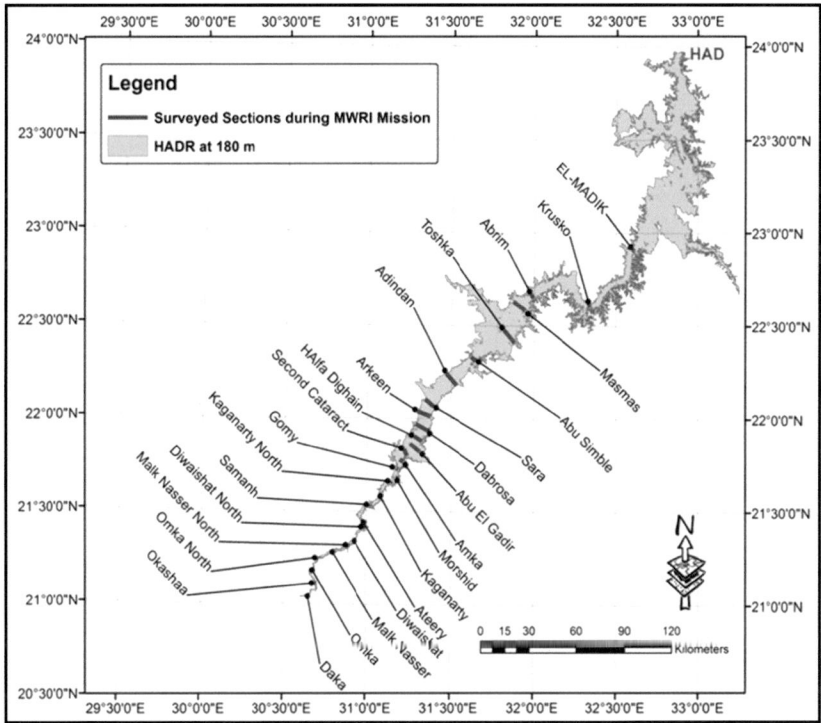

Figure 4.5: Location of the sections surveyed by the MWRI mission in 2007

The data were analyzed with the Depth Sounding software, and then reviewed and cleaned in the post-processing stage. The data were then imported into a GIS system, plotted and examined. Geographic coordinates of data were converted into decimal degrees with a precision of six decimal places. Depths were corrected for the submergence of the transducer and for the elevation of the reservoir surface below the actual water surface elevation measured in known cross stations on the HADR. The data files containing a sequential identification number, longitude, latitude, and depth for each point were then prepared for transfer to the GIS software (MWRI, 2003).

The bathymetric survey for the HADR has been conducted regularly by a mission from the MWRI according to political and finical conditions. The number of sections increased from 18 sections in 1999 to about 29 sections in 2006. Table 4.3 demonstrates the cross sections names, locations, and distances from the HAD. Past missions realized that the sedimentation rate was considered extremely high on Lake Nubia within the Sudan border. Therefore, the mission has usually conducted a comprehensive survey for the entire Lake Nubia, and for the sections that undergoes high erosion or sedimentation rates in Lake Nasser within the Egyptian border, as shown in Figure 4.6.

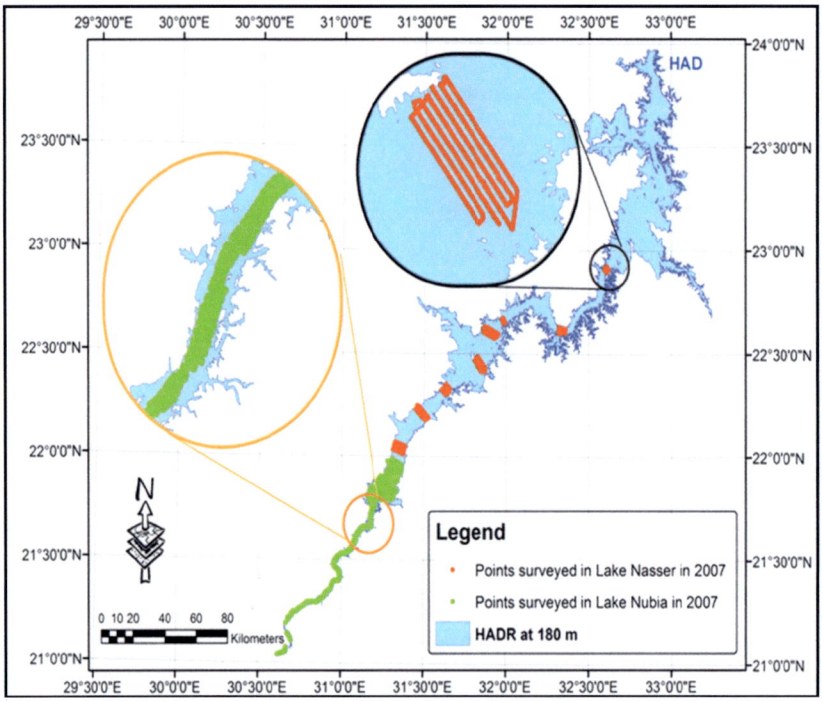

Figure 4.6: Points surveyed in 2007 on Lake Nasser and Lake Nubia

Table 4.3: The HADR cross-sections' names, codes, West and East head geographical coordinates, and distances upstream from the HAD in Kilometers (HDA-MWRI, 2009).

Location	Section No	Section Code	Section Name	Distance US HAD	West Head of Cross Sec.		East Head of Cross Sec.	
					N	E	N	E
Lake Nubia	1	23	Daka	487.5	21.052	30.647	21.046	30.650
	2	A-A	Okashaa	477	21.110	30.693	21.109	30.698
	3	19	Okma	466	21.193	30.668	21.186	30.670
	4	B-B	North Okma	456	21.248	30.709	21.243	30.715
	5	16	Malk Nasser	448	21.289	30.785	21.279	30.788
	6	C-C	North Malk Nasser	438	21.319	30.889	21.313	30.896
	7	13	Diwaishat	431	21.352	30.920	21.344	30.930
	8	H-H	North Diwaishat	422	21.415	30.986	21.413	30.996
	9	10	Ateery	415.5	21.449	30.972	21.457	30.979
	10	8	Samanh	403.5	21.533	31.019	21.524	31.021
	11	6	Kaganarty	394	21.593	31.073	21.588	31.085
	12	W-W	North Kaganarty	384	21.663	31.139	21.657	31.140
	13	3	Morshid	378.5	21.678	31.176	21.678	31.187
	14	D	Gomy	372	21.729	31.177	21.724	31.192
	15	28	Madik Amka	368	21.764	31.185	21.774	31.204
	16	27	Amka	364	21.793	31.167	21.773	31.207
	17	26	Second Cataract	357	21.838	31.220	21.797	31.237
	18	25	Abd El kadir	352	21.868	31.251	21.821	31.315
	19	24	Halfa Dighaim	347	21.905	31.280	21.879	31.327
	20	22	Dabrosa	337.5	21.974	31.282	21.933	31.367
lake Nasser	21	21	Arkeen	331.1	22.037	31.302	22.012	31.377
	22	20	Sara	325	22.106	31.340	22.068	31.400
	23	9	Adindan	307	22.241	31.455	22.178	31.522
	24	8	Abu Simbel	282	22.334	31.623	22.317	31.642
	25	T-T	Toshka	256	22.479	31.816	22.436	31.851
	26	7	Masmas	237	22.622	31.861	22.583	31.945
	27	6	Apreem	214	22.673	31.979	22.650	31.993
	28	K-K	Korosko	182.5	22.622	32.337	22.605	32.335
	29	M-M	El Madik	130	22.917	32.599	22.906	32.609

In January 2007, I shared in the mission survey of the HADR and spent 30 days on the mission ship collecting the required data (MWRI, 2007). In this study, the data collected for the missions executed in 1999, 2003, 2006, 2007, 2008, 2009 and 2010 are available in spreadsheet format. To determine the lakebed altitudes above mean sea level, the water depths were subtracted from the lake surface water level measured during this phase. The surveyed points reached up to 300000 points for one mission. Within Lake Nasser, five kilometers around each section were surveyed except in 2010 when the bathometric survey covered the entire areas between the surveyed sections. The bathymetric survey covered some parts of Lake Nubia in 1999, and the entire lake from 2007 to 2009. In 2010, the mission only reached to Diwaishat Section due to low water depths (MWRI, 2011).

4.1.4.3 Satellite images

As, the bathymetric survey points alone are not adequate to define the lakebed altitudes, historical satellite imagery is essential to define the bed altitudes. The Landsat imagery were used for this research because the images have been taken regularly each 16 days since July 1972 as stated on the Landsat mission website (USGS, 2013). The purpose of the Landsat program is to provide scientists, application engineers, land planners, environmental and nature conservers, and others with a continuing stream of remote sensing data for monitoring and managing the Earth's resources. It was the first Earth-observing satellite to be launched with the express intent to study and monitor our planet's landmasses (Baumann, 2010). Furthermore, Landsat images are used successfully in catchment management, flood delineation and for land use and vegetation mapping. The price of Landsat image was reduced from $5000 to $600, which led to a large increase of Landsat data users. In October 2008, United States Geological Survey (USGS) made all Landsat 7 data free to the public, leading to a 60 - fold increase of data downloads. Since the end of May 2003, the satellite Landsat 7 suffered the loss of its scan line corrector, causing missing strips in the images.

In this study, Landsat images were used in identifying the border line of the HADR at different water levels. The Landsat satellite captures the HADR area with its huge inaccessible areas in nine different image frames as shown in Figure 4.7. Some images were covered with clouds or had other problems. Thus, it was difficult to find complete sets on different dates. Therefore, all the available Landsat images (over 500 images) were downloaded for the period from 1972 to 2003 from the Landsat images' library of the USGS to pick the appropriate sets with similar water levels. Fifty-seven images with Mobile Mapping Systems MMS sensors were downloaded from Landsat 1, Landsat 2, and Landsat 3 taken between 1972 to 1979 as listed in Appendix E. Figure 4.7 shows the path and rows for the downloaded images. Seventy-five images with MMS sensors were downloaded from Landsat 4 and Landsat 5 taken between 1983 to 1990 (Appendix E). Figure 4.8 shows the path and rows for the downloaded images. Moreover, 246 images with TM sensors were downloaded from Landsat 4 and Landsat 5 taken between 1984 to 2003 (Appendix E). Figure 4.9 shows the path and rows for these downloaded images. In addition, 135 images with ETM+ sensors were downloaded from Landsat 7 taken between 1999 to 2003 (Appendix E). Figure 4.10 shows the path and rows for these downloaded images. The images were downloaded as tar compressed files.

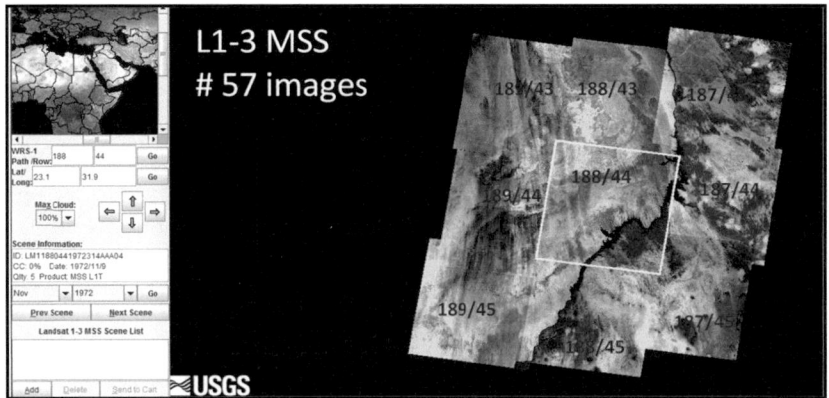

Figure 4.7: Path and row of the downloaded image sets from Landsat 1,2,3 MSS

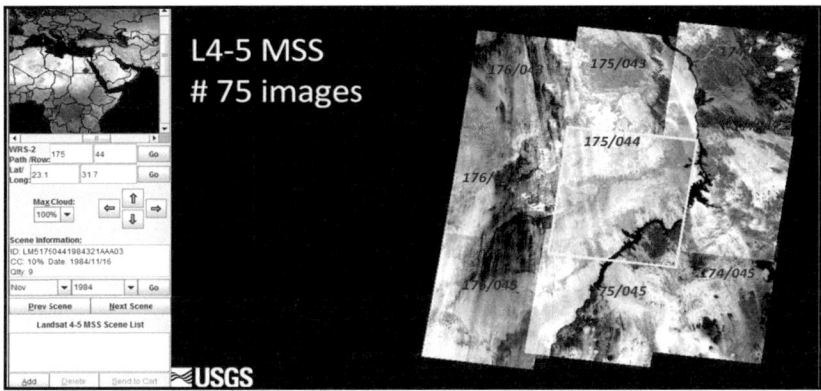

Figure 4.8: Path and row of the downloaded image sets from Landsat 4,5 MSS

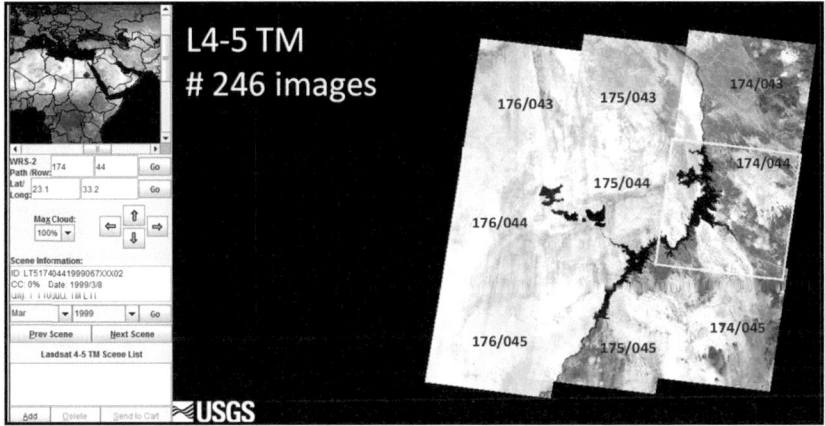

Figure 4.9: Path and row of the downloaded image sets from Landsat 4,5 TM

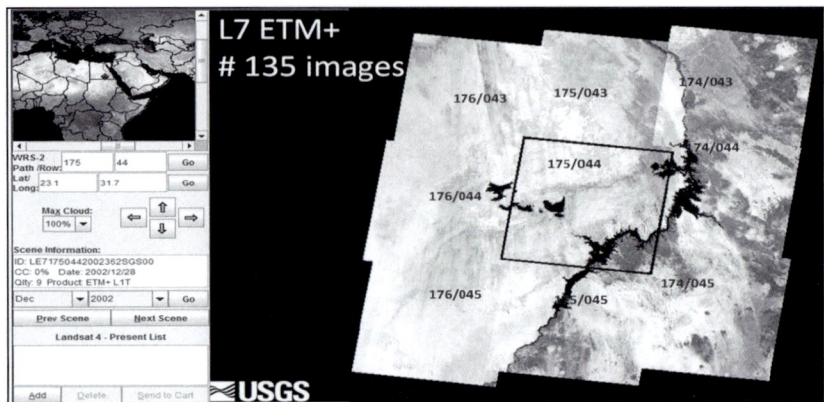

Figure 4.10: Path and row of the downloaded image sets from Landsat 7 ETM+

4.1.4.4 Aerial photos:

Since 1969, the former Nile morphology has changed due to submergence after the operation of the HAD. However, this morphology is essential to define the bottoms of the HADR khors due to the lack of old topographic maps before the HAD construction. Therefore, mono aerial photos are needed to delineate the original Nile morphology. Two old aerial photos acquired in 1955 were used to simulate the Nile at surface water level of 110 m AMSL. Figure 4.11 shows the used aerial photos overlain on the current the HADR morphology.

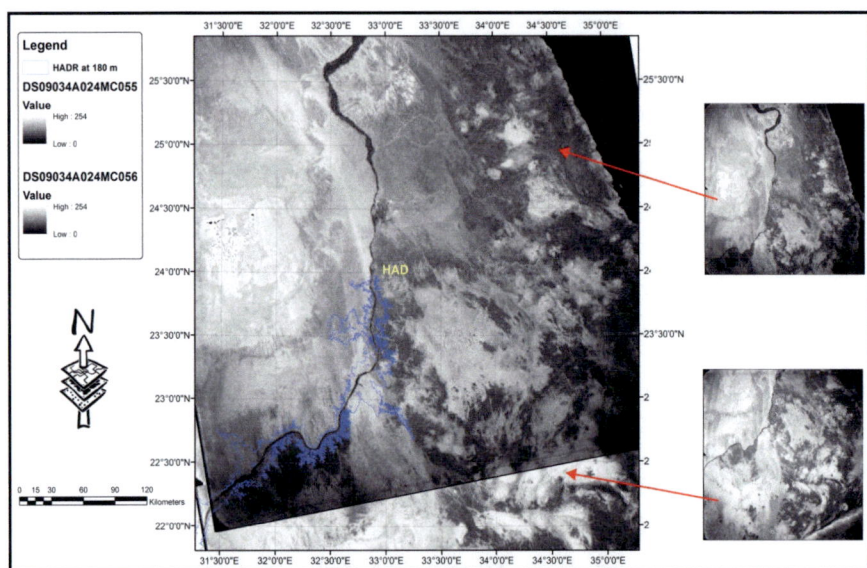

Figure 4.11: Aerial Photos acquired for the Nile south Egypt before the HAD construction overlaid by the HADR current morphology at 180 m AMSL

4.1.5 Bed soil sedimentary data

The sediment load brought into the river is relatively large and reaches from 50 to 228 million tons (Abu El-Ata 1978). Particles of the main river's sediment load include sand, silt and clay, in addition to a small percentage of small-size gravel. During the mission of the bathymetric survey, the MWRI also monitors the HADR bed soil at the cross sections listed in Table 4.3. Samples of the bed soil have been collected to determine properties of the deposited sediment for these cross-sections. The samples have been usually collected from east, west and middle of the sections. Sieve analysis and sedimentation analysis were applied to the samples to determine their categorizes and distributions. Three main categorizes have been usually defined, clay and silt, sand, and gravel. The data included the MWRI mission results in 2007 (Appendix F).

4.2 Methods

This research intends to simulate the HADR current hydrological and meteorological features to investigate the HADR water losses problems. The simulation models were developed using several created databases and geo-databases that provide data about the HADR current hydrological and meteorological conditions. The created databases were used in generating a Digital Elevation Model (DEM) for the HADR which was used in simulating the HADR's hydrological characteristics. Moreover, each khor was modeled to investigate alternatives for reducing the evaporation losses from the HADR. The methods for creating the databases and the simulation models are explained below.

4.2.1 Softwares

EXCEL, SPSS, ARCGIS, ERDAS imagine and ENVI EX were used to carry out the research methodology. Description of these tools is summarized below.

4.2.1.1 Spreadsheet-based software

Numerous spreadsheet/graphics/database programs are on the market. *LOTUS 1-2-3*, *QUATTRO PRO* and *EXCELL* are among the more popular packages. In this study, *EXCEL 2007* was used in building the databases, and in estimating the hydrological characteristics for the HADR and its khors.

4.2.1.2 Geographic information systems

A geographic information system *(GIS)* is a set of computer-based tools for storing, processing, combining, manipulating, analyzing, and displaying data that are spatially referenced to the earth. There are some popular software packages such as: *ArcGIS, ARC/INFO* and *GRASS*. ARCGIS 10 with its spatial module was used to generate the above mentioned shape files, interpolating the point coverages to raster data for remote areas and generating the DEM for the study area.

4.2.1.3 SPSS statistical packages

SPSS Statistics is a software package used for statistical analysis. *SPSS* Statistics (originally, Statistical Package for the Social Sciences, later modified to read Statistical Product and Service Solutions) was released in its first version in 1968 after being developed by Norman H. Nie, Dale H. Bent, and C. Hadlai Hull. SPSS is among the most widely used programs for statistical analysis in social science. It is used by market researchers, health researchers, survey companies, governments, education researchers, marketing organizations, and others. For modeling the HADR and the hydrological characteristics of the khors, regression models were constructed using the curve estimate option within the analysis menu.

4.2.1.4 ERDAS Imagine

Earth Resource Data Analysis System (*ERDAS*) *Imagine* software has tools for image mapping, visualization, enhancement, geocorrection, and reprojection, including remote sensing analysis and spatial modeling. For this study, pre-processing Landsat imagery was conducted in *ERDAS Imagine* 13 with basic image manipulation tools, reprojection and geocorrection processing. The extracted water areas from *ENVI EX* were edited using *ERDAS* Imagine to correct the misclassified pixels.

4.2.1.5 ENVI EX

ENvironmental for **V**isualizing **I**mages (*ENVI*) *EX* is a powerful viewer with a dynamic display that allows for rapid viewing and manipulation of remotely sensed images, vectors, and annotations. *ENVI EX* works seamlessly with ESRI layers and feature classes. Feature Extraction is a module for extracting information from high-resolution panchromatic or multispectral imagery based on spatial, spectral, and texture characteristics. One can extract multiple features at a time, including vehicles, buildings, roads, bridges, rivers, lakes, and fields. Feature Extraction works well with Digital Globe image data in an optimized, user-friendly, and reproducible fashion so users can spend less time with processing details and more time interpreting results. Feature Extraction in *ENVI EX* uses an object-based approach to classify imagery. In this study, Feature Extraction workflow was used to define the water bodies from Landsat imagery.

4.2.2 Creation of databases

For creating the databases, EXCEL has been used. For establishing the Geodatabases, ARCGIS software has been applied to generate and manipulate data. The steps for developing the databases and the subsequent analysis are described below.

4.2.2.1 HADR Meteorological database (HADRMTDB)

For this study, the **H**igh **A**swan **D**am **R**eservoir **Met**eorological **Dat**abase (HADRMTDB) was created to record the measured data at the six stations since their operation till 2010. Figure 4.12 shows the framework for creating the database and interpolating the data. The database was build for many parameters, such as wind speed, wind direction, air temperature, barometric pressure and relative humidity measured at two meters and at four meters above the ground, in addition to water temperature measured at water surface level, and at various water depths. Moreover, average net radiation measured at two meters at some stations, and the estimated evaporation rate were recorded. The database was stored in EXCEL files through which the averages of the daily, monthly and annually evaporation losses rates were computed. The analysis of each parameter was conducted using the SPSS statistical package to determine their trends. The evaporation rates vary from one station to another according to their meteorological characteristics. A GIS shape file was created for the six meteorological stations as point coverage. The attributes of this shape file included the mean daily evaporation rates (mm/day), the mean annual evaporation rates (mm), the mean air temperature, the mean relative humidity, the mean water temperature, and the mean wind speed. This shape file was used as geo-database for the mentioned parameters estimated at the different meteorological stations.

Since only a limited number of stations were available as data points, interpolation is essential to predict unknown values for any other geographic point data. Interpolation can estimate values for cells in a raster from a limited number of sample data points providing predictions for each location in the study area as shown in Figure 4.13. The unknown values can be predicted with a mathematical formula that uses the values of nearby known points (Olea, 1999). Many mathematical formulas are available such as **I**nverse **D**istance **W**eighted (IDW), Kriging, and Natural neighbour. The IDW formula uses a method of interpolation that estimates cell values by averaging the values of sample data points in the neighborhood of each dealing out cell. The nearer the point is to the center of the cell being estimated, the more influence (weight) it has in the averaging process. (NAOUM and TSANIS, 2004, Li and Heap, 2011) showed that IDW can provide reliable estimates regardless of the number of gages or the cell size used in the interpolation. In this study, each of the attributes of the meteorological shape file was interpolated using ARCGIS Spatial Analyst modules to produce their distribution along the HADR according to IDW formula. The output of these interpolation were grid coverages for the mean daily evaporation rates (mm/day), the mean annual evaporation rates (mm), the mean air temperature, the mean water temperature, the mean relative humidity, and the mean wind speed.

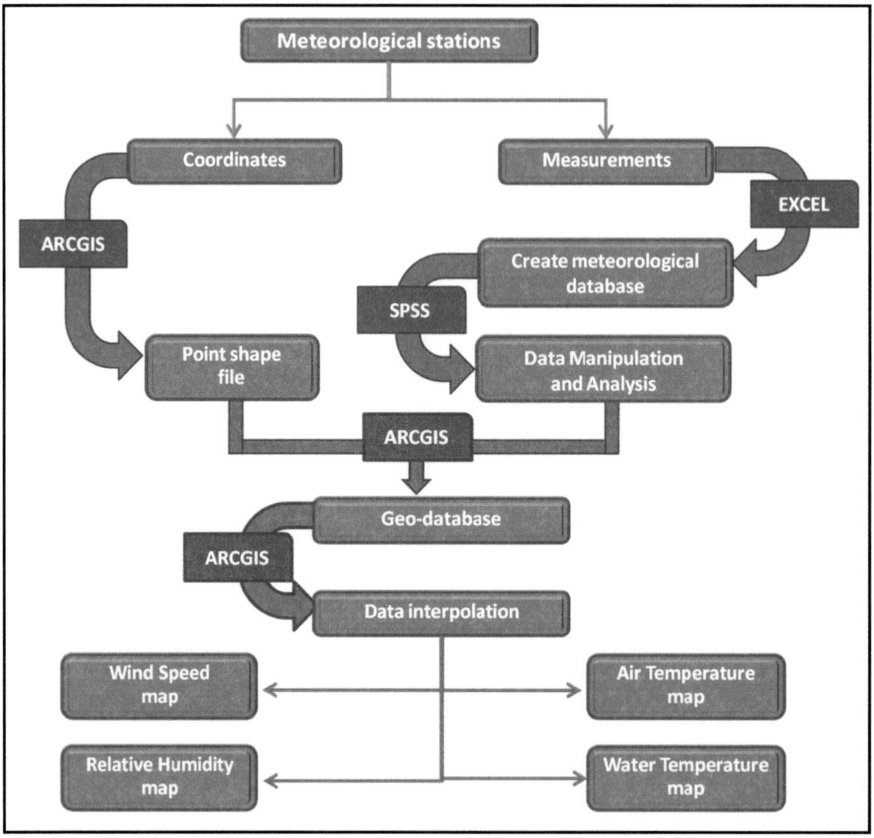

Figure 4.12: General methodology flowchart for creating database and geodatabase for the meteorological data.

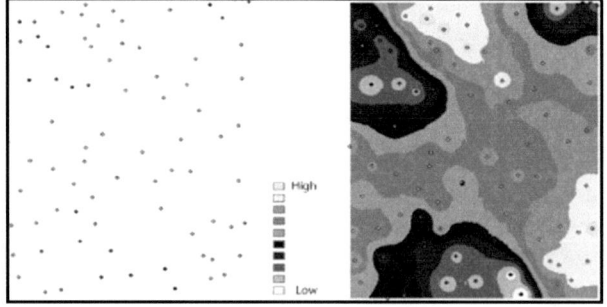

Figure 4.13: Grid coverage interpolated from point coverage. (ESRI, 2011)

4.2.2.2 HADR Hydrological database (HADRHYDB)

In this study, the **H**igh **A**swan **D**am **R**eservoir **Hy**drological **D**ata**b**ase (HADRHYDB) was created to record the lake mean surface water level and the lake flow, measured daily at Aswan station since 1964. Figure 4.14 shows the framework for creating the database and interpolating the data. More than 16000 water level records, covering the period from 1964 to July 2010, were saved as EXCEL files. Moreover, the mean surface water level measured at Wadi Halfa between August 1968 to July 2007, were stored in the database. The monthly average discharge measured at Dongola was also recorded. The saved data were then analyzed using the SPSS statistical package to determine the main characteristics of each station.

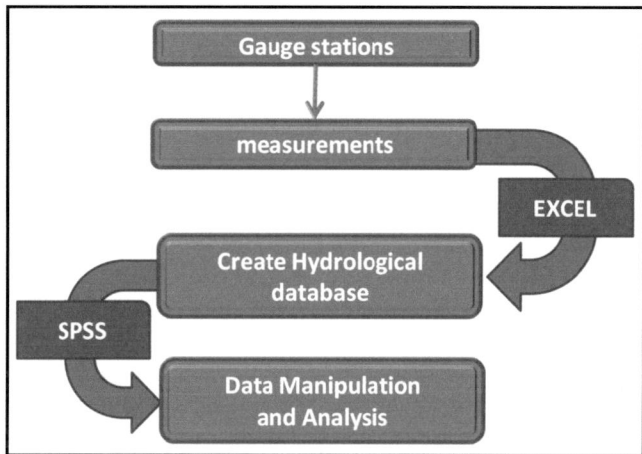

Figure 4.14: General methodology flowchart for creating hydrological database for the HADR.

4.2.2.3 Contour lines for the HADR

A **D**igital **E**levation **M**odel (DEM) for the HADR area was necessary to simulate the topographic features of the area. Until now, there was no DEM for this area available. Thus contour lines of the study area were required. Figure 4.15 shows a general flowchart of the used data to create the contour lines coverage. As stated before, the latest topographic maps of the HADR and the surrounding area were produced in the 1960s with a scale of 1:100,000 before the construction of the HAD. Moreover, these old maps did not include the HADR bed altitudes below 160 m AMSL. In addition, the HADR morphology has changed due to sediment and erosion events since the construction of the HAD. Therefore, in this study, these topographic maps were used to delineate the contour lines for the regions outside the lake edges at the maximum water level of 181.6 m AMSL. The contour lines below this level were extracted from satellite images covering the lake during the past years as listed in Appendix E.

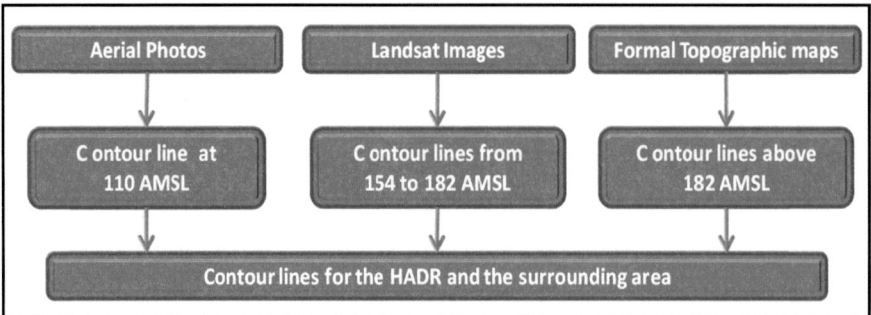

Figure 4.15: General flowchart of the used data to create the contour lines coverage

The historical satellite images of the HADR, provided by the Landsat satellites from 1973 to 2003, were utilized in constructing new contour lines for the HADR at different altitudes. To pick the appropriate sets from the downloaded images, the surface water level for each image was identified using the HADRHYDB. Images with similar water levels were selected to identify the lake morphology at different elevations, and considered one imagery set. Each set consists of nine images covering the HADR and the surrounding area. The lake sides have a smooth incline, thus a step of 1.5 m is adequate to embrace various surface areas. To preprocess the downloaded images, the tar compressed files were extracted to uncompress the bands of each image. The different bands were then stacked to have image with different available bands using the image processing software ERDAS IMAGINE. The images were geographically rectified to Universal Transverse Mercator (UTM) projections. The images of each imagery set were co-registered to ensure that their features match and were mosaiced into one image. To identify the water area, the mosaiced images had to be classified into two classes, water and land. Band four is the near infrared band with wave length ranges between 0.77 and 0.90. Consequently, it emphasizes biomass content and shorelines as stated on the USGS website (USGS, 2013). Therefore, band four was used in classifying the images

The image processing software ENVI EX provides the Feature Extraction Workflow to extract information from high-resolution panchromatic or multispectral imagery based on spatial, spectral, and texture characteristics. Feature Extraction workflow uses an object-based approach to classify imagery. An object is a region of interest with spatial, spectral, and/or texture characteristics that define the region. Traditional remote sensing classification techniques are pixel-based. For example, the supervised classification, produced by ERDAS Imagine, depends on the spectral reflectance of pixels, while the Object-based classification with its knowledge base includes other attributes such as shape, texture, relation to neighboring objects and the spectral information provided in an image. Feature Extraction Workflow consists of two primary steps: Find Objects and Extract Features. The Find Objects task is divided into four stages: Segment,

Merge, Refine, and Compute Attributes. Then, the Extract Features task can be performed. This task consists of supervised or rule-based classification and exporting classification results to shape files and/ or raster images as shown in Figure 4.16 (ENVI EX, 2013).

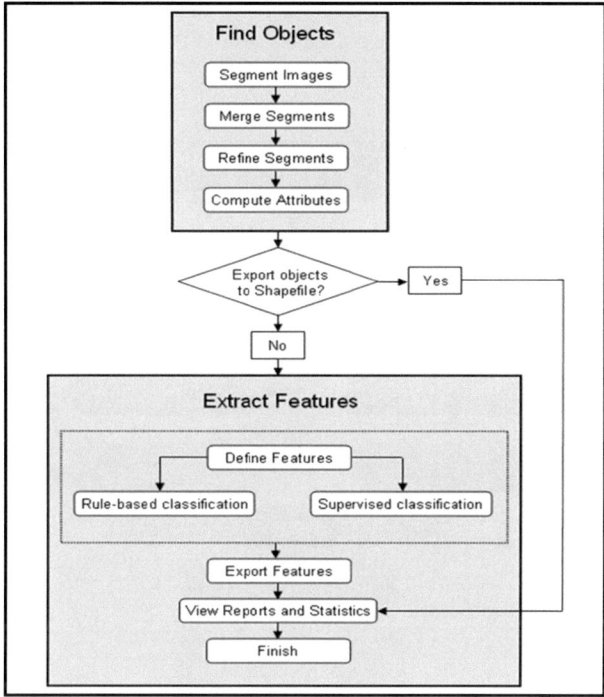

Figure 4.16: Methodology flow chart of the object oriented classification using ENVI EX image processing software - (ENVI EX, 2013)

In this study, a pilot area was tested using the near infra red band (band four) to select the appropriate method for classifying the images. Figure 4.17 shows an overall workflow of the data processing methodology applied to the pilot area. Supervised and unsupervised classifications were generated using ERDAS Imagine based on the pixel values. Moreover, object oriented classification was produced using the feature extraction module under the EVNI EX image processing software. The results of the pilot area showed that when using the feature extraction module, a higher classification accuracy was achieved. Therefore, mosaiced images of the HADR were imported into the image processing software ENVI EX to extract the water area.

Figure 4.18 shows an overall workflow of the data processing methodology applied to the Landsat images. Rule based classification was used according to the mean attributes (average value of the pixels comprising the region in band 4). Finally the classification outputs were exported to raster images. The raster images produced from the ENVI EX were imported to ERDAS IMAGINE software, in order to manually edit the water areas to correct the errors occurring around the lake shore lines. The modified layers were converted into polygon shape files to delineate the borders of the water areas. These polygons were converted into polylines representing contour lines of individual water levels using ARCGIS. The contour lines were combined into a new shape file, simulating the contour lines between 154.5 m to 181.6 m AMSL.

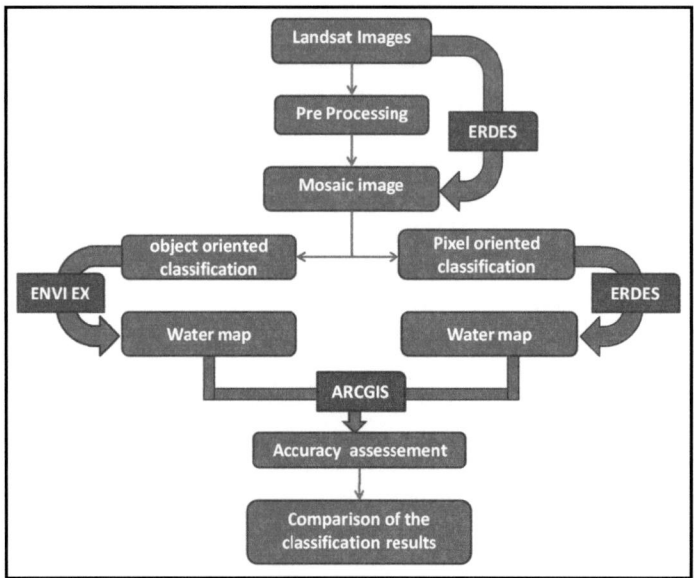

Figure 4.17: An overall workflow of the data processing methodology applied to the pilot area.

For the altitudes below 154 m AMSL, the formal Nile morphology before the HAD construction was extracted from old aerial photo taken in 1955 to simulate the contour line of 110 m AMSL. A similar approach was applied to these aerial photos to extract the water bodies. Figure 4.19 shows an overall workflow of the data processing methodology applied to the Aerial photos. The two Aerial images were mosaiced into one image. The mosaiced image was imported into ENVI EX to extract the water area. The water area was exported to ERDAS Imagine to edit the misclassified areas. The water class was converted into a polygon shape file to delineate the border of the water area using ARCGIS. This polygon was transformed to polyline shape file representing contour line of 110 m AMSL using ARCGIS. The generated contour line of 110 m was attached to the generated contour lines from the Landsat imagery.

Material and Methods

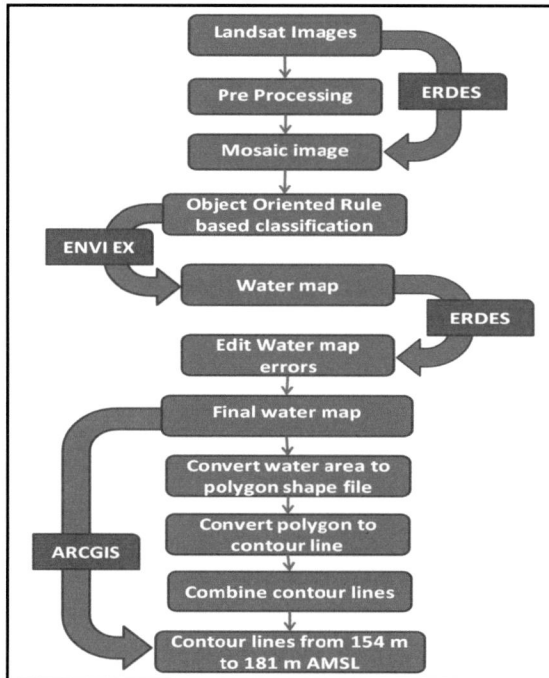

Figure 4.18: Overall workflow of the data processing methodology applied to Landsat images.

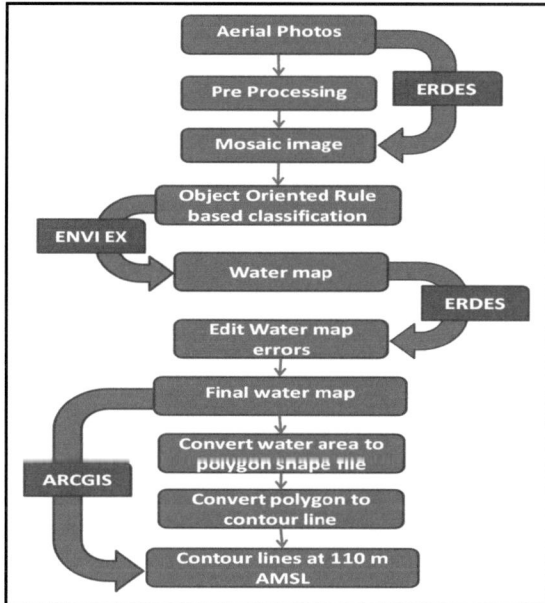

Figure 4.19: Overall workflow of the data processing methodology applied to Landsat images.

4.2.2.4 Spot heights for the HADR

The HADR bathymetric survey data were used to identify the altitudes of the HADR bed. Figure 4.20 shows an overall workflow of the bathymetric survey measurements processing methodology. The water depths were saved as text files including the surveyed point coordinates and the water depths. The text files were converted to EXCEL files. The water depths were subtracted from the lake surface water level, measured during this phase to determine the lakebed altitudes above mean sea level. The data were checked to eliminate the wrong altitudes and coordinates outside the HADR region. The data were created as point shape file using ARCGIS10. The shape files were utilized to create spot heights for the lake altitudes for 1999 and 2003, as well as from 2006 to 2010.

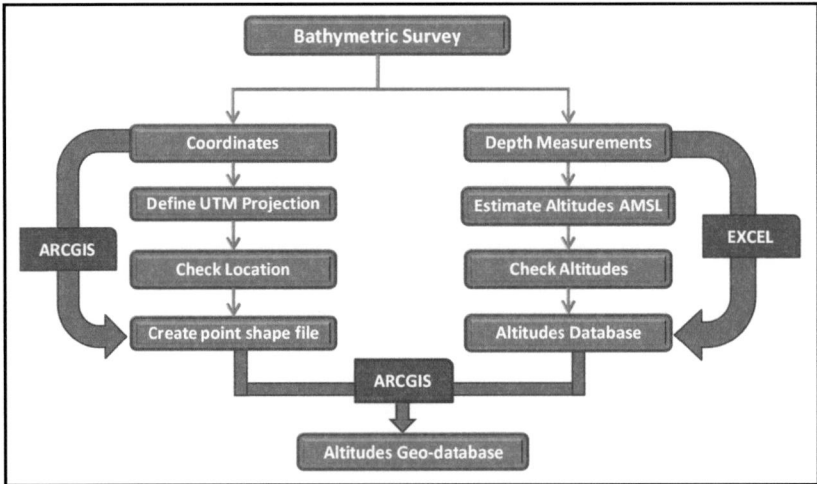

Figure 4.20: Overall workflow of the bathymetric survey measurements processing methodology

4.2.2.5 HADR Seepage Data Base (HADRSPDB)

As stated before, piezometers are used to study and follow up the behavior of seepage from the HADR through the banks. In this study, the piezometers' data were saved as EXCEL files for the period from 1964 to 2009 forming the High Aswan Dam Reservoir Seepage Database (HADRSPDB). The amount of lateral seepage flow from the reservoir is calculated by Darcy's law according to equation 4.1.

$$Q = K\,I\,A = K\,I\,H\,L \qquad \text{Equation 4.1}$$

Where:
    $Q =$ Discharge in m3/ day.
    $A =$ Area of flow in m2.

I = Hydraulic gradient.
K = Coefficient of Permeability in m/ day.
H = Thickness of the seepage face in m.
L = Length of seepage face in m.

The seepage per unit area depends on two factors, the coefficient of permeability of the formation and the hydraulic gradient. The first factor was obtained from field tests. Results of the field tests were presented in Table 4.4. The second factor was calculated from the piezometeric readings.

Table 4.4: Average Coefficients of Permeability Test

Section	K (m/day)	
	West	East
Garf Husein	7.09×10^{-1}	3.55×10^{-1}
Afia	1.66×10^{-1}	---
Tushka	1.66×10^{-1}	8.64×10^{-2}
Adindan	8.64×10^{-1}	1.73×10^{-1}

4.2.2.6 HADR Bed Soil Sedimentary Categories Database (HADRBSDB)

In this study, HADR Bed Soil sedimentary categories Database (HADRBSDB) was created to record the results of MWRI missions to analyse and identify the HADR bed soil sedimentary categories. The average of the percentage of clay and silt, sand, and gravel were stored in EXCEL files. A point shape file was created including the soil classification data for each section along the HADR using ARCGIS 10.

4.2.3 HADR model development

To study the morphological and hydrological characteristics of the HADR, a Digital Elevation Model (DEM) was developed to model the topography of the HADR bed. Regression models were then established to simulate the HADR's characteristics as described below.

4.2.3.1 HADR Digital Elevation Model (HADRDEM)

To establish an actual Digital Elevation Model for the High Aswan Dam Reservoir (HADRDEM), the generated contour lines from the satellite images, spot heights computed from the lake bathometry survey in 2010, and the old topographic maps were used. Figure 4.21 shows overall workflow methodology for creating digital elevation models.

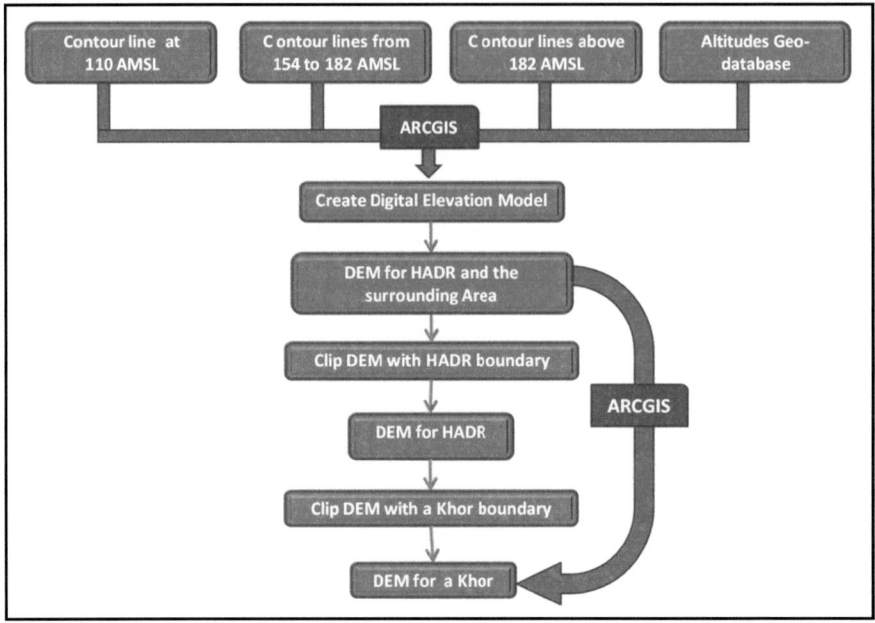

Figure 4.21: Overall methodology workflow for creating digital elevation models.

The new DEM was generated using the Topo to Raster tool in the ARCGIS 10 software with a spatial resolution of 25 m for those laying areas between 21° and 24°N and between 31°30`and 33°30`E. The Topo to Raster tool is an interpolation method particularly used for the creation of a hydrological accurate DEM, based on the ANUDEM program developed by (Hutchinson, 1988). Topo to Raster interpolates elevation values for a raster while imposing constraints to ensure a connected drainage structure, and correct representation of ridges and streams from input contour data. Initially, Topo to Raster uses information inherent to the contours and the spot heights to build a drainage model. The areas of the steepest slope were identified by classifying areas of local maximum curvature in each contour. Consequently a network of streams and ridges is created to ensure appropriate hydro-geomorphic properties of the output DEM. Thus, the general morphology of the surface has been determined. Then, the contour and the spot height data have been used to interpolate the elevation values at each cell (Hutchinson, 1988). In this study, the HADR boundary at 182 m AMSL was used to extract the generated the HADRDEM. To study their hydrological characteristics at maximum water level of 182 m AMSL, several khors were extracted as well.

The bathymetric surveys usually covered the entire Lake Nubia. Therefore, the point shape files generated for years 1999, 2003, 2006, 2007, 2008, 2009, and 2010 and the boundary of Lake Nubia at 182 m AMSL were used for generating DEMs for the bed of Lake Nubia during those years. These DEMs were used in monitoring the sediment events within Lake Nubia because this part of the HADR usually undergoes high sedimentation rates. The annual sediment amount for those particular years was computed by estimating the difference between the altitude values in the generated DEMs, as shown in Figure 4.22. The volume of sediment and erosion events was estimated using EXCEL.

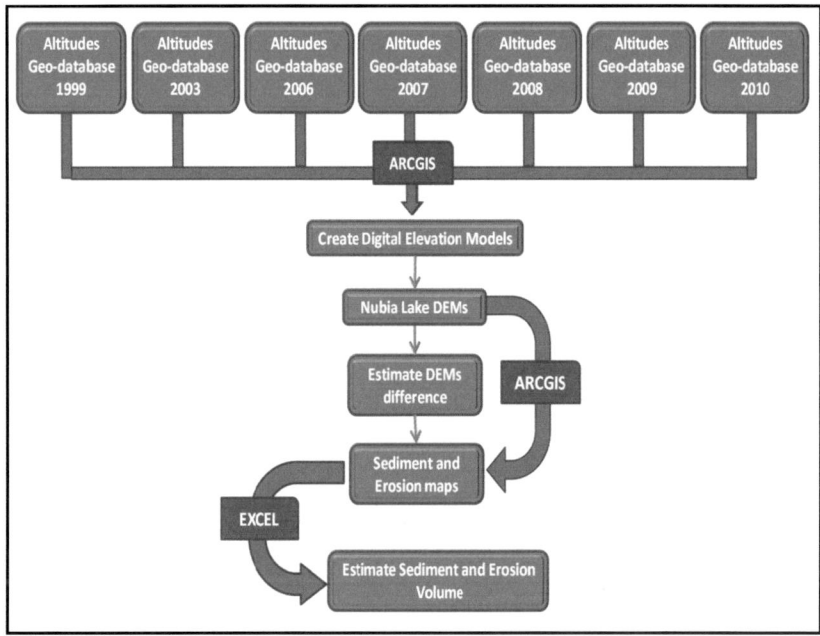

Figure 4.22: Overall methodology workflow for estimating sediment and erosion events.

4.2.3.2 HADR mathematical models

In this study, HADRDEM was isolated by using the HADR boundary at 182 m to determine the number of pixels for each level within the HADR area. The HADRDEM was produced as a floating number format. To estimate the HADR area and volume, sheets of one meter thickness were assumed to facilitate the calculations. Therefore, the HADRDEM was transformed to unsigned 16 bit format, in order to measure the altitudes to the nearest meter. Consequently, the number of pixels for each altitude with one meter step, were estimated from the histogram field in the attributes of the HADRDEM. The number of pixels corresponding to each altitude was exported from the HADRDEM as EXCEL files to estimate the hydrological and morphological characteristics for the HADR and the studied khors. Figure 4.23 shows a schematic example of a cross section through a reservoir.

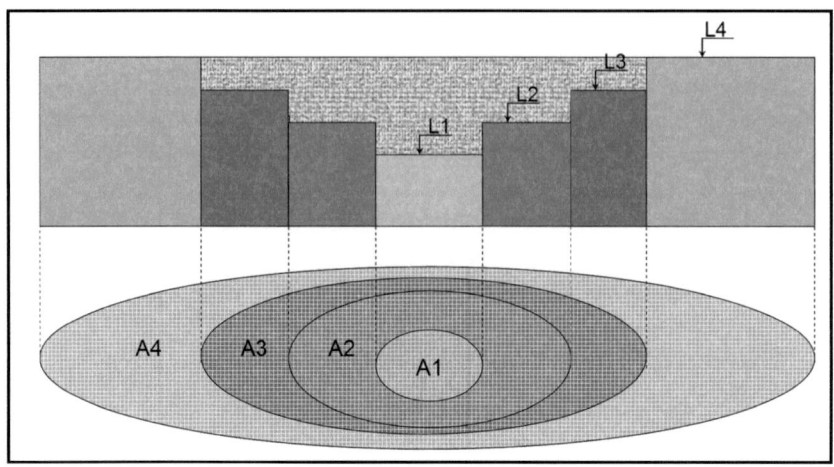

Figure 4.23: A scheme for estimating the reservoir area and volume from the DEM, where L is the altitude and A is the lakebed area at this altitude.

L is the altitude in one meter steps, N is the number of pixels at this altitude, A is the lakebed area at the same altitude, AW is the water surface area at this level, h is the HADR depth. The pixel area is equal to 25 m × 25 m. A_1 is equal to the number of pixel at altitude L_1 (N_1) multiplied by the pixel area. A_2 is equal to the number of pixel at altitude L_2 (N_2) multiplied by the pixel area. The water surface area at altitude L_1 (AW_1) is equal to A_1. The water surface area at altitude L_2 (AW_2) is equal to the sum of A_1 and A_2. The water surface area at altitude L_3 (AW_3) is equal to the sum of A_1, A_2 and A_3 as shown in Equitation 4.2 and Equation 4.3.

$$A_h = 25 \times 25 \times N_h \qquad \text{Equation 4.2}$$

$$AW_h = A_1 + A_2 + A_3 + \cdots \ldots + A_h \qquad \text{Equation 4.3}$$

The volume at altitude L_1 is equal to zero. The volume up to altitude L_2 (V_2) is equal to the water surface area at altitude L_1 (AW_1) multiplied by one meter, where the difference between L_1 and L_2 equal one meter. The volume up to altitude L_3 (V_3) equal to the water surface area at altitude L_2 (AW_2) multiplied by one meter plus V_2. The volume up to altitude L_4 (V_4) is equal to the water surface area at altitude L_3 multiplied by one meter plus V_3, as shown in Equation 4.4. Area and volume were computed at each altitude for the HADR and the studied khors.

$$V_m = (AW_{m-1} \times 1) + (AW_{m-2} \times 1) + \ldots \ldots + (AW_2 \times 1) + (AW_1 \times 1)$$
$$\text{Equation 4.4}$$

To model the HADR and the investigated khors, the EXCEL files for estimating the area and the volume were exported to the SPSS statistical package to build regression models for the relationship between three pairs: area and altitude; volume and altitude; in addition to area and volume. Mathematical models were developed to reveal the relationship between surface water level, lake surface area and lake water volume. The available types of regression models including linear, logarithmic, inverse, quadratic, cubic, compound, power, s, growth, and exponential equations were investigated to detect the most proper curve fit for each relationship. The selected equations were evaluated to check the model reliability. Similar mathematical models were built for the investigated khors to model their hydrological characteristics up to an elevation of 200 m AMSL using the DEM for the whole area. These models were used to investigate the possibility of eliminating these khors by determining proper location for building dams. The criteria for evaluating different alternatives were based on the dam cross section, dam length, dam depth, and the surface area eliminated downstream the dam. Moreover, the mathematical models were used for estimating the required filling materials and the availability of sufficient filling material around the eliminated khors. The HADR was divided into two divisions at the El-Madik section. HADRDEM was used as well to generate mathematical models to simulate the lake characteristics upstream and downstream of the proposed new dam at El-Madik section.

Chapter 5

Results and Discussion

This chapter presents the main results of this study. The databases generated for HADR, which include data on meteorology, hydrology, seepage, and bed soil sediment, are described and analyzed, moreover, the geo-databases created for the HADR topographic characteristics are presented. The Digital Elevation Model created for the HADR is analyzed and verified, and the mathematical models generated to describe the hydrological characteristics of the HADR and their verification are also addressed. The results of the digital elevation models and the mathematical models developed for this study are used to explore the HADR, its khors, and their respective sediment status. Several scenarios are analysed for eliminating khors, controlling water levels, and lowering the lakebed level. The effects of each scenario on the characteristics of the HADR and its environment are presented. Several combinations of scenarios are discussed. Finally, the measures proposed to reduce evaporation losses are presented and evaluated.

5.1 Produced databases

For this study, many databases were created in order to simulate the HADR and the surrounding environments. The developed databases were the HADR Meteorological Database, the HADR Hydrological database, the Contour Lines for the HADR, Spot heights for the HADR and the HADR Seepage Database. The characteristics of each database are provided below.

5.1.1 HADR Meteorological Database (HADRMTDB)

For this study, the HADRMTDB was created to record the measured data of the six meteorological stations along the HADR measured by ARHDA (Figure 3.40). As stated in 4.1.1.1, the measured data were wind speed, wind direction, air temperature, relative humidity measured at two meters, relative humidity measured at four meters, water temperature measured at water surface level, water temperature measured at various water depths, average net radiation measured at two meters at some stations and the calculated evaporation rate. These measured data were collected since the first operation of each station until 2010. The analysis of these measured data is provided below.

5.1.1.1 Air temperature

In this study, daily and monthly average air temperatures measured at two and four meters were computed using the hourly measured data recorded in the

HADRMTDB. Figure 5.1 shows the air temperature data measured at two meters above surface water level at the Aswan station before constructing the HAD from 1953 to 1963 and at the Wadi Halfa station from 1959 to 1963. The air temperature at Aswan station is higher than that at Wadi Halfa station with a mean annual difference of about 3%. Figure 5.2 shows the data measured at each station after operating the HAD at two meters above surface water level at Aswan station from 1974 to 2009, at El-Alaky station from 1995 to 2007, at Abu Simbel station from 1995 to 2007, at Kalabsha station from 2004 to 2008, at Amada station from 2004 to 2008 and at Toshka station from 2004 to 2006. The highest air temperatures were measured at Aswan station. Compared to Aswan station, the mean air temperature at the other stations were less by about 15% at El-Alaky station, 7% at Abu Simbel station, 22% at Kalabsha station, 11% at Amada station and 12% at Toshka station.

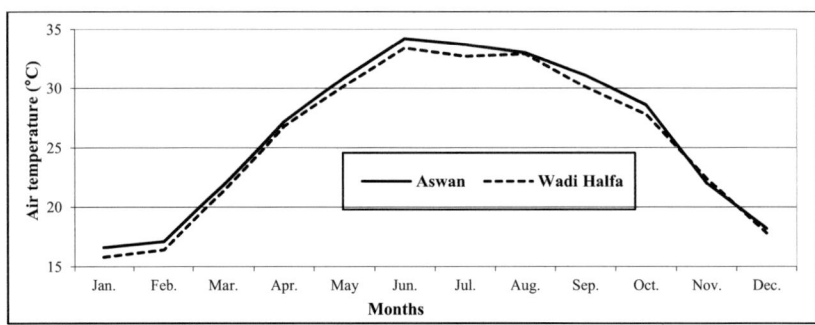

Figure 5.1: Monthly variation of the mean values of air temperature in °C before constructing the HAD at Aswan station from 1953 to 1963 and at Wadi Halfa station from 1959 to 1963 (based on the data measured by ARHDA – section 4.1.1.1)

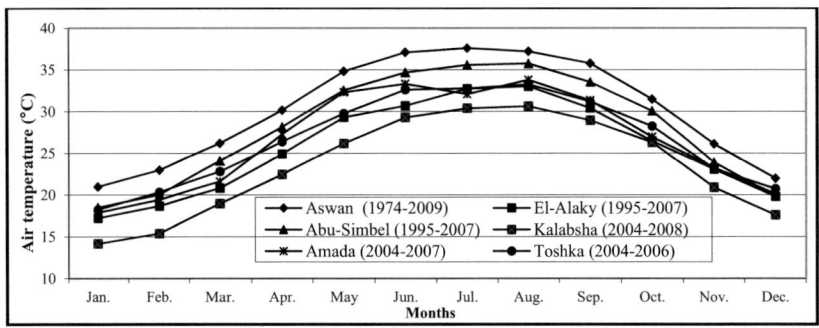

Figure 5.2: Monthly variation of the mean values of air temperature in °C after constructing the HAD, at Aswan (1977-2010), El-Alaky (1995-2007), Abu Simbel (1995-2007), Kalabsha (2004-2008), Amada (2004-2007) and Toshka (2004-2006) (based on the data measured by HDA – section 4.1.1.1)

The monthly averages of the air temperature, measured at two and four meters above surface water level at the different stations of the HADR, were analyzed using the SPSS statistical software. Figure 5.3 shows the histogram for the monthly means of temperatures along the HADR measured at two and four meters above surface water level at all the meteorological stations. For the last four decades, the monthly mean air temperatures were about 28.5°C and 27.2°C measured at two and four meters above surface water level, respectively. The temperatures that most frequently calculated were 37°C and 35°C at two and four meters above surface water level, respectively.

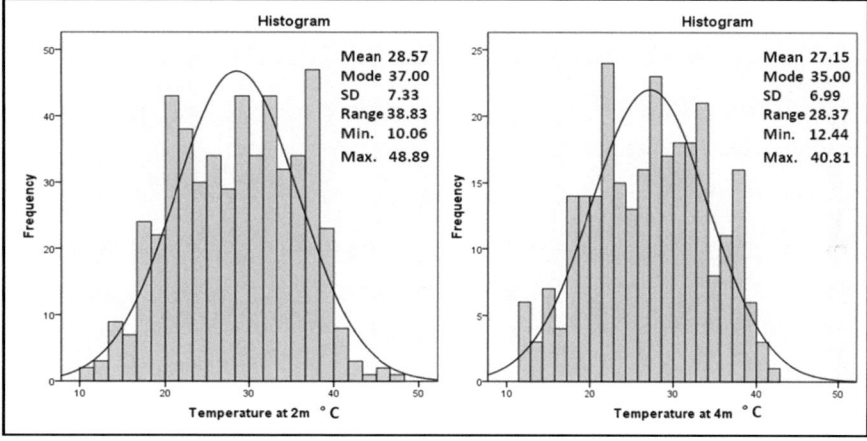

Figure 5.3: Histogram of the monthly mean air temperature in °C measured at two and four meters above surface water level at all the meteorological stations along HADR

Table 5.1 shows the average, median, maximum and minimum monthly air temperature measured at two and four meters above surface water level at all the meteorological stations along the HADR. June, July and August were the hottest months over the years with a mean air temperature of about 35°C. The coldest months were December, January and February where the mean temperature was about 20°C. The highest air temperature of 48°C was recorded in August 2004, while the lowest one of 11°C was recorded in February 2007. Table 5.2 shows the average, median, maximum and minimum monthly air temperature measurements by the different meteorological stations along the HADR in different decades. The highest mean air temperature of 30°C was computed at Aswan station. The highest air temperature of 48°C was measured at Aswan station, while the lowest one of 11°C was measured at El-Alaky station. The mean air temperature was increased by about 1°C per decade at Aswan, Abu-Simbel and El-Alaky stations. A summary of each station is provided in Appendix G.

To generate a geo-database of the air temperature which is presented in table 5.2, the mean air temperature for each station measured at two meters above surface water level, was used. A point shape file was created using the mean air temperature values. The created point coverage was interpolated to determine the mean air temperature at remote areas on the HADR as shown in Figure 5.4.

Table 5.1: The mean, median, maximum and minimum monthly air temperatures in °C measured at two and four meters above surface water level at all the meteorological stations on the HADR

Month	Air temperature at 2 m (°C)				Air temperature at 4 m (°C)			
	Mean	Median	Maximum	Minimum	Mean	Median	Maximum	Minimum
January	18.97	19.49	26.23	10.98	17.39	18.13	23.73	12.44
February	20.75	20.70	40.72	10.06	19.16	19.83	25.76	12.51
March	24.55	24.38	43.04	13.94	22.95	22.34	30.16	17.42
April	28.98	29.99	39.50	17.81	27.27	27.00	33.56	20.91
May	32.68	33.49	41.29	19.68	31.19	30.88	39.81	24.83
June	35.02	35.17	47.42	24.98	32.82	32.49	37.68	27.80
July	35.59	37.01	45.72	27.65	34.04	33.50	40.45	28.32
August	35.67	36.60	48.89	28.83	34.43	33.92	40.81	29.13
September	33.38	34.09	40.58	23.66	32.48	31.86	38.35	26.28
October	29.76	29.72	36.24	20.70	29.02	28.87	36.55	24.11
November	24.42	24.66	32.37	17.57	22.99	22.13	29.12	17.52
December	20.79	21.33	28.41	13.57	19.50	21.18	24.01	13.45
Annual	28.57	28.83	48.89	10.06	27.15	27.84	40.81	12.44

Table 5.2: The mean, median, maximum and minimum monthly air temperatures in °C measured at two and four meters above surface water level at different meteorological stations on the HADR

Station Name	Decade	Temperature at 2 m (°C)				Temperature at 4 m (°C)			
		Mean	Median	Maximum	Minimum	Mean	Median	Maximum	Minimum
Abu-Simbel	1990s	28.77	29.9	37.37	19.65	28.31	30.58	38.14	19.89
	2000s	29.49	30.35	38.12	14.89	28.9	30.92	38.81	14.77
	mean	28.54	30.09	38.12	14.89	28.96	30.92	38.81	14.77
Aswan	1970s	28.75	29.32	38.6	13.98	no data	no data	no data	no data
	1980s	31.36	32.6	40.24	17.6	no data	no data	no data	no data
	1990s	32.1	34.59	40.9	15.94	31.3	32.42	40.45	14.14
	2000s	33.49	35.27	48.89	13.57	32.04	33.16	40.81	13.45
	mean	30.48	31.24	48.89	13.57	29.36	30.2	40.81	13.45
El-Alaky	1990s	28.25	29.16	36.52	19.51	no data	no data	no data	no data
	2000s	29.19	29.98	37.42	10.06	28.08	24.95	33.74	12.44
	mean	25.87	25.31	37.42	10.06	28.08	24.95	33.74	12.44
Kalabsha	2000s	23.91	23.84	34.35	12.4	24.08	24.95	33.74	12.44
Amada	2000s	28.21	28.92	38.16	16.79	28.38	29.23	38.58	16.59
Toshka	2000s	25.79	25.74	33.23	17.35	no data	no data	no data	no data
All stations		28.57	28.83	48.89	10.06	27.15	27.84	40.81	12.44

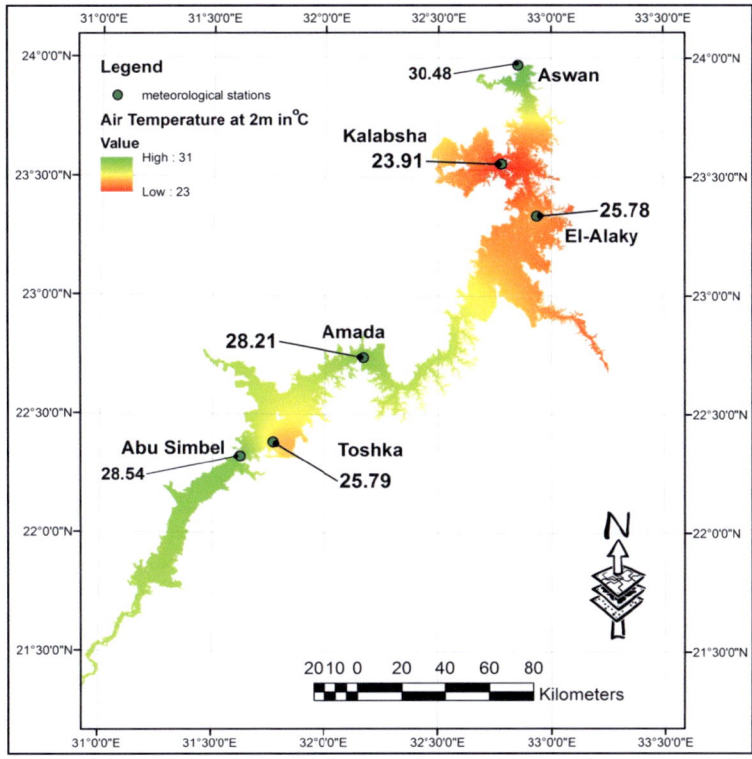

Figure 5.4: Distribution of mean air temperature of the HADR

5.1.1.2 Relative humidity

In this study, daily and monthly averages of relative humidity measured at two and four meters were computed using the hourly measured data recorded in the HADRMTDB. Figure 5.5 shows data on relative humidity measured at two meters above surface water level at Aswan station before constructing the HAD namely from 1953 to 1963 and at Wadi Halfa station from 1959 to 1963. Relative humidity values at Wadi Halfa were higher than those at Aswan station with a mean annual difference of about 26%. Figure 5.6 shows the relative humidity data measured after operating the HAD at two meters above surface water level at Aswan station from 1974 to 2009, at El-Alaky station from 1995 to 2007, at Abu Simbel station from 1995 to 2007, at Kalabsha station from 2004 to 2008, at Amada station in 2000 and at Toshka station from 2004 to 2006. The lowest mean relative humidity was calculated at Aswan station. Compared to Aswan station, the mean relative humidity values were higher than the other stations by about 50% at El-Alaky station, 32% at Abu Simbel station, 24% at Kalabsha station, 31% at Amada station and 36% at Toshka station.

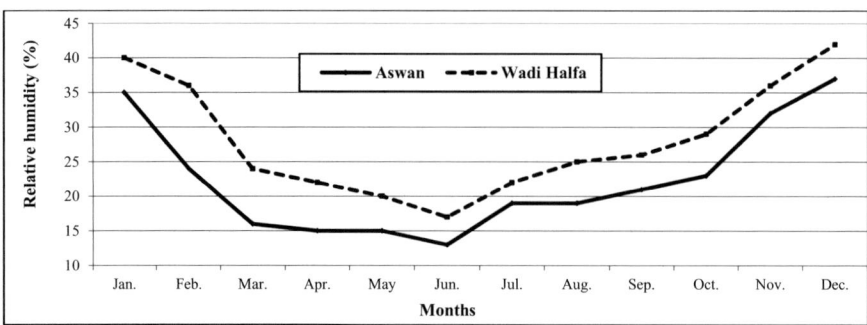

Figure 5.5: Monthly variation of the mean values of relative humidity before constructing the HAD at Aswan station (1953-1963) and at Wadi Halfa station (1959-1963)

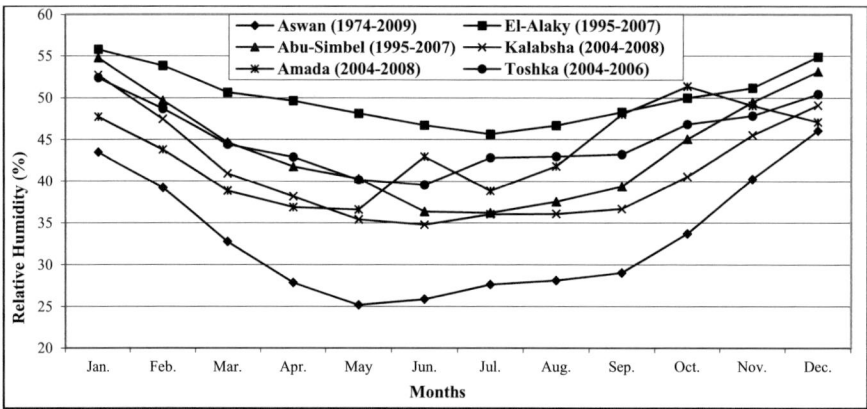

Figure 5.6: Monthly variation of the mean values of relative humidity after constructing the HAD at different meteorological stations

The monthly averages of relative humidity, measured at two and four meters above surface water level at the different stations of the HADR, were analyzed using the SPSS statistical software. Figure 5.7 shows the histogram for the monthly mean relative humidity along the HADR measured at two and four meters above surface water level at all the meteorological stations. For the last four decades, the monthly mean relative humidity was about 40% and 37% measured at two and four meters above surface water level, respectively. The most frequently calculated mean relative humidity was 50% and 34% at two and four meters above surface water level, respectively. Table 5.3 shows the average, median, maximum and minimum monthly relative humidity measured at two and four meters above surface water level at all the meteorological stations along the HADR. June had the lowest mean relative humidity of about 33%, while January had the highest mean relative humidity of about 51%. The highest relative humidity of 72% was recorded in December 1975, while the lowest one of 17% was recorded in May 1984.

Results and Discussion

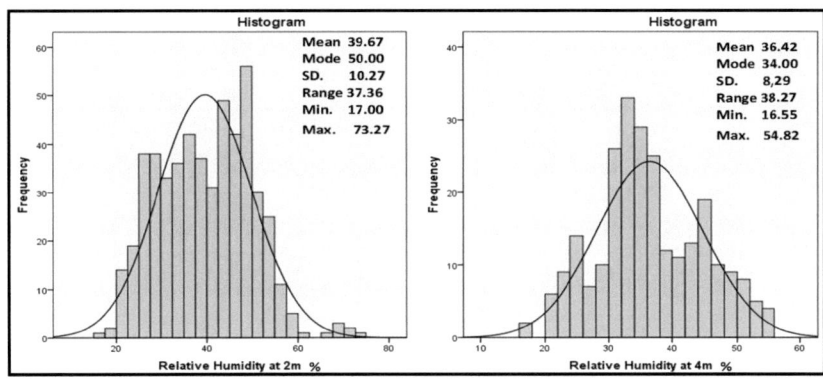

Figure 5.7: Histogram of the monthly mean relative humidity measured at two and four meters above surface water level of all the meteorological stations along the HADR

Table 5.3: The mean, median, maximum and minimum monthly relative humidity measured at two and four meters above surface water level at all the meteorological stations on the HADR

Month	Relative humidity at 2 m (%)				Relative humidity at 4 m (%)			
	Mean	Median	Maximum	Minimum	Mean	Median	Maximum	Minimum
January	50.73	51.27	69.00	39.00	49.95	51.71	54.82	38.32
February	44.96	44.52	69.34	33.00	42.32	43.11	49.69	32.69
March	39.92	39.31	61.84	23.00	35.53	36.70	45.29	26.69
April	36.20	36.60	52.93	22.00	31.06	33.86	40.03	20.49
May	33.85	34.29	49.58	17.00	28.93	30.87	36.50	20.46
June	33.29	32.86	50.23	19.00	28.87	29.08	33.50	17.40
July	34.42	34.39	51.39	21.00	30.77	32.33	36.92	21.65
August	34.56	32.48	57.29	21.74	31.10	31.84	37.42	16.55
September	35.22	31.17	50.70	25.00	32.49	34.62	38.46	24.86
October	39.85	37.94	59.48	28.82	36.73	37.69	43.47	28.46
November	45.10	44.00	69.30	23.28	43.17	43.62	48.91	36.57
December	49.31	49.64	72.42	19.38	47.72	48.51	54.63	36.12
Annual	39.61	39.77	72.42	17.00	36.42	35.15	54.82	16.55

Table 5.4 shows the average, median, maximum and minimum monthly relative humidity at the different meteorological stations along the HADR in different decades. The highest mean relative humidity of 50% was computed for El-Alaky station. The lowest mean relative humidity of about 35% was computed at Aswan station. Both the highest and lowest relative humidity values were measured at Aswan station. A summary of each station is provided in Appendix G. To generate a geo-database for the relative humidity, the mean values for each station measured at two meters above surface water level, which are listed in table 5.4, were used. A point shape file was created using the mean relative humidity values. The created point coverage was interpolated to determine the mean relative humidity in remote areas of the lake as shown in Figure 5.8.

Table 5.4: The mean, median, maximum and minimum monthly relative humidity measured at two and four meters above surface water level at different meteorological stations of the HADR

Station Name	Decade	Relative humidity at 2 m (%)				Relative humidity at 4 m (%)			
		Mean	Median	Maximum	Minimum	Mean	Median	Maximum	Minimum
Abu-Simbel	1990s	45.88	44.51	57.56	36.77	42.26	40.3	54.63	33.37
	2000s	42.83	42.22	57.11	23.14	40.69	38.83	54.82	30.20
	mean	43.24	42.28	57.56	23.14	41.02	39.56	54.82	30.20
Aswan	1970s	40.48	38.00	72.42	19.38	no data	no data	no data	no data
	1980s	32.20	30.00	51.00	17.00	no data	no data	no data	no data
	1990s	35.21	32.13	50.23	27.42	32.57	28.85	49.06	23.65
	2000s	34.32	32.55	51.5	21.45	30.94	29.12	49.32	16.55
	mean	35.20	32.55	72.42	17.00	31.16	29.12	49.32	16.55
El-Alaky	1990s	50.33	49.52	55.65	46.6	no data	no data	no data	no data
	2000s	49.21	48.57	73.27	41.23	38.14	35.08	52.36	28.95
	mean	45.81	48.81	73.27	41.23	38.14	35.08	52.36	28.95
Kalabsha	2000s	40.81	38.48	53.67	31.09	38.14	35.08	52.36	28.95
Amada	2000s	41.58	40.82	51.21	33.62	35.50	32.81	46.92	28.21
Toshka	2000s	45.51	44.80	53.51	38.64	no data	no data	no data	no data
All stations		39.67	39.93	73.27	17.00	36.42	35.15	54.82	16.55

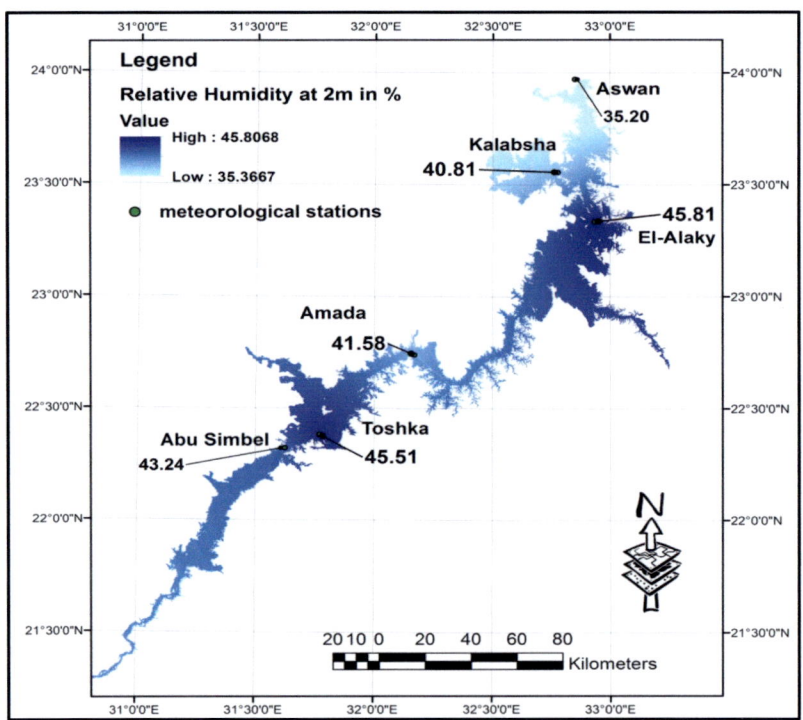

Figure 5.8: Distribution of mean relative humidity of the HADR

5.1.1.3 Wind speed

In this study, daily and monthly average wind speeds measured at two and four meters above surface water level were computed using hourly measured data recorded in the HADRMTDB. Figure 5.9 shows the mean wind speed data measured after operating the HAD at two meters above surface water level at Aswan station from 1977 to 2010, at El-Alaky station from 1995 to 2007, at Abu Simbel station from 1995 to 2007, at Kalabsha station from 2004 to 2008, at Amada station in 2000 and at Toshka station from 2004 to 2006. The least wind speed mean was calculated at El-Alaky station. Compared to El-Alaky station, the mean wind speed at other stations was higher by about 35% at Aswan station, 53% at Abu Simbel station, 27% at Kalabsha station, 25% at Amada station and 14% at Toshka station. The monthly averages of wind speed, measured at two and four meters above surface water level at the different stations of the HADR, were analyzed using the SPSS statistical software. Figure 5.10 shows the histogram for the monthly mean wind speeds along the HADR measured at two and four meters above surface water level at all the meteorological stations.

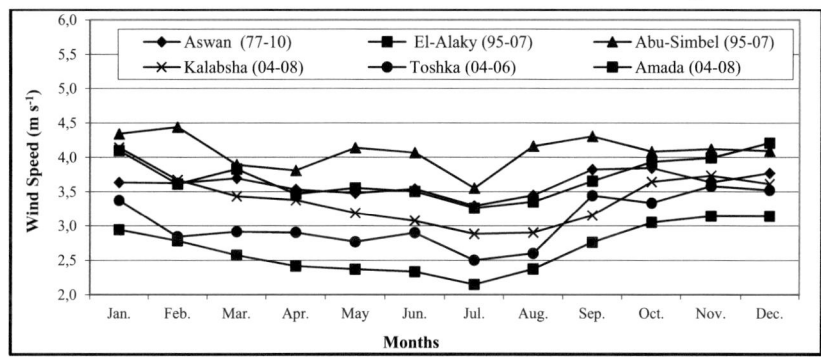

Figure 5.9: Monthly variations of wind speed at different meteorological stations of the HADR

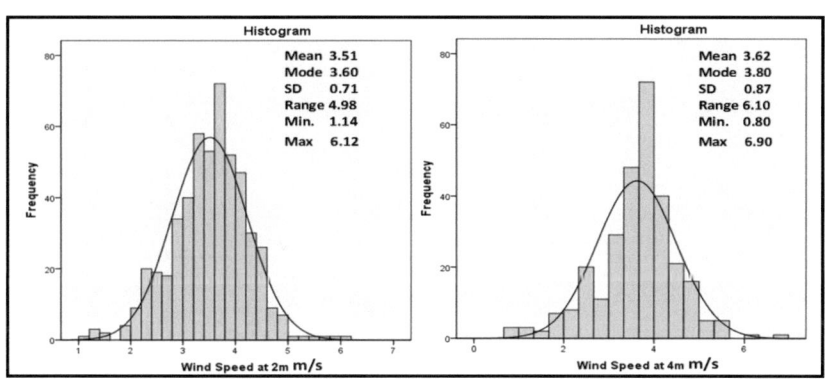

Figure 5.10: Histogram of the monthly mean wind speeds measured at two and four meters above surface water level at all the meteorological stations along the HADR in m/s

For the last four decades, the monthly mean wind speeds were about 3.51 m s^{-1} and 3.62 m s^{-1} measured at two and four meters above surface water level, respectively. The wind speeds that have been recorded most frequently were 3.6 m s^{-1} and 3.8 m s^{-1} at two and four meters above surface water level, respectively. Table 5.5 shows the average, median, maximum and minimum monthly wind speeds measured at two and four meters above surface water level at all the meteorological stations along the HADR. July had the least wind speed of about 3 m s^{-1}, while October had the highest mean wind speed of about 4 m s^{-1}. The lowest wind speed of 1.14 m s^{-1} was recorded in December 2008, while the highest one of 6.12 m s^{-1} was recorded in October 2003.

Table 5.6 shows the average, median, maximum and minimum monthly wind speeds measured at the different meteorological stations along the HADR in different decades. The highest mean wind speed of about 3.9 m s^{-1} was computed at Abu-Simbel station. The lowest mean wind speed of about 2.6 m s^{-1} was computed at El-Alaky station. The highest wind speed was measured at Aswan station while the lowest one was measured at Kalabsha station. A summary of each station is provided in Appendix G. To generate a geo-database for wind speed, the mean values for each station measured at two meters above surface water level, which are listed in table 5.6, were used. A point shape file was created using the mean wind speed values. The created point coverage was interpolated to determine the mean wind speed at remote sites (Figure 5.11).

Table 5.5: The mean, median, maximum and minimum monthly wind speeds measured at two and four meters above surface water level at all the meteorological stations on the HADR

Month	Wind speed at 2 m (m s^{-1})				Wind speed at 4 m (m s^{-1})			
	Mean	Median	Maximum	Minimum	Mean	Median	Maximum	Minimum
January	3.79	3.80	5.40	2.25	3.84	3.85	5.16	0.99
February	3.51	3.66	5.08	1.52	3.55	3.71	5.37	0.86
March	3.44	3.45	4.97	1.25	3.53	3.76	4.53	1.11
April	3.30	3.30	4.36	1.29	3.42	3.63	4.70	1.07
May	3.30	3.36	4.46	2.11	3.43	3.74	4.39	0.80
June	3.36	3.49	4.33	2.03	3.56	3.71	6.90	1.22
July	3.08	3.16	4.05	1.54	3.18	3.33	3.79	1.52
August	3.30	3.28	4.51	2.11	3.45	3.52	4.12	1.85
September	3.69	3.71	4.88	1.83	3.80	3.93	4.72	1.78
October	3.85	3.82	6.12	1.99	4.03	4.21	5.35	1.92
November	3.80	3.80	5.89	1.40	4.03	4.11	6.20	2.37
December	3.82	3.87	5.80	1.14	3.75	3.78	5.53	2.16
Annual	3.51	3.57	6.12	1.14	3.62	3.72	6.90	0.80

Results and Discussion

Table 5.6: The average, median, maximum and minimum monthly wind speed measured at two and four meters above surface water level at different meteorological stations of the HADR

Station Name	Decade	Wind speed at 2 m (m/s)				Wind speed at 4 m (m/s)			
		Mean	Median	Maximum	Minimum	Mean	Median	Maximum	Minimum
Abu-Simbel	1990s	4.15	4.07	5.08	3.08	3.68	3.69	4.32	2.93
	2000s	3.87	3.92	5.42	2.37	3.46	3.77	4.94	1.49
	mean	3.91	3.95	5.42	2.37	3.49	3.76	4.94	1.49
Aswan	1970s	3.98	4.00	4.97	3.07	no data	no data	no data	no data
	1980s	3.75	3.70	4.90	2.89	no data	no data	no data	no data
	1990s	3.61	3.53	4.16	3.16	3.85	3.77	4.38	3.46
	2000s	3.38	3.36	6.12	1.25	3.56	3.67	4.75	1.86
	mean	3.70	3.70	6.12	1.25	3.60	3.70	4.75	1.86
El-Alaky	1990s	2.55	2.47	3.22	2.06	2.81	2.72	3.54	2.28
	2000s	2.67	2.51	4.33	1.92	3.70	3.77	6.90	0.80
	mean	2.66	2.51	4.33	1.92	3.56	3.67	6.90	0.80
Kalabsha	2000s	3.38	3.31	4.66	1.14	3.93	3.83	5.47	2.35
Amada	2000s	3.27	3.02	3.72	2.84	3.34	3.17	3.84	2.98
Toshka	2000s	3.03	3.02	3.91	1.54	no data	no data	no data	no data
All stations		3.51	3.57	6.12	1.14	3.62	3.72	6.90	0.80

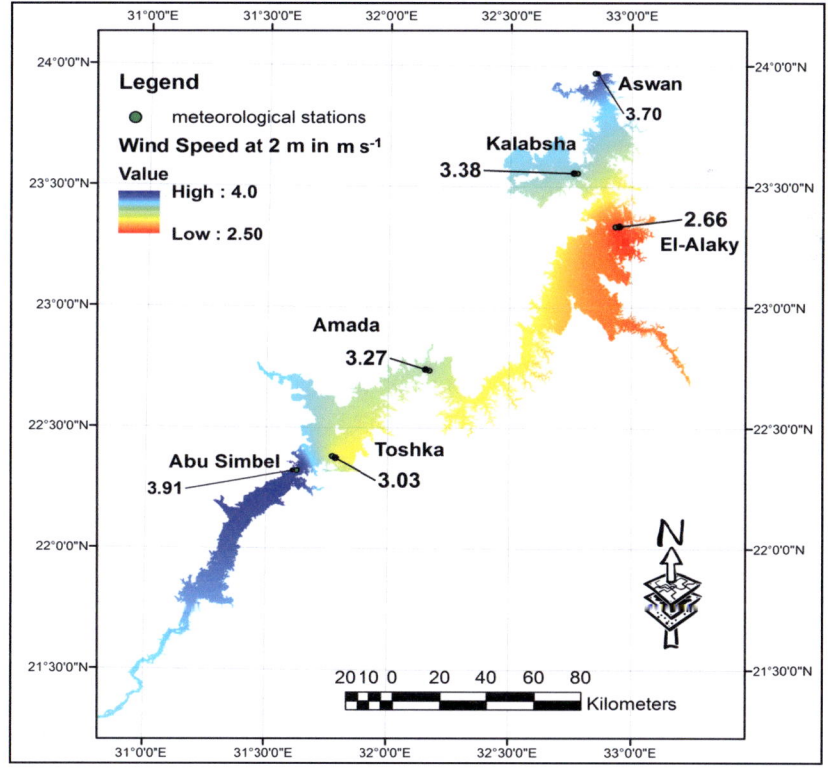

Figure 5.11: Distribution of mean wind speeds of the HADR

5.1.1.4 Water temperature

In this study, daily and monthly average water temperatures measured at the surface of water were computed using the hourly measured data recorded in the HADRMTDB. Figure 5.12 shows the mean water temperature data measured after operating the HAD at Aswan station from 2000 to 2009, at El-Alaky station from 2000 to 2007, at Abu Simbel station from 2000 to 2007, at Kalabsha station from 2004 to 2008, at Amada station in 2000 and at Toshka station from 2004 to 2006. The least water temperature mean was calculated at Aswan station. Compared to Aswan station, the mean wind speeds at other stations were higher by about 14% at El-Alaky station, 12% at both Abu Simbel and Kalabsha station, about 9% at Amada station and about 13% at Toshka station.

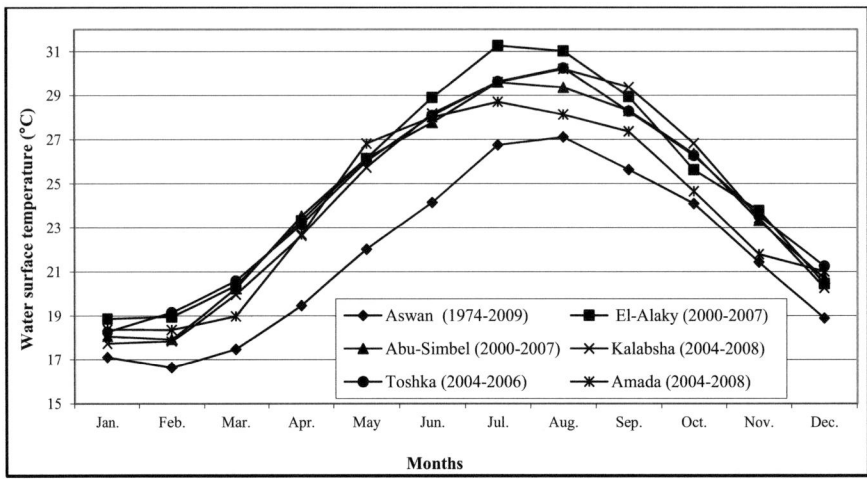

Figure 5.12: Monthly variations of water surface temperature at different meteorological stations of the HADR °C

The monthly averages of the water temperature, measured at the surface of water at the different stations of the HADR, were computed using the statistical software SPSS. Figure 5.13 shows the histogram of the monthly means of water temperature along the HADR measured at all the meteorological stations. For the last four decades, the monthly means of water temperature was about 23.93°C. The water temperature that most frequently recorded was 25°C. Table 5.7 shows the average, median, maximum and minimum monthly water temperature measured at all the meteorological stations along the HADR. January and February had the lowest water temperature of about 18°C, while July and August had the highest mean water temperature of about 30°C. The lowest water temperature of 16°C was recorded in February 2000, while the highest one of 34°C was measured in September 2008.

Results and Discussion

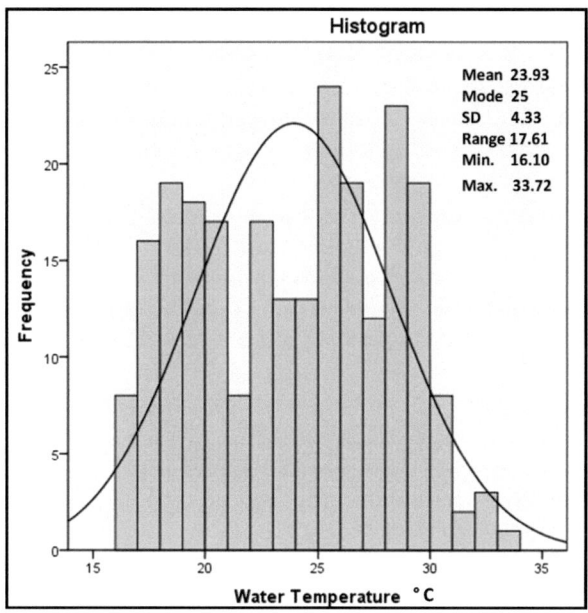

Figure 5.13: Histogram of the monthly mean water temperature measured at all the meteorological stations along the HADR °C

Table 5.7: The mean, median, maximum and minimum monthly water temperature measured at all the meteorological stations on the HADR

Month	Water temperature (°C)			
	Mean	Median	Maximum	Minimum
January	17.96	17.99	20.01	16.41
February	17.77	17.91	19.57	16.10
March	19.53	19.75	21.51	16.53
April	22.31	22.70	25.00	19.29
May	24.97	25.52	26.90	21.26
June	27.56	27.94	31.56	24.31
July	29.57	29.52	32.64	27.50
August	29.25	29.54	32.89	25.87
September	28.00	28.09	33.72	23.19
October	25.80	25.95	29.26	21.11
November	23.09	23.02	24.81	20.68
December	20.20	20.00	22.10	18.70
Annual	23.93	24.25	33.72	16.10

Table 5.8 shows the average, median, maximum and minimum monthly water temperature at the different meteorological stations along the HADR in different decades. The highest mean water temperature of about 25°C was computed at Kalabsha station. The lowest mean water temperature of about 22°C was computed at Aswan station. The highest water temperature of 34°C was measured at Kalabsha station while the lowest one of 16°C was measured at Aswan station. A summary of each station is provided in Appendix G. To generate a geodatabase for the water temperature, the mean values for each station, which are listed in table 5.8, were used. A point shape file was created using the mean surface water temperature values. The created point coverage was interpolated to determine the mean water temperature in remote areas of the lake as shown in Figure 5.14.

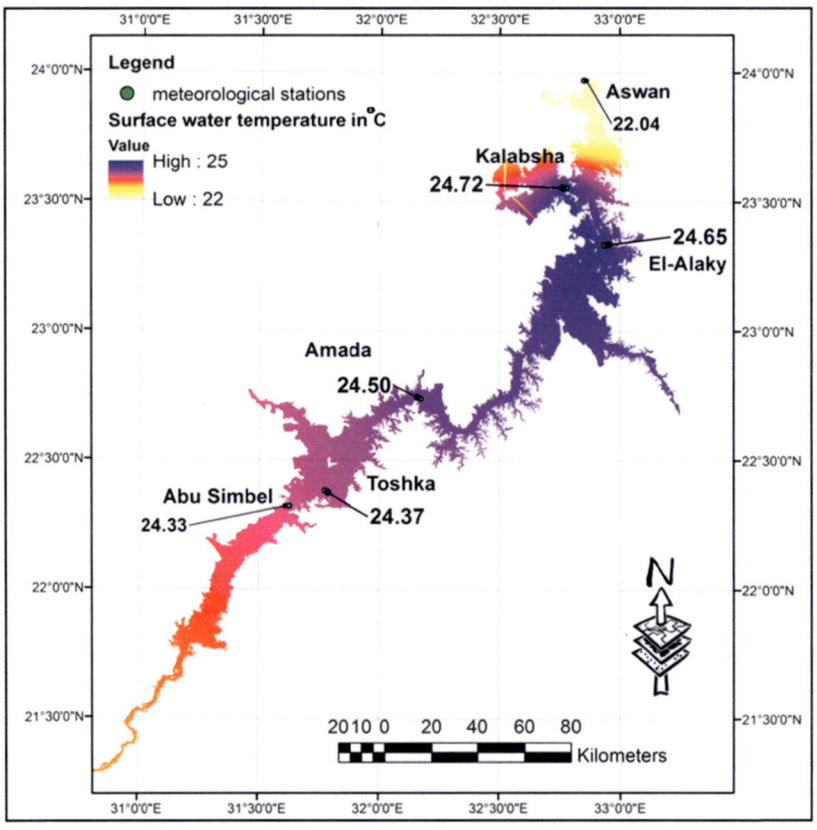

Figure 5.14: Distribution of mean surface water temperatures on the HADR

Table 5.8: The mean, median, maximum and minimum monthly water temperature measured at different meteorological stations of the HADR

Station Name	Decade	Water temperature (°C)			
		Mean	Median	Maximum	Minimum
Abu-Simbel	1990s	24.35	25.16	30.73	17.12
	2000s	24.32	24.96	30.93	16.76
	mean	24.33	25.00	30.93	16.76
Aswan	1990s	21.55	21.21	27.50	16.10
	2000s	22.17	22.49	28.39	16.41
	mean	22.04	21.89	28.39	16.10
El-Alaky	1990s	24.96	24.42	32.41	19.57
	2000s	24.50	24.42	32.89	17.56
	mean	24.65	24.42	32.89	17.56
Kalabsha	2000s	24.72	25.56	33.72	16.76
Amada	2000s	24.51	24.20	29.40	18.30
Toshka	2000s	24.37	24.33	30.24	18.27
All stations		23.93	24.25	33.72	16.10

5.1.1.5 Evaporation rates

In this study, the daily evaporation value was obtained from the sum of the calculated 24 hour measured data of the meteorological stations of the HADR and recorded in the HADRMTDB. The monthly and annual means of evaporation rates were also computed. Figure 5.15 shows the monthly means of the evaporation rates measured at Aswan station from 1995 to 2009, at El−Alaky station from 1999 to 2007, at Abu Simbel station from 1995 to 2007, at Kalabsha station from 2004 to 2008, at Amada station from 2004 to 2008 and at Toshka station from 2004 to 2006. The highest mean evaporation rates were calculated at Amada station. Compared to Amada station, the mean evaporation rate at other stations was less by about 25% at El-Alaky station 4% at Abu Simbel, 9% at Kalabsha station, 13% at Aswan station and 31% at Toshka station.

The monthly averages of the evaporation rate, measured at the different stations of the HADR, were analyzed using the statistical software SPSS. Figure 5.16 shows the histogram of the monthly means of evaporation rates along the HADR measured at all the meteorological stations. For the 1990s and 2000s, the monthly mean evaporation rate was about 7.4 mm/day. The most frequent evaporation rate was 6 mm/day. Table 5.9 shows the average, median, maximum and minimum monthly means of evaporation rate measured at all the meteorological stations along the HADR. August had the highest mean evaporation rate of about 10 mm/day, while February had the least one of 5 mm/day. The highest evaporation rate was recorded in August 2001 with 14.5 mm/day, while the lowest one was recorded in December 2008 with 1.9 mm/day.

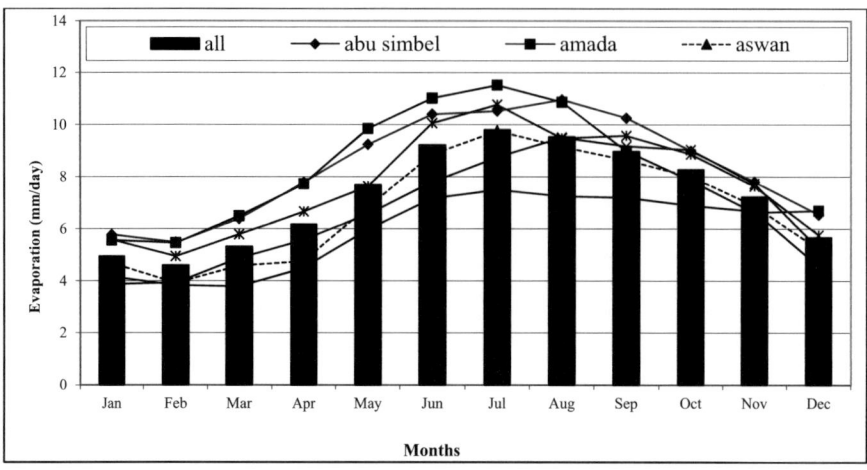

Figure 5.15: Monthly means of variations of the evaporation rate from the meteorological stations along the HADR in mm/day for the 2010s

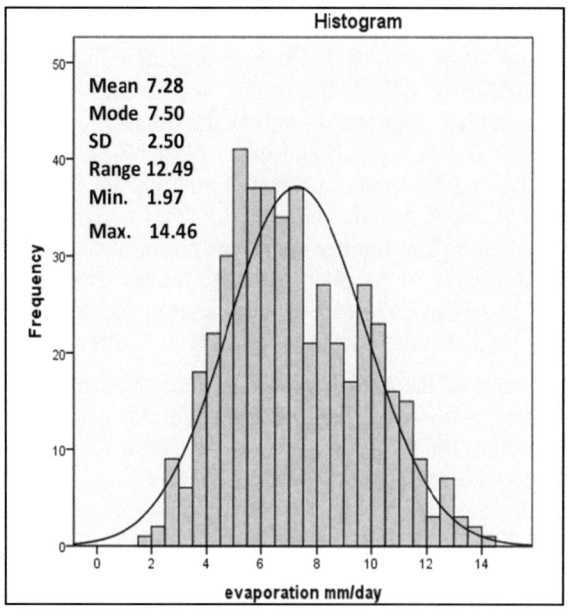

Figure 5.16: Histogram of the monthly means of evaporation rate measured at all the meteorological stations along the HADR °C

Table 5.10 shows the average, median, maximum and minimum monthly means of evaporation rates at the different meteorological stations along the HADR for the different decades. The highest mean evaporation rate of 8.5 mm/day was computed at Amada station, while the lowest one of 6.10 mm/day was computed at El-Alaky station. In the 2000s, the mean evaporation rate had increased by about 1 mm/day equivalent to an increase of 12%, where the mean evaporation rate had increased by about 14% at Abu-Simbel station, by about 9% at Aswan station and by about 16% at El-Alaky station. A summary of each station is provided in Appendix G. To generate a geo-database for evaporation rates, the mean evaporation rates for each station, which are listed in table 5.10, were used. A point shape file was created using the mean values of the evaporation rates. The created point coverage was interpolated to determine the mean evaporation rate of remote lake areas as shown in Figure 5.17.

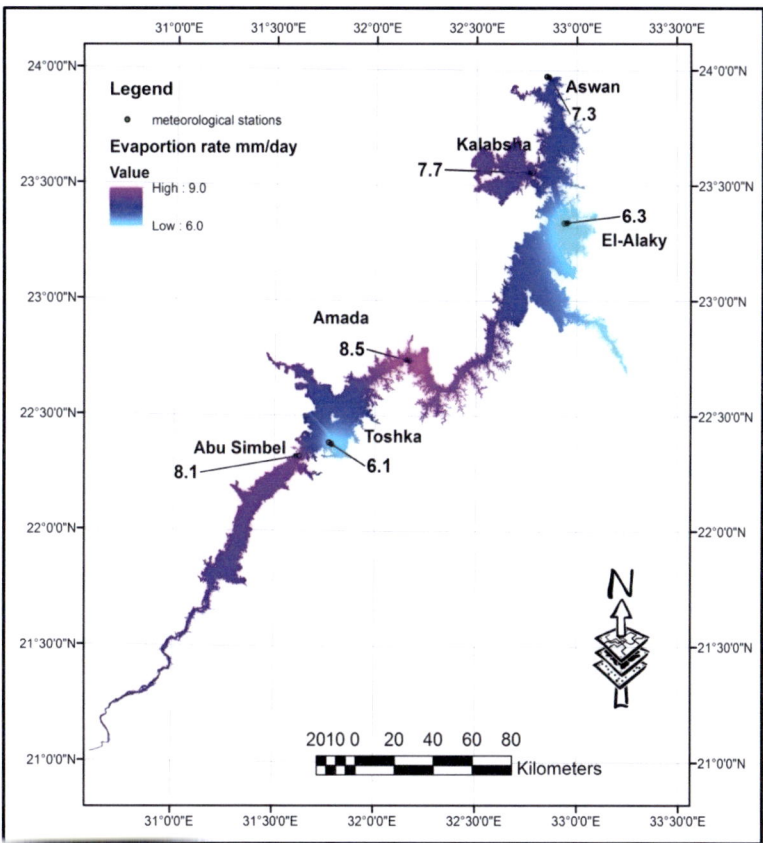

Figure 5.17: Distribution of mean evaporation rate of the HADR

Table 5.9: The average, median, maximum and minimum monthly means of evaporation rates measured at all the meteorological stations along the HADR

Month	Evaporation rate mm/day			
	Mean	Median	Maximum	Minimum
January	4.94	5.13	7.79	2.64
February	4.87	5.01	7.86	2.50
March	5.41	5.38	7.00	3.05
April	6.33	6.34	10.13	3.83
May	7.38	7.36	11.37	4.15
June	8.89	9.23	12.99	5.04
July	9.05	9.51	12.79	4.12
August	9.91	9.90	14.46	6.29
September	9.50	9.63	13.66	6.00
October	8.65	8.49	12.57	3.87
November	6.91	6.83	11.25	2.93
December	5.48	5.44	7.39	1.97
Annual	7.28	7.00	14.46	1.97

Table 5.10: The average, median, maximum and minimum monthly means of evaporation rates measured at the various meteorological stations of the HADR for the different decades

Station Name	Decade	Evaporation mm/day			
		Mean	Median	Maximum	Minimum
Abu-Simbel	1990s	7.46	7.06	13.40	4.12
	2000s	8.52	7.90	14.46	3.90
	mean	8.10	7.44	14.46	3.90
Aswan	1990s	6.96	7.12	11.89	4.29
	2000s	7.56	6.78	11.40	2.23
	mean	7.29	6.93	11.89	2.23
El-Alaky	1990s	5.64	5.95	7.46	3.84
	2000s	6.59	6.06	12.52	2.50
	mean	6.28	6.03	12.52	2.50
Kalabsha	2000s	7.73	8.05	11.69	1.97
Amada	2000s	8.50	8.10	12.79	2.88
Toshka	2000s	6.03	6.46	8.54	3.79
All stations	1990s	6.69	6.58	13.40	3.84
	2000s	7.49	7.22	14.46	1.97
	mean	7.28	7.00	14.46	1.97

5.1.1.6 Forecasted evaporation rates

As stated in chapter 1.1.2, MWRI used two climate models namely ECHAM5, HadCM3 to forecast the expected evaporation rates until the end of century (LNFDC, 2008). The models calculations were based on mean annual evaporation rate of 2700 mm and lake's surface area of 6500 km^2 resulting in total evaporation losses of 17.60 km^3. The Climate Model ECHAM5 predicted that the mean annual evaporation losses will increase by about 0.47 km^3, 0.88 km^3, and 1.66 km^3 for the years 2030, 2050, and 2100, respectively. The Climate Model HadCM3 predicted that the annual evaporation losses will increase by about 0.52 km^3, 0.8 km^3, and 1.46 km^3. These predictions were converted into percentages in order to utilize them in computing the saved evaporation losses with respect to the suggested alternatives in Chapter 6 as presented in Table 5.11.

Table 5.11: Forecasted additional Losses in the evaporation rates until end of the century based on the climate models simulated by MWRI - (LNFDC, 2008).

Model Name	Evaporation additional losses					
	2030		2050		2100	
	(km^3)	%	(km^3)	%	(km^3)	%
ECHAM5	0.47	2.7	0.880	5.0	1.66	9.4
HadCM3	0.52	3.0	0.800	4.5	1.46	8.3

5.1.2 Hydrological Database (HADRHYDB)

For this study, the HADRHYDB was created to record lake surface water level and the lake flow, measured daily at Aswan gauge station from 1964 to 2010 and the monthly mean lake surface water level at Wadi-Halfa gauge station from 1968 to 2002. The monthly mean water level was computed at Aswan gauge station to compare the surface water level data with Wadi-Halfa data. The analysis of these measured data is provided below.

5.1.2.1 Aswan gauge station:

More than 16800 records of daily surface water levels covering the period from 1964 to July 2010 were stored in the HADRHYDB, as shown in Figure 5.18. Based on the HADRHYDB, the lowest water level of 150.62 after operating the HAD was recorded on July 21, 1988. The highest level of 181.60 was measured for the period from November 10 to 14, 1999. Using the SPSS, these records were analyzed. Figure 5.19 shows the histogram for the daily water level. The mean surface water level was about 168 m AMSL. The most frequently computed surface water level was 178 m AMSL. The minimum water level was about 112 m AMSL for the reservoir filling time, while 181.60 m AMSL was the maximum water level. Table 5.12 shows the number of records for each month, in addition to average, median, maximum, minimum, range and standard deviation of the daily surface water level measured at the Aswan gauge station for each month. October to January had the highest mean surface water level of 170 m AMSL (flood season), while July and August had the lowest mean surface water level of 164 m AMSL. The maximum surface water levels were also recorded for the flood season from October till January. The minimum surface water levels were recorded in August.

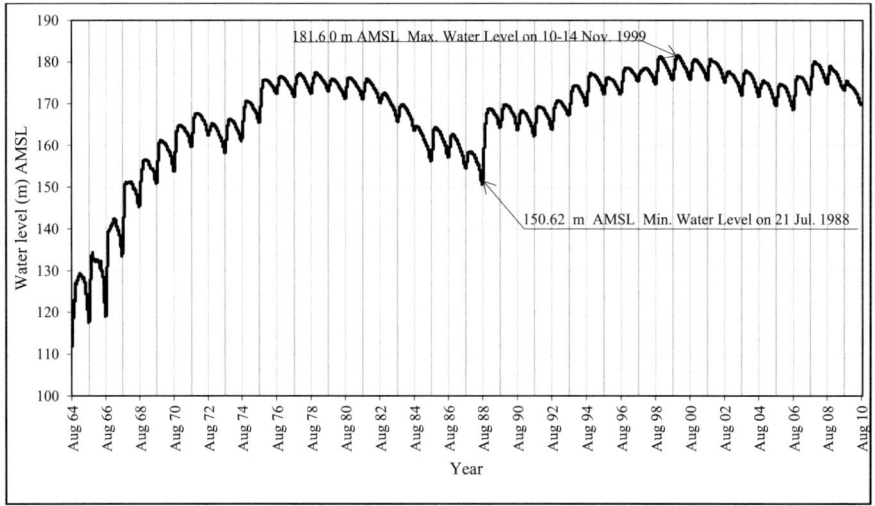

Figure 5.18: Surface water level in m AMSL measured at HAD gauge station from 1964 to 2010

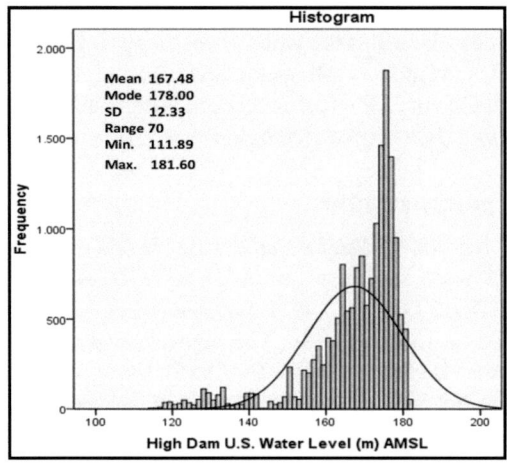

Figure 5.19: Histogram of the mean daily surface water levels measured at Aswan gauge station in m AMSL

Table 5.12: The number, mean, median, maximum, minimum, range and standard division of the daily surface water levels measured at Aswan gauge station of each month

Month	High Dam U.S. Water Level (m) AMSL						
	N	Mean	Median	Minimum	Maximum	Range	Std. Deviation
January	1426	169.17	174.11	128.72	180.93	52.21	11.52
February	1300	168.81	173.71	128.75	180.25	51.50	11.46
March	1425	168.14	173.07	128.08	179.66	51.58	11.57
April	1380	167.45	172.39	127.36	178.90	51.54	11.63
May	1426	166.78	171.56	124.47	178.41	53.94	11.80
June	1380	165.58	170.35	120.71	177.52	56.81	12.25
July	1426	164.20	169.48	117.57	176.45	58.88	13.08
August	1426	164.03	169.09	111.89	178.94	67.05	14.31
September	1380	167.61	172.01	118.72	180.66	61.94	12.56
October	1426	169.22	174.05	122.68	181.39	58.71	12.08
November	1380	169.52	174.36	126.76	181.60	54.84	11.86
December	1426	169.38	174.31	127.78	181.38	53.60	11.71
Annual	16801	167.48	171.42	111.89	181.60	69.71	12.33

Table 5.13 shows the mean, median, maximum and minimum daily surface water levels measured at Aswan gauge station for each decade. Figure 5.20 shows the histograms for the daily surface water levels measured at Aswan gauge station for the last five decades. The highest mean water level was calculated for the 1990s. The lowest mean water level was calculated for the 1960s, during the construction of the HAD, with a maximum water level of about 161 m AMSL and a mode of 150 m AMSL. For the 1970s, the HAD was fully operating and the mean water level increased to about 169 m AMSL with a mode of 177 m AMSL. For this time, the surface water level reached a maximum water level of over 177 m AMSL while the minimum surface water level was about 154 m AMSL. For the 1980s, the mean water level was relatively low and reached only 166 m AMSL because of sequence droughts. For this time, the water level did not rise above 176 m AMSL and did not fall below 150 m AMSL with a mode of 170 m AMSL. In the 1990s, the mean water level increased to about 173 m AMSL because of sequence high floods. For this time, the water level reached minim water level of about 162 m AMSL and a maximum water level of 181.6 m AMSL which is the maximum allowed water level in the HADR, with a mode of 178 m AMSL. For the 2000s the mean water level was only about 176 m AMSL because of many high floods. For this decade, the HADR did not reach above 181 m AMSL and below 168.6 m AMSL.

Figure 5.21 shows the average, maximum and minimum surface water levels at Aswan gauge station for the different months for the last five decades. The water year starts usually in August when the water level starts to increase for the flood months up to December. From January to July, the surface water level usually decreases after the end of the flood season. The minimum water level for operating the turbines in the HAD is 155. The 1980s experienced many droughts where the minimum water level reached less than 155 m AMSL causing the turbines to stop. The 1990s and 2000s witnessed high floods and had almost no variation of the mean and maximum surface water levels for each month. For the last five decades, July had the lowest water levels. For the 1990s the lowest surface water levels was about 162 m AMSL. For the 2000s, the lowest water level was about 168 m AMSL.

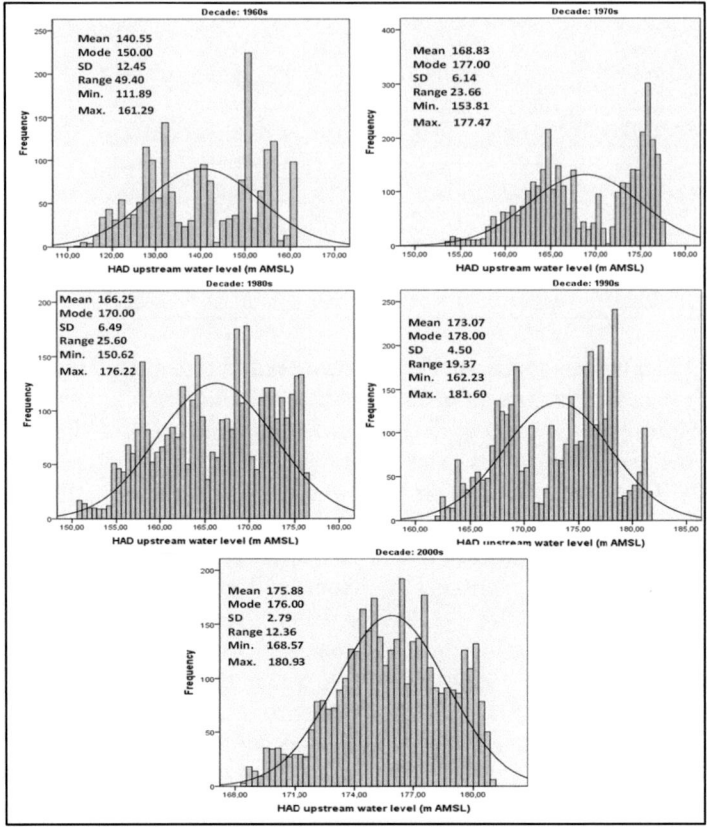

Figure 5.20: Histograms of the daily surface water levels measured at Aswan gauge station for the last five decades

Table 5.13: The mean, median, maximum and minimum daily surface water levels measured at Aswan gauge station of each decade

Decade	HAD upstream water level (m) AMSL			
	Mean	Median	Maximum	Minimum
1960s	140.55	140.57	161.29	111.89
1970s	168.83	168.20	177.47	153.81
1980s	166.25	167.08	176.22	150.62
1990s	173.07	174.17	181.60	162.23
2000s	175.88	175.98	180.93	168.57
Mean	167.48	171.42	181.60	111.89

Results and Discussion

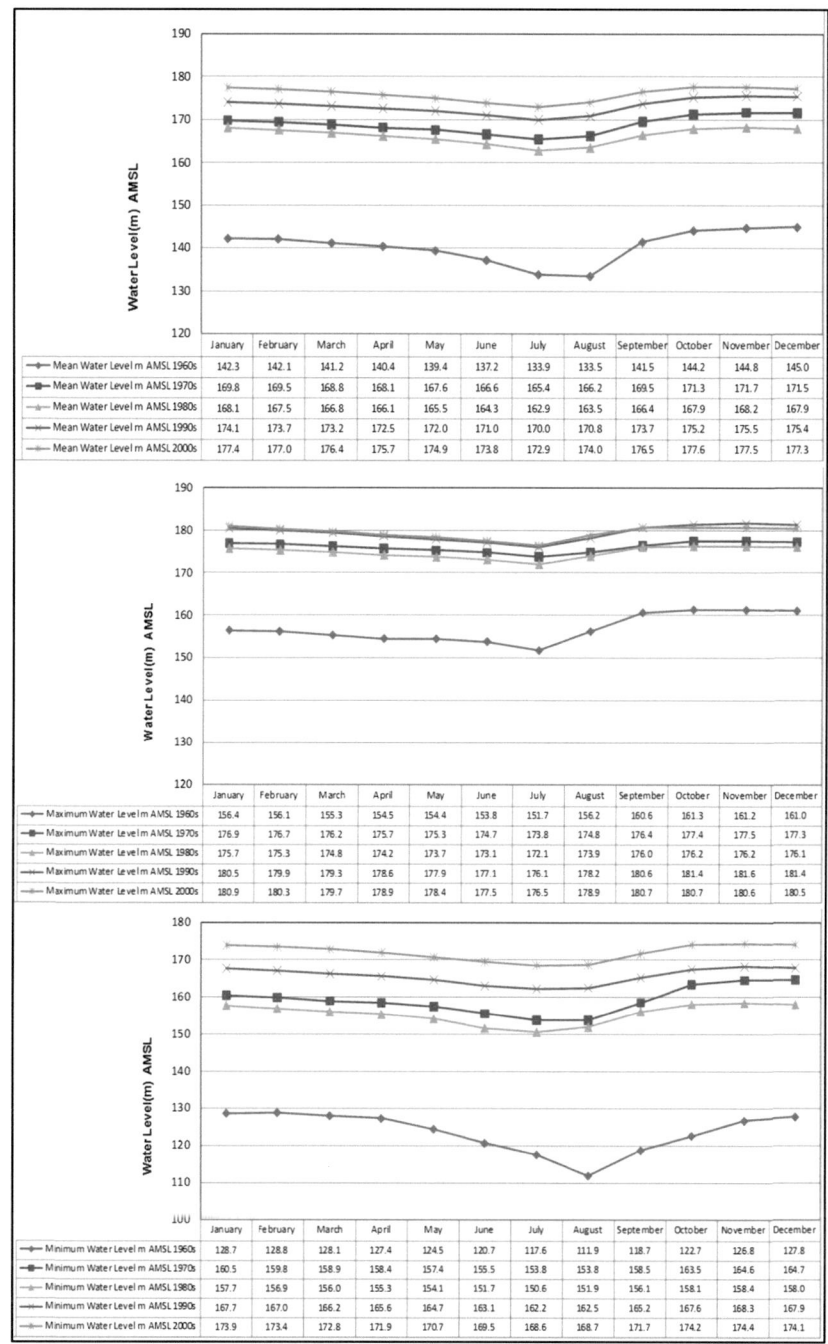

Figure 5.21: Monthly mean, maximum and minimum surface water levels at Aswan gauge station for the last five decades

5.1.2.2 Wadi Halfa gauge station:

More than 400 records of monthly mean surface water level records covering the period from 1968 to 2002 were stored in the HADRHYDB as shown in Figure 5.22. Based on the HADRHYDB, the lowest water level of 151.62 m AMSL was recorded in July 1988. The highest level of 181.60 m AMSL was measured in November 1999. The difference between the mean surface water levels of the two stations was analyzed using SPSS. Figure 5.23 shows the histogram of the differences in the monthly mean surface water levels between Aswan and Wadi Halfa gauge stations. The mean difference was about 11 cm. The most frequent calculated difference was only 2 cm. The maximum difference was about 90 cm for a high flood season.

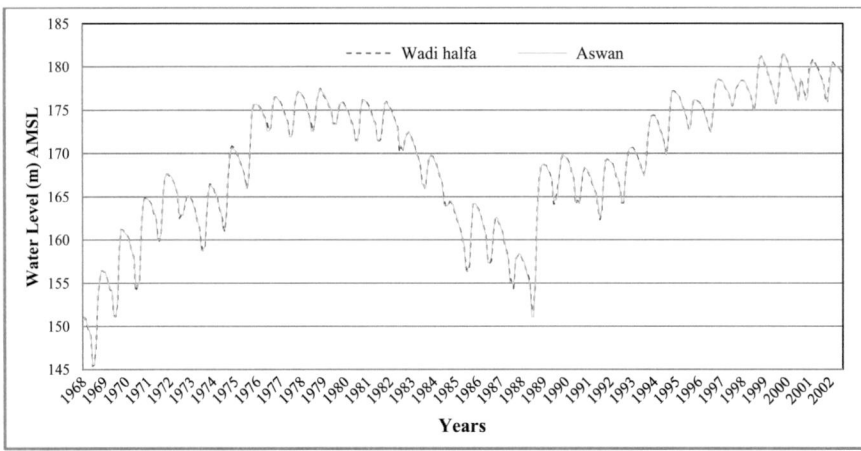

Figure 5.22: Monthly mean surface water levels measured at Wadi Halfa and Aswan gauge stations

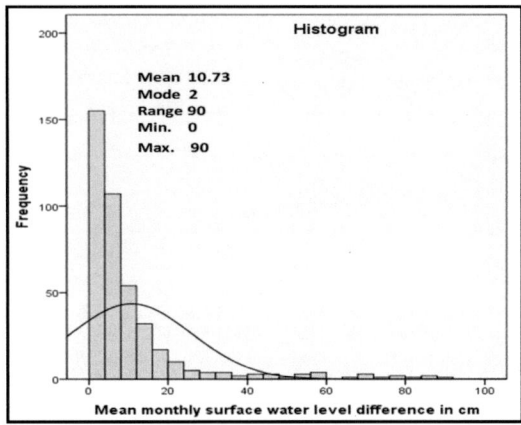

Figure 5.23: Histogram of the differences in the monthly mean surface water levels between Aswan and Wadi Halfa gauge stations

5.1.3 Topographic maps

In this study, contour lines simulating the HADR's bed were produced using the Landsat satellite images and aerial photos. Furthermore, spot heights simulating the altitudes of the HADR's bed were produced using bathymetric survey results. Details of the generated layers are provided below.

5.1.3.1 Contour lines for the HADR

The historical satellite images over the HADR acquired by the Landsat satellites were utilized for constructing new contour lines for the HADR at different elevations. Over 500 Landsat images were downloaded for the period from 1972 to 2003. Using the HADRHYDB, the altitude for each image was identified as shown in table 5.14. The images with similar water levels were selected to identify the lake morphology at different elevations. Since the HADR sides have a smooth incline, a step of 1.5 m was adequate to embrace various surface areas. Eighteen sets were used consisting of 162 Landsat images taken for the period from 1984 to 2005 as listed in Appendix E. The smallest surface area was obtained from the imagery set of August 1987 at a water level of about 154.5 m AMSL, while the largest surface area was obtained from the imagery set of November 1999 when the lake reached its peak water level of 181.6 m AMSL. The aerial photos were used to simulate the HADR at 110 m AMSL. Each imagery set was mosaiced together into one image to extract the water area boundary and to convert them into contour lines. The contour lines were attached together into a new shape file, simulating the contour lines between 154.5 m to 181.6 m AMSL and the contour line of 110 m AMSL as shown in Figure 5.24.

Table 5.14: List of utilized image sets with acquisition dates and corresponding water levels of the HADR in meters AMSL

No.	Acquisition dates	Water level (m) AMSL	No.	Acquisition dates	Water level (m) AMSL
1	Aug-87	154.50	10	Jan-98	169.50
2	Mar-88	156.50	11	Jul-05	171.00
3	Jan-88	158.00	12	Jul-03	172.00
4	Mar-95	159.50	13	Aug-03	174.00
5	May-85	161.00	14	Jul-98	175.50
6	Dec-86	162.50	15	Jun-02	177.00
7	Sep-84	164.50	16	Mar-99	179.00
8	Jun-84	166.00	17	Jan-99	180.00
9	Apr-90	167.50	18	Nov-99	181.60

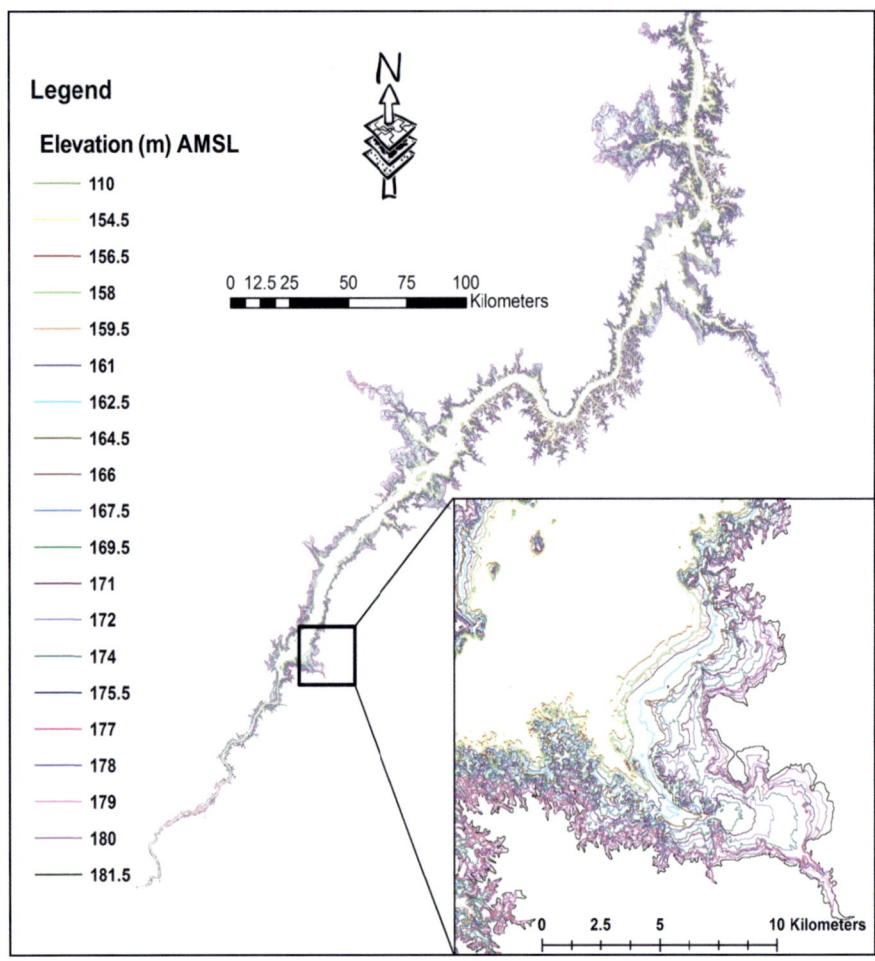

Figure 5.24: Contour lines extracted from satelitte images and areial photos

5.1.3.2 Spot heights for the HADR

In this study, the data collected for the missions executed in 1999, 2003, 2006, 2007, 2008, 2009 and 2010 were used to generate point shape files. The surveyed points reached up to 300000 points for one mission. The generated points are shown in Figure 5.25. From 2006, the entire Lake Nubia was surveyed to investigate sediment problems. Lake Nasser was only surveyed in certain sections except the conduct survey in 2010 where the lake was completely surveyed up to the El-Madik section. Some parts of Lake Nubia were not surveyed due to navigation problems.

Results and Discussion 179

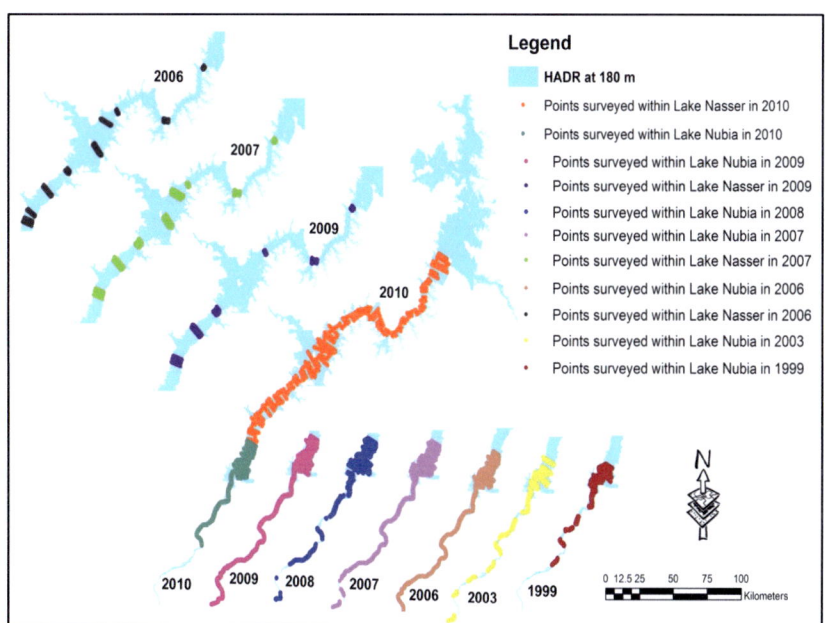

Figure 5.25: Points surveyed on Lake Nasser and Lake Nubia between 1999 and 2010

5.1.4 HADR Seepage Database (HADRSPDB)

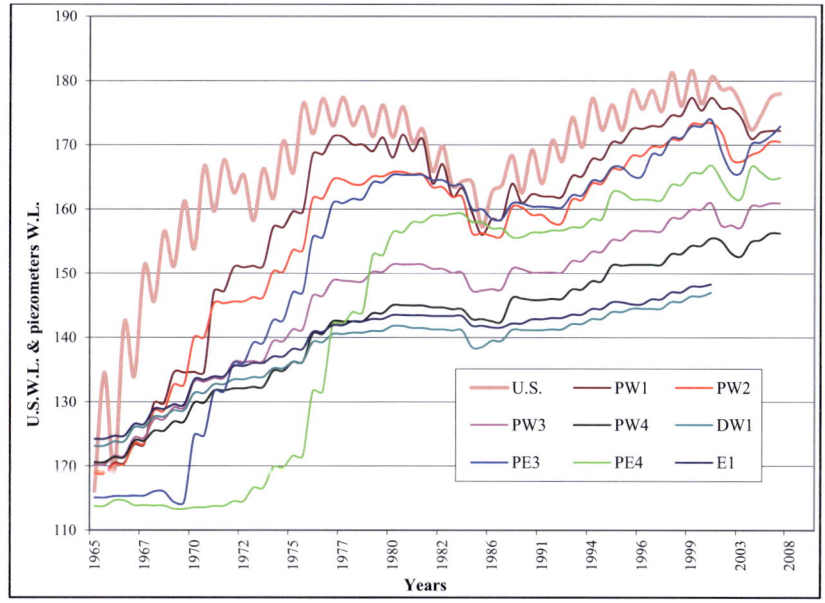

Figure 5.26: Observations of upstream water level and Piezometers water level at Adindan Sector 316 km upstream the HAD from 1965 to 2009

In this study, the piezometer data were stored in an excel database. Figure 5.26 shows piezometer water level measured from 1965 to 2009 and the corresponding upstream (U.S.) water levels in meters AMSL. DW represents deep western piezometers, PW the shallow western piezometers, E the deep eastern piezometers and PE the shallow eastern piezometers. The piezometers data were used to estimate the amount of lateral seepage flow from the reservoir according to Darcy's law as shown in Table 5.15. The maximum seepage amount reached about 110 hm^3 in the water year 1999/2000 where the HADR reached its maximum allowed surface water level. The mean seepage losses were about 55 hm^3.

Table 5.15: Annual average water losses by seepage (hm^3) in the HADR

Water Year	Upstream water level (m) AMSL		Seepage losses	Water Year	Upstream water level (m) AMSL		Seepage losses
	Max.	Min.	hm^3		Max.	Min.	hm^3
1965/1966	132.9	119	24	1987/1988	158.5	150.6	38
1966/1967	142.5	133.5	26	1988/1989	168.8	164.3	40
1967/1968	151.2	145.3	30	1989/1990	169.8	163.8	39
1968/1969	156.6	150.9	26	1990/1991	168.4	162.2	40
1969/1970	161.3	153.8	36	1991/1992	169.4	163.8	27
1970/1971	164.9	159.7	40	1992/1993	170.8	167.2	24
1971/1972	167.6	162.5	50	1993/1994	174.3	169.5	31
1972/1973	165.3	158.2	56	1994/1995	177.3	172.3	38
1973/1974	166.3	161	55	1995/1996	176.3	172.3	45
1974/1975	170.6	165.6	62	1996/1997	178.6	175.4	62
1975/1976	175.7	172.4	68	1997/1998	178.5	174.7	51
1976/1977	176.6	171.7	99	1998/1999	181.3	175.7	94
1977/1978	177.2	172.4	82	1999/2000	181.6	175.8	109
1978/1979	177.5	173	85	2000/2001	180.63	175.7	88
1979/1980	176	171.2	100	2001/2002	180.68	175.14	98
1980/1981	176.2	171.1	79	2002/2003	177.69	172.02	47
1981/1982	176	170.2	84	2003/2004	177.91	171.7	43
1982/1983	172.6	165.6	79	2004/2005	175.56	169.57	48
1983/1984	169.9	163.6	38	2005/2006	174.72	168.57	53
1984/1985	164.7	156.1	33	2006/2007	176.53	168.69	56
1985/1986	164.3	157.1	28	2007/2008	180.11	173.46	55
1986/1987	162.7	154.5	36	2008/2009	179	173.3	54
Average (hm^3)							55

5.1.5 HADR Bed Soil Sedimentary Categories Database (HADRBSDB)

In this study, the HADR Bed Soil Sediment Categories Database (HADRBSDB) was created to record the sediment categories of the HADR bed soil data collected by the MWRI mission in 2007. Figure 5.27 shows the distribution of bed soil sedimentary along the HADR cross sections in 2007. Figure 5.28 shows a chart with percentages of the HADR's bed soil categories in 2007.

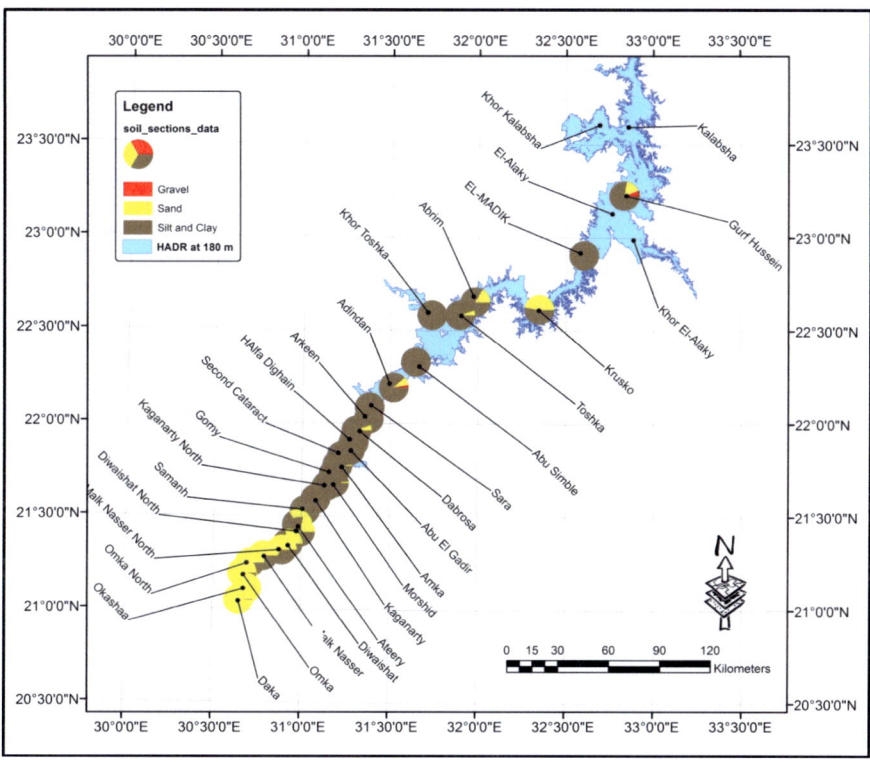

Figure 5.27: Distribution of bed soil sedimentary along the HADR's cross sections in 2007

The section from km 410 up to the HADR end has undergone higher sedimentation of sand while the other part had more silt and clay sedimentary particularly in the section from km 350 to km 400 upstream the HAD. However, strong winds from the east desert caused some sand sedimentary from km 100 to km 350 upstream from the HAD (HDA-MWRI, 2009, NWS, 2012). Only miniature percentage of gravel was available in few sections. The HADR usually had high silty and clay sediments in the cross section from the HAD to Ateery cross section, at 415 km upstream the HAD. The sandy sediments increased from the Ateery cross section to the Daka cross section, at 478.5 km upstream the HAD, to reach about 100%.

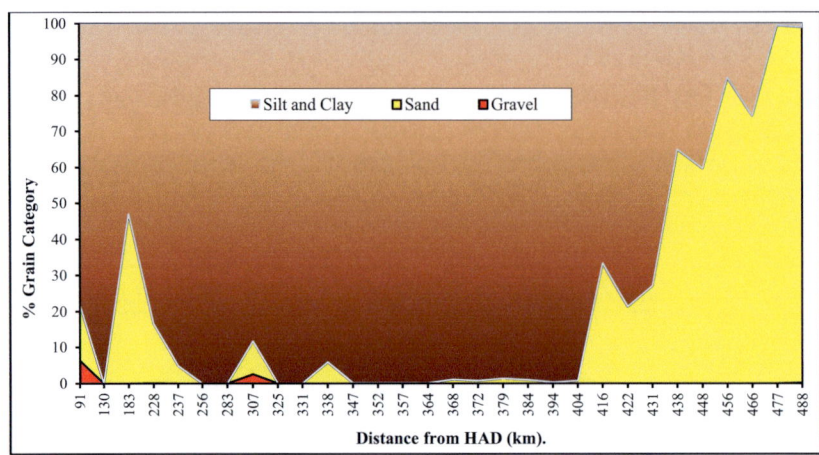

Figure 5.28: Chart of percentages of the HADR's bed soil sediment textures in 2007

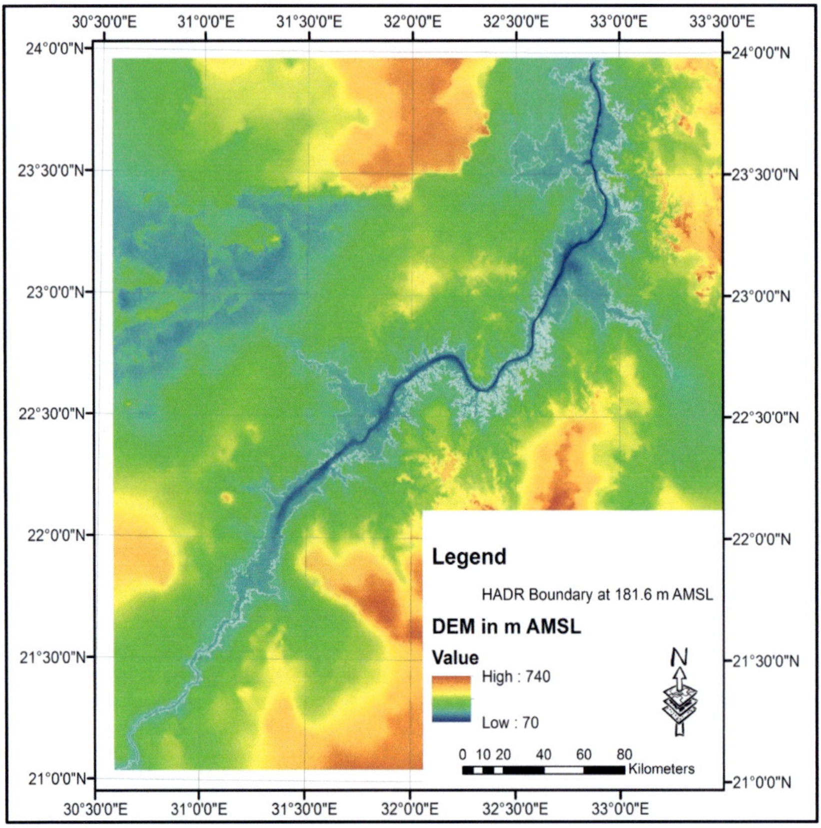

Figure 5.29: Digital elevation model of the entire area of the HADR in m AMSL and the HADR boundary at 181.6 m AMSL.

5.2 Modeling HADR

In this study, a digital elevation model (DEM) was developed to model the topography of the HADR bed in order to study the morphological and hydrological characteristics of the HADR. The DEM was used in establishing regression models to simulate the HADR's characteristics as provided below. The created models were also verified to investigate their effectiveness.

5.2.1 HADR Digital Elevation Model (HADRDEM)

In this research, a DEM was created using the contour lines extracted from satellite images and aerial photos, spot heights measured from the bathymetric survey in 2007 and old topographic maps. It was generated using ARCGIS with 25 m spatial resolution for the area extends between 21° and 24° north to 31°30`and 33°30`E as shown in Figure 5.29. To study the lake hydrological features, the HADR Digital Elevation Model (HADRDEM) was isolated using the lake boundary at 181.60 m AMSL as demonstrated in Figure 5.30. The HADR was divided into two parts at El−Madik section to study the hydrological characteristics of each division About 44 khors were also isolated to study their hydrological features.

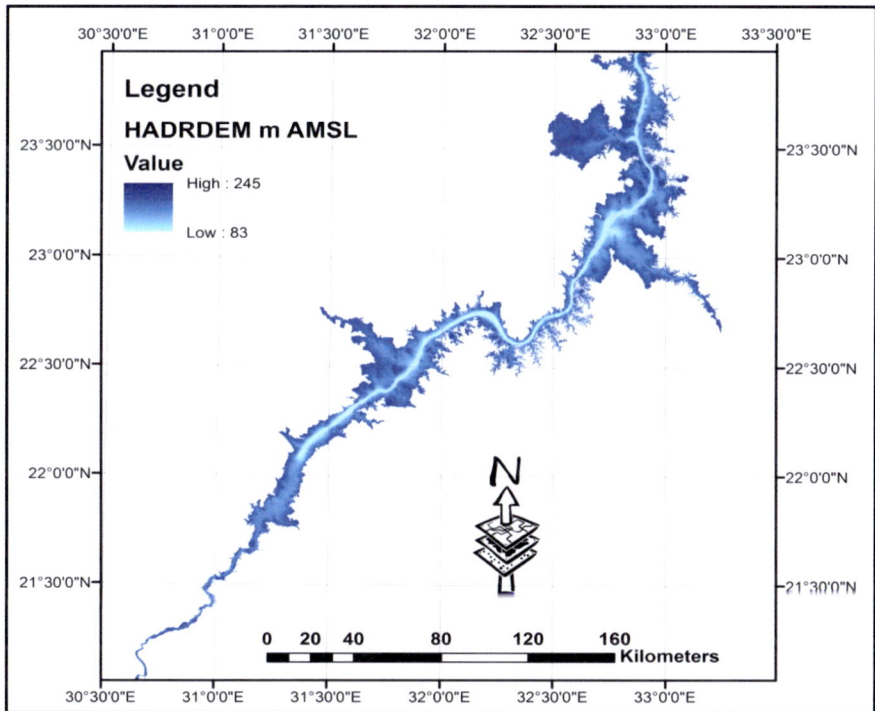

Figure 5.30: The isolated HADR digital elevation model (HADRDEM) in m AMSL.

5.2.2 HADR mathematical models

In this research, the HADRDEM was used initially to build mathematical models for the HADR's hydrological characteristics. The number of pixels at each elevation was imported from HADRDEM to excel files in order to estimate the lake surface area and water volume at the various HADR altitudes with one meter step. Elevations, computed surface areas and volumes were analyzed using SPSS to develop the proper equation type between them. Three mathematical models were built to expose the relationship between the different hydrological characteristics: water level (WL) versus surface area (A); water level versus lake volume (V); and surface area versus lake volume. Figure 5.31 shows the optimum equation to embody the relationship between water level and the surface area, which is a power equation (Equation 5.1). Figure 5.32 shows the optimum equation to represent the relationship between water level and the water volume, which is a power equation (Equation 5.2). Figure 5.33 shows the optimum equation to represent the relationship between water volume and the surface area, which is a linear equation (Equation 5.3).

$A = 5.0833 \times 10^{-12} \times (WL)^{6.6853}$ \hfill Equation 5.1

$V = 2.4585 \times 10^{-16} \times (WL)^{7.87324}$ \hfill Equation 5.2

$A = 39.902\, V + 541.027$ \hfill Equation 5.3

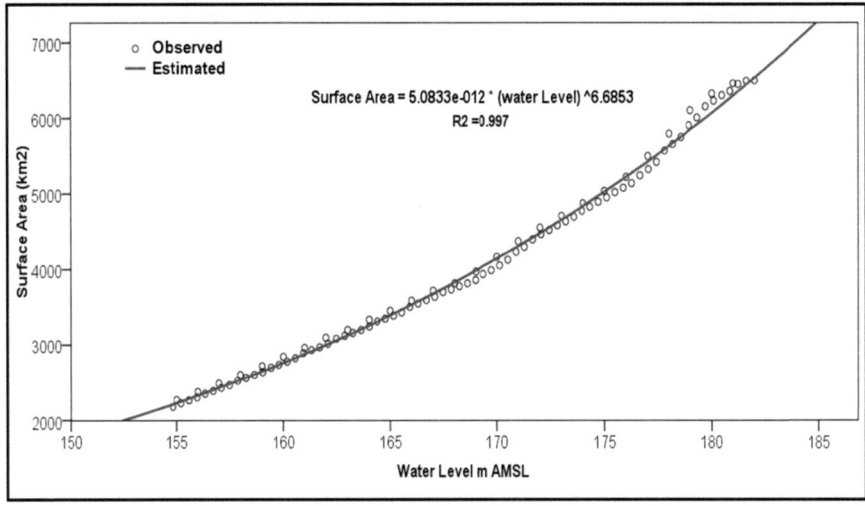

Figure 5. 31: Evaluation of the relationship between surface area in km² and water level in meter AMSL and the optimum equation

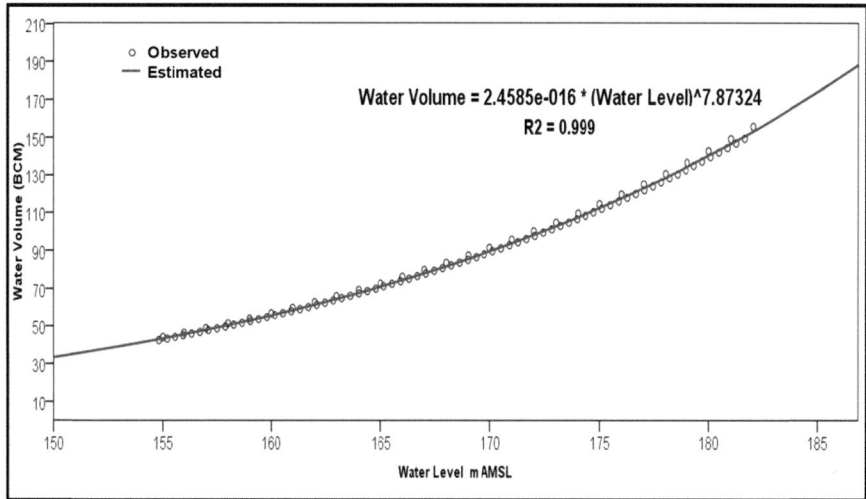

Figure 5.32: Evaluation of the relationship between water volume in km³ and water level in meter AMSL and the optimum equation

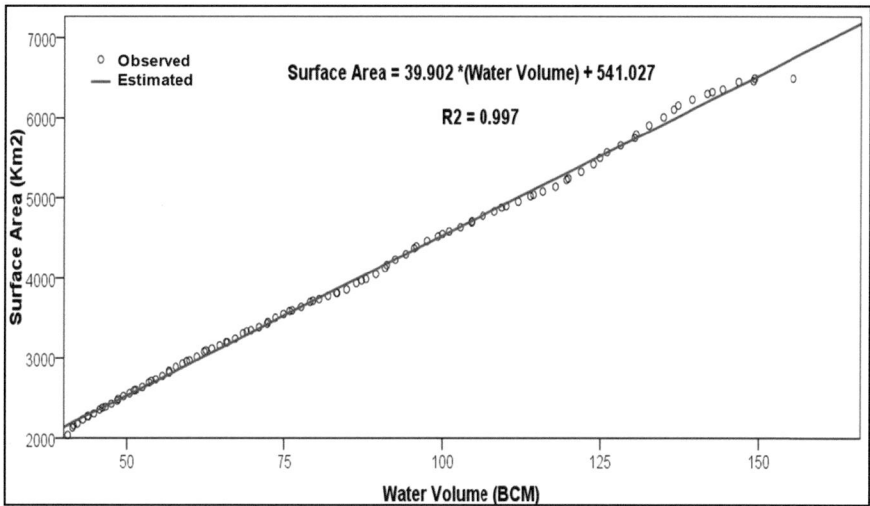

Figure 5.33: Evaluation of the relationship between water volume in km³ and surface area in km² and the optimum equation

5.2.3 HADRDEM and mathematical model verification

The HADRDEM was evaluated to investigate its quality and significance in estimating the surface area and the water volume of the HADR. Figure 5.34 shows the HADR altitudes and the corresponding surface area calculated by the HADRDEM and the original surface area as functions of HADR level given by the NWS (Table 3.9). The differences between both areas are average to about ±100 km² due to the morphological changes of the HADR for the last four decades. Figure 5.35 shows a scatter diagram of the surface area calculated by the HADRDEM and by the NWS. The points are highly correlated and the relationship between them can be presented with a line inclined at 45°.

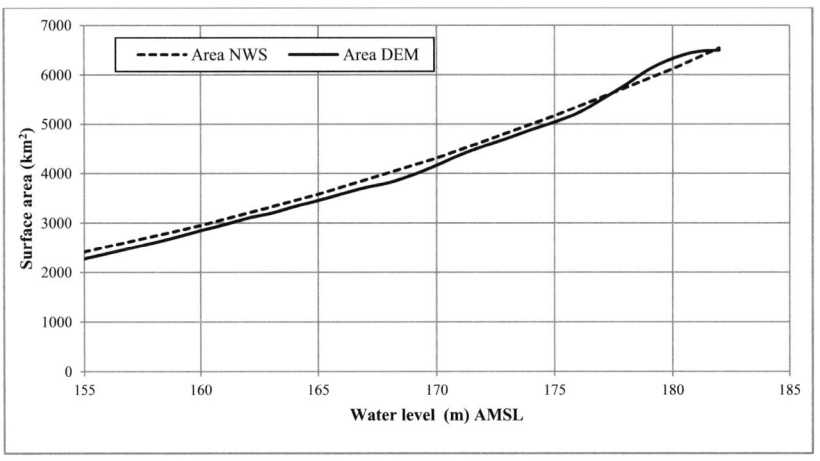

Figure 5.34: The HADR altitudes (m) AMSL and the corresponding surface area calculated by the HADRDEM and by the NWS in km²

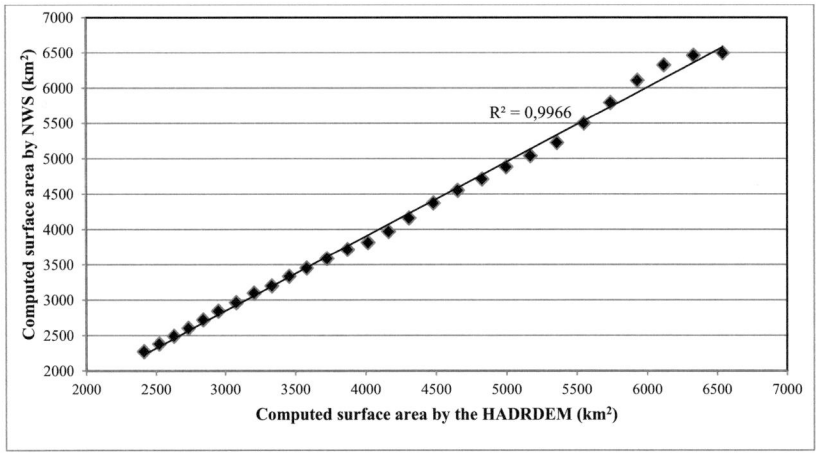

Figure 5.35: Scatter diagram between the HADR surface area calculated by the HADRDEM and the HADR surface area calculated by the NWS in km²

Figure 5.36 shows the HADR's altitudes and the corresponding water volume calculated by the HADRDEM and the original water volume as functions of HADR level given by the NWS (Table 3.9). Both lines are parallel and have a difference of about 7 km³ between both volumes due to the sediment events over the last four decades. Using the HADRHYDB, the water levels were usually below 177 m AMSL till 1996 and caused sediment loads that decreased the surface area. On the other hand, the water levels raised above 177 m AMSL since 1996 for the high flood years, which caused erosion to some parts of the HADR banks and increased consequently the surface area. Figure 5.37 shows a scatter diagram between the water volume calculated by the HADRDEM and by the NWS. The points are highly correlated and the relationship between them can be presented with a line.

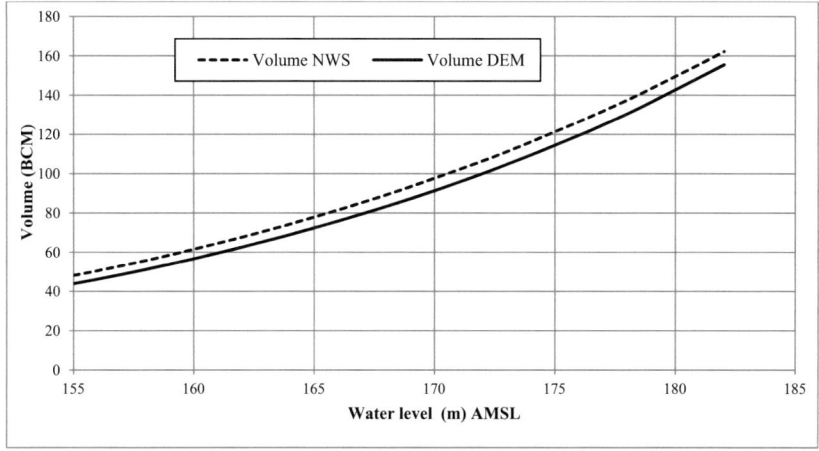

Figure 5.36: The HADR's altitudes and the correspondence water volume calculated by the HADR-DEM and by the NWS in km³

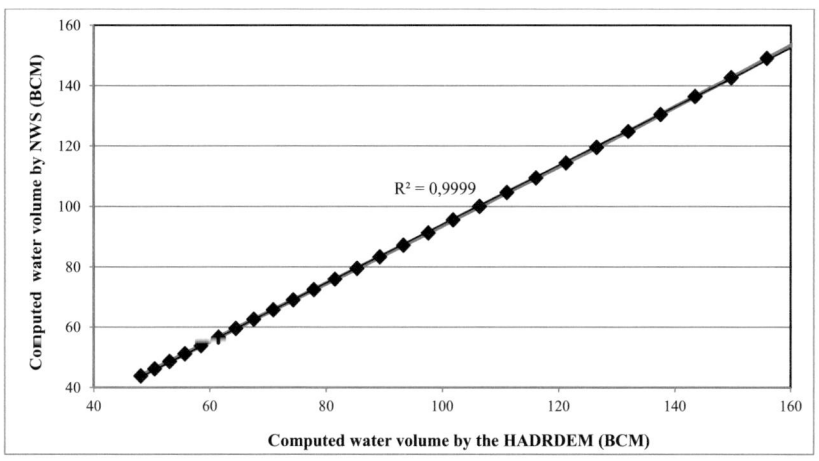

Figure 5.37: Scatter diagram of the HADR water volume calculated by the HADRDEM and the HADR water volume calculated by the NWS in km³.

The resulting mathematical models were also evaluated to investigate their significance in estimating the surface area and the water volume of the HADR with respect to the water levels. Figure 5.38 shows a scatter diagram depicting the surface area calculated by the HADRDEM and by the resulting mathematical model (Equation 5.1). The points are highly correlated and the relationship between them can be presented with a line inclined at 45°. Figure 5.39 shows a scatter diagram between the water volume calculated by the HADRDEM and by the resulting mathematical model (Equation 5.2). The points are highly correlated and the relationship between them can be presented with a line inclined at 45°.

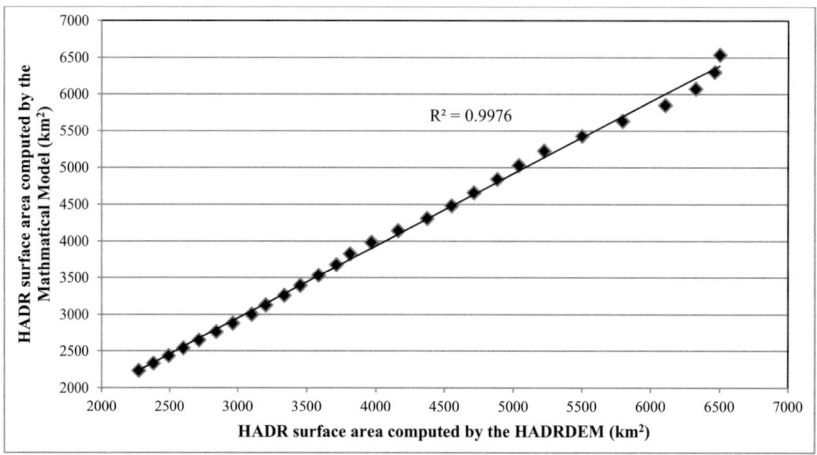

Figure 5.38: Scatter diagram of the HADR's surface area calculated by the HADRDEM in km^2 and calculated by the HADR mathematical model −Equation 5.1 in km^2

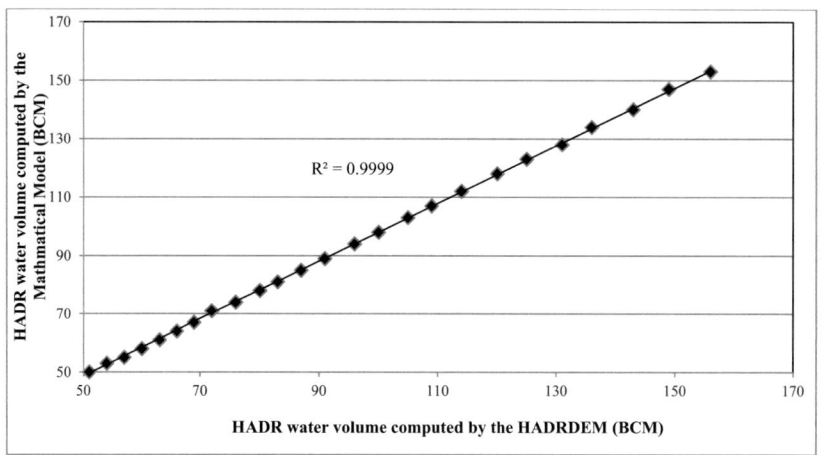

Figure 5.39: Scatter diagram of the HADR water volume calculated by the HADRDEM in km^3 and the HADR water volume calculated by mathematical model−Equation 5.2 in km^3

5.3 Exploring the HADR

In order to study the possible alternatives of reducing evaporation losses on the HADR, the HADRDEM was used to explore the HADR hydrological and meteorological characteristics. The most important khors and cross sections were studied. Furthermore, the positions suffering from high sediment rates were examined.

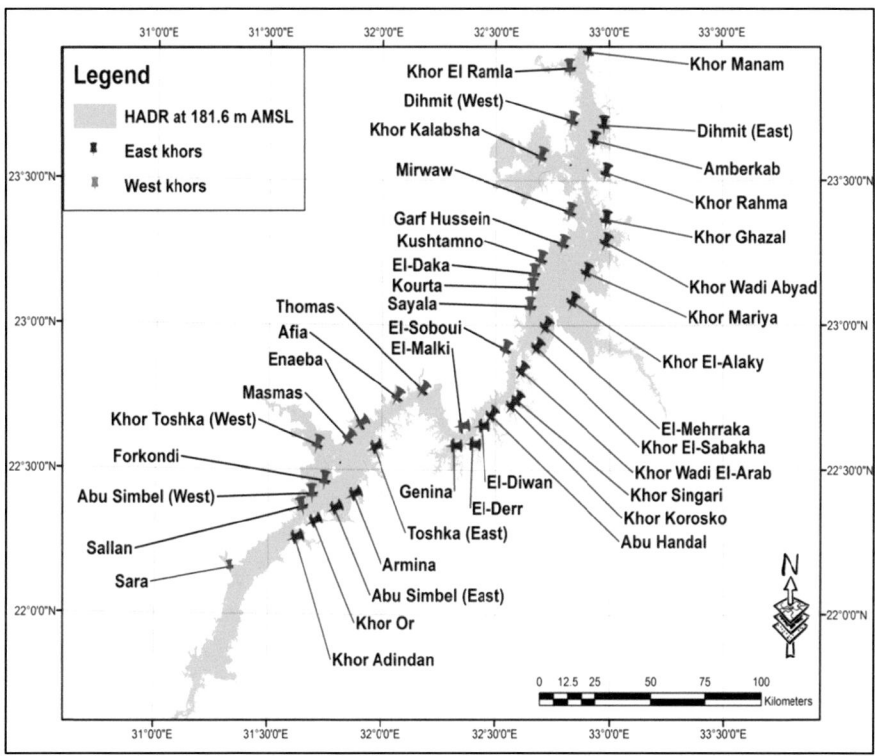

Figure 5.40: The main khors along the HADR - modified after (FAO, 2012)

5.3.1 Exploring the HADR Khors

In this research study, the developed HADRDEM was used to investigate and explore 42 khors along the HADR. Twenty-two khors on the eastern side of the reservoir were studied and twenty khors on the western side, as shown in Figure 5.40. Each khor was separated from the HADRDEM to study its hydrological characteristics as shown in Figure 5.41. Table 5.16 shows the list of the studied khors, including the volume of each khor and the surface area of the HADR at maximum water level of 181.6 m AMSL. Figure 5.42 shows the distribution of the khors' surface area and water volume. Khor Kalabsha has the largest surface area of over 700 km^2 with a mean depth of 12 m while Khor Sallano has the least surface area of only 3 km^2 and mean depth of 18 m. Khor El-Alaky has the largest water volume and the second largest surface area with a mean depth of 19 m. Khor Toshka west is the third largest khor in surface area and volume with approximately 300 km^2 surface area and approximately 4 km^3 water volume with a mean depth of over 14 m at 181.6 m AMSL. Most of the khors have a small surface area and large volume as they are deep. The total area of the HADR khors is approximately 3000 km^2, representing 46% of the HADR surface area. The khors have 44 km^3 of water volume, representing 28% of the HADR water volume with a mean depth of 14 m.

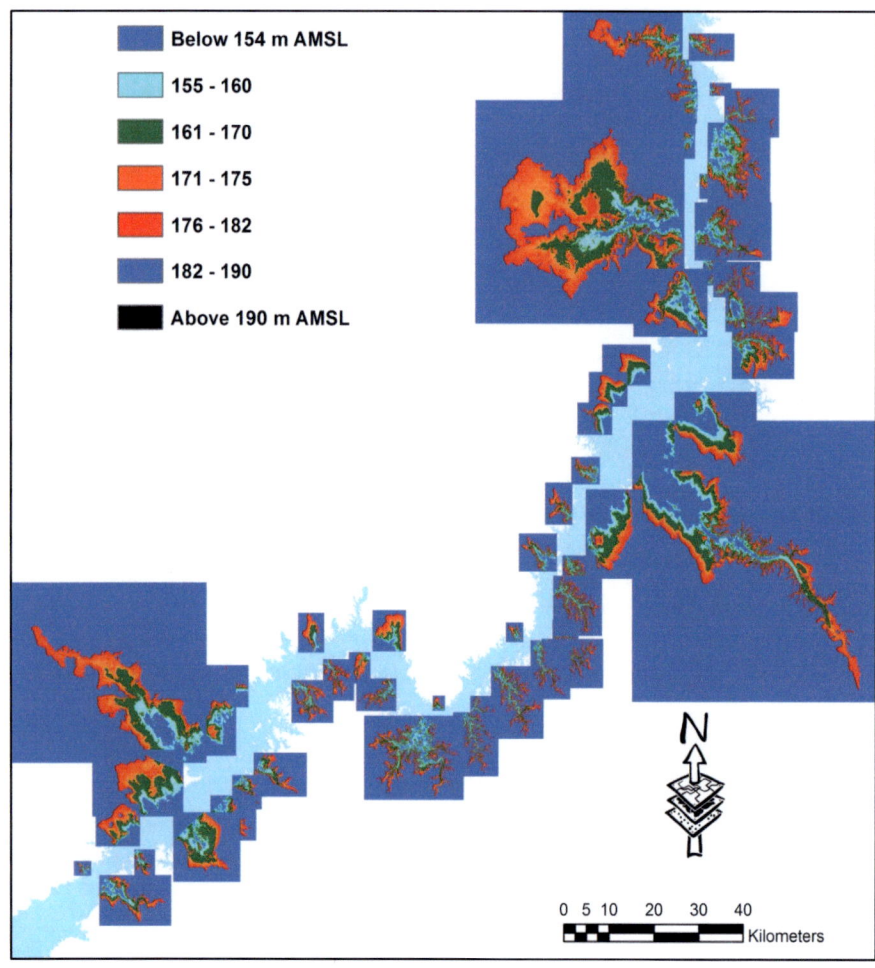

Figure 5.41: Isolated DEMs for the studied khors deduced from the digital elevation model of the HADR (HADRDEM)

Table 5.16: List of the studied khors of the HADR, their surface areas in km^2 and their volume in km^3 and corresponding mean water depths in m

No	Location	Name	Surface Area (km^2)	Volume (km^3)	Mean Depth (m)
1	west	Elramla	84.39	0.94	11.2
2	west	Dihmit	45.26	0.58	12.8
3	west	Khor Kalabsha	708.01	8.46	9.0
4	west	Minrwaw	91.50	1.58	17.3
5	west	Garf Hussein	30.41	0.45	14.7
6	west	Kushtamno	25.57	0.34	13.5
7	west	Eldaka	19.40	0.35	18.1
8	west	Kourta	13.68	0.24	17.6
9	west	Sayala	19.39	0.36	18.7
10	west	Elsoboui	4.71	0.08	16.0
11	west	Elmaliki	3.72	0.06	15.4
12	west	Thomas	34.98	0.62	17.6
13	west	Afia	16.70	0.23	13.6
14	west	Naeba	30.37	0.49	16.1
15	west	Masmas	53.66	1.22	22.8
16	west	Khor Toshka west	306.26	4.38	14.3
17	west	Forkondi	122.47	1.87	15.3
18	west	Abu Simbel West	43.60	0.54	12.4
19	west	Sallano	3.08	0.06	18.0
20	west	Sara	39.4	0.51	16.5
21	east	Khor Manam	14.56	0.16	11.2
22	east	Dihmit	19.08	0.18	9.2
23	east	Amberkab	105.23	1.89	18.0
24	east	Khor Rahma	57.86	0.93	16.0
25	east	Khor Ghazal	15.35	0.15	10.0
26	east	Khor Wadi Abyad	45.41	0.65	14.4
27	east	Khor Mariya	60.80	0.78	12.9
28	east	Khor El-Alaky	508.13	9.40	18.5
29	east	Elmehrraka	96.73	1.68	17.3
30	east	Khor Elsabakha	31.12	0.34	11.0
31	east	Khor Wadi Elarab	16.63	0.22	13.5
32	east	Khor Singari	25.23	0.29	11.3
33	east	Khor Korosko	24.32	0.28	11.5
34	east	Abu Handal	14.21	0.18	12.8
35	east	Eldiwan	11.62	0.12	10.6
36	east	Elderr	17.43	0.16	9.2
37	east	Genina	89.50	1.20	13.8
38	east	Toshka East	12.81	0.20	15.7
39	east	Armina	31.50	0.43	13.6
40	east	Abu Simbel East	79.12	1.22	15.4
41	east	Khor Or	8.93	0.12	13.9
42	east	Khor Adindan	42.38	0.64	15.1
Total			2985.07	44.09	14.44

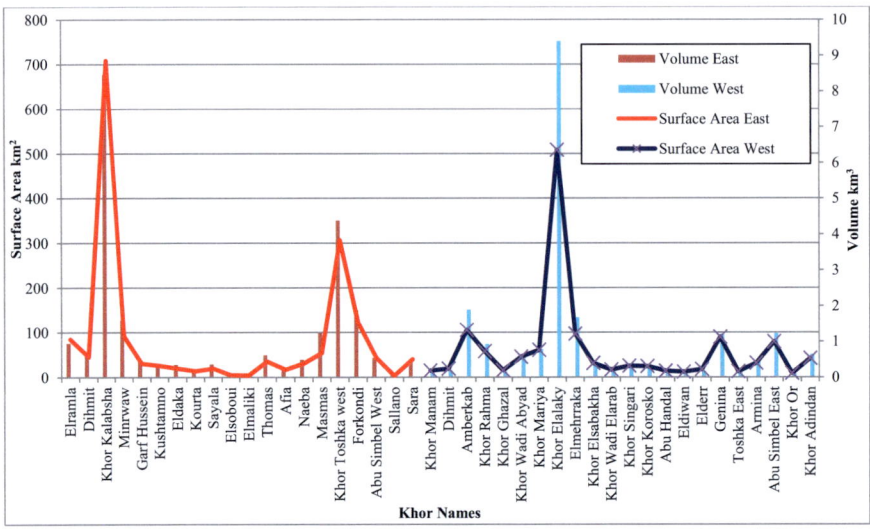

Figure 5.42: Volumes of individual HADR khors in km³ and surface areas in km²

5.3.2 Exploring the HADR cross sections

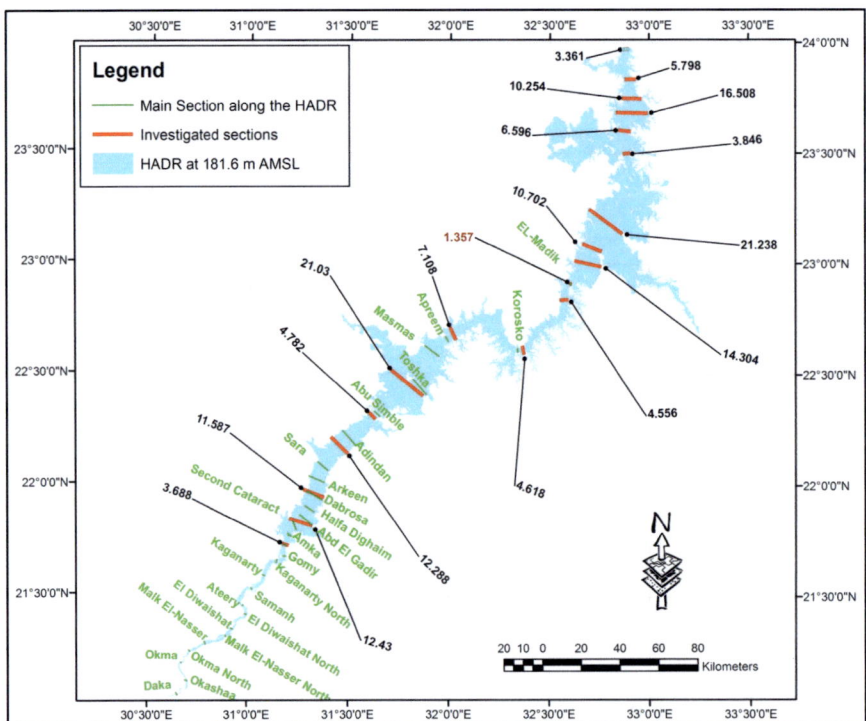

Figure 5.43: The width of the investigated sections and main sections along the HADR

In this study, 19 different sections at narrow and wide locations along the lake were selected to investigate the widths of the HADR as shown in Figure 5.43. The narrowest cross section is located at the El-Madik section at 130 km downstream the HAD. Its width is approximately 1350 m where the HADR reaches its maximum water level of 181.6m AMSL. The longest cross section is located north of the Khor El-Alaky mouth, at 90 km downstream the HAD. Its width is more than 21 km at the altitude of 181.6 m AMSL. The average width of the HADR is approximately 9 km at the same altitude. At 500 m downstream the HAD, the HADR's width is approximately 3.3 km at 181.6 m AMSL. South of Khor Kalabsha, the HADR's width is approximately 3.8 km at 50 km downstream the HAD and increases to over 4.5 km at 500 m south of the El-Madik section at 181.6 m AMSL.

5.3.3 Exploring the HADR sediment conditions:

In this research, the generated DEM's of the Lake Nubia for the years 1999, 2003, 2006, 2007, 2009 and 2010 were compared with each other to monitor the changes in the HADR bed due to sediment and erosion events. Figure 5.44 shows the longitudinal section along the HADR's deepest points from 1964 to 2010. List of sections names and locations are given in Table 4.3. The section for 1964 was taken from the main Nile river regime before constructing the HAD. The section for 1977 was taken from the old survey of the HADR. A new delta has developed between the Samanh section at approximately 403 km upstream the HAD and the Gomy section at 378 km upstream the HAD (NWS, 2012, HDA-MWRI, 2009, Ahmed and Ismail, 2008). Figure 5.45 shows the distribution of changes in the altitudes of the HADR due to a successive sedimentation and erosional events during the past 50 years.

During the filling period of the HADR from 1964 to 1977, successive sediment events had occurred, particularly in Lake Nubia while the El-Madik section in Lake Nasser suffered from minor erosion problems. On the other hand, Lake Nasser slightly suffered from sediment problems particularly between Sara section and Apreem section, while the distance between Korosko and El-Madik did not experience any changes in the HADR's bed altitudes during the past decade. During the filling period from 1964 to 1977, the Madik Amka section experienced the largest increase of the amount of sediment (43 m). The Second Cataract section experienced the greatest increase in the amount of sediment which reached over 60 m during the past five decades. During the period from 1977 to 1999, the largest sediment layers were found between the Samanh and Abd El kadir sections. They had a mean sediment depth of 20 m, equivalent to approximately one meter of sediment per year. During the past decade, the Second Cataract section suffered from a high sedimentation rate that reached over 12 m, accounting for 1.2 meter per year. The sections between Second Cataract and Dabarosa also suffered from sediment problems. In the other direction, the sections north of Second Cataract all the way up to Diwaishat suffered from erosion problems. The highest erosion rate was measured at Amka section and reached approximately 80 cm per year. Therefore, the area between Second Cataract section and Dabarosa section was selected to study the changes in the lakebed in the past decade.

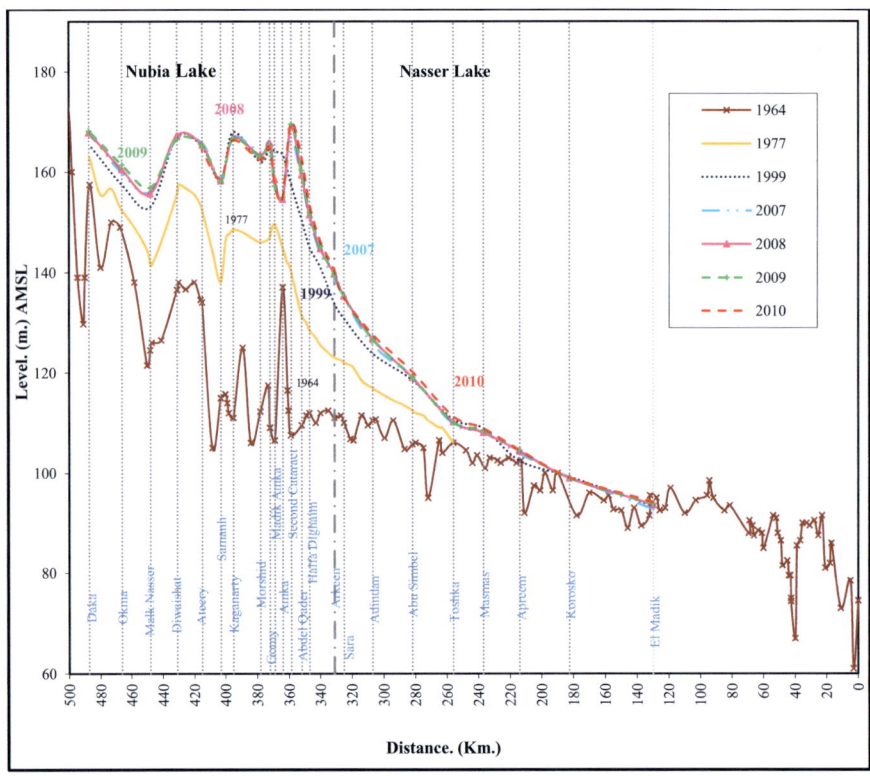

Figure 5.44: A longitudinal section along the deepest points of the HADR from 1964 to 2010

Figure 5.46, Figure 5.47, Figure 5.48, and Figure 5.49 show the HADRDEMs of Lake Nubia in the years 1999 and 2003 as well as from 2006 to 2010 and the corresponding changes in the lakebed's altitudes. Figure 5.50 shows the changes in the lakebed's altitudes cross section due to sediment events from 1999 to 2010 at the cross sections Abd El-kader, Halfa-Dighaim, and Dabrosa. The sediment deposits increased from Section Daka to Section Second Cataract and then decreased till Section Dabarosa. This pattern is due to the increase in cross section width (12.4 km) at Section Second Cataract, which follows the narrow sections upstream, which have a maximum width of 3.7 km (Figure 5.43). The accumulated sediments are decreasing gradually to Section Dabrosa, where the width of cross sections decreases slightly.

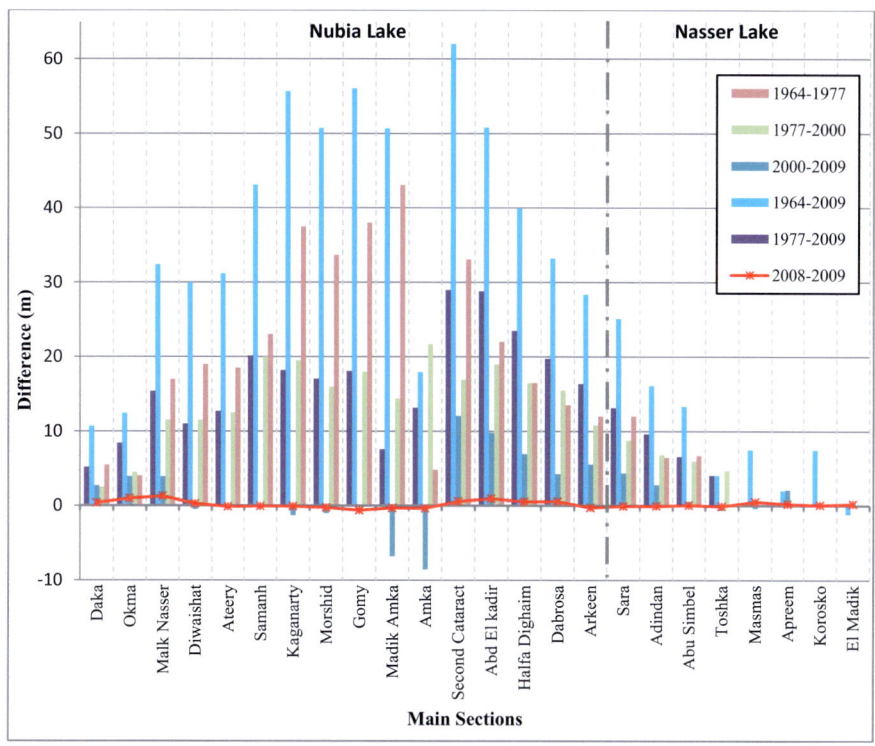

Figure 5.45: Sedimentation and erosion of lake Nubia and Lake Nasser reflected by changes of the altitudes from 1964 to 2009

The volumes of sediment and erosion from 1964 to 2010 are given in Appendix F. The total erosion volume was only approximately 126 hm^3 while the total volume of sediment reached over 6.6 km^3. The total change in Lake Nubia was approximately 5.5 km^3 in the past five decades, while the total change in Lake Nasser was approximately 1.1 km^3. The changes in the HADR's volume were usually positive due to the small volumes caused by erosion in the different years, which account for only 2% of the change in the HADR's bed. The annual mean total sediment volume was approximately 140 hm^3 while the annual mean total volume caused by erosion was approximately 4 hm^3. The sections between Second Cataract and Dabarosa experienced the greatest changes in sediment volume during the past years.

The sedimentation accumulated at major sections in Lake Nubia from 1977 to 2010 due to successive sediment and erosion events are shown in Figure 5.51. Section Daka suffered from successive erosion events till 1996 and then had successive sediment events with minor changes. Section Okma suffered from successive sediment events with slight changes till 2006. This period from 2006 to 2008 was characterized by low erosion events; after 2008, more sediment was deposited than eroded. Section Ateery, section Amka, and section Dabarosa suffered from successive moderate sediment events till 2010. Section Second Cataract experienced successive and extremely high sediment, while section Halfa Dighaim experienced successive high sediment events. The sediment deposits increased from Section Daka to Section Second Cataract and then decreased till Section Dabarosa. Section Second Cataract had the largest amount of accumulated sediments with 800 hm^3, which account for 17% of the total sediments deposited in Lake Nubia till 2010. Furthermore, the 2.8 km^3 of sediment deposited between Sections Amka to Dabrosa, a distance of approximately 26 km, account for approximately 60% of the total sediments in Lake Nubia, which equals 50% of the total sediments in the HADR during the period from 1977 to 2010.

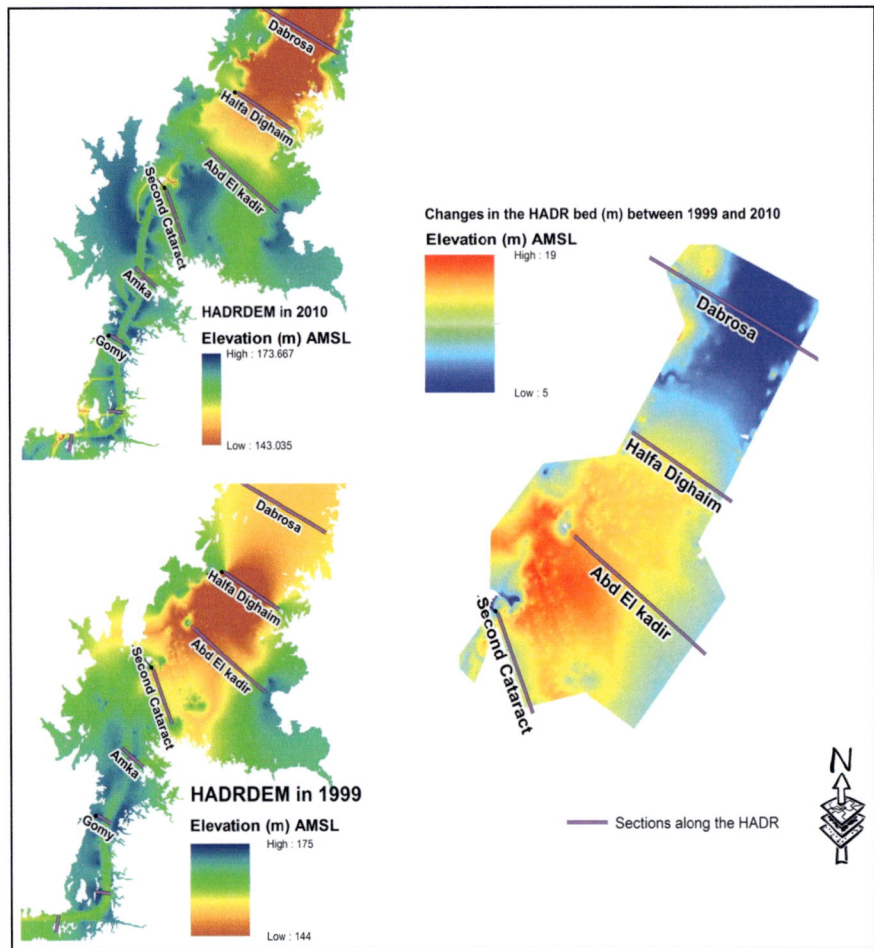

Figure 5.46: The HADR digital elevation models (HADRDEM) for 1999, 2010 and the corresponding changes of the lakebed

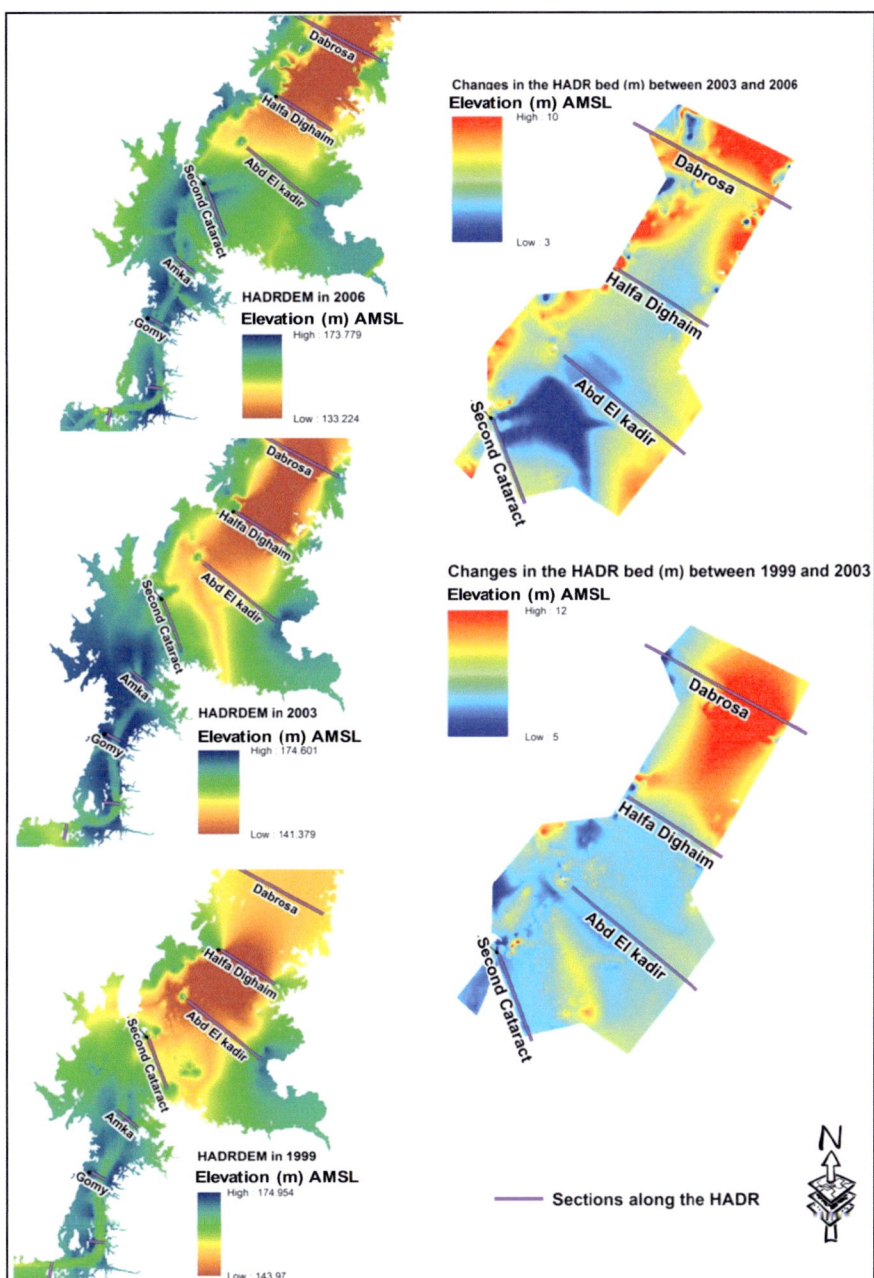

Figure 5.47: The HADR digital elevation models (HADRDEM) for 1999, 2003, 2006, and the corresponding changes of the lakebed

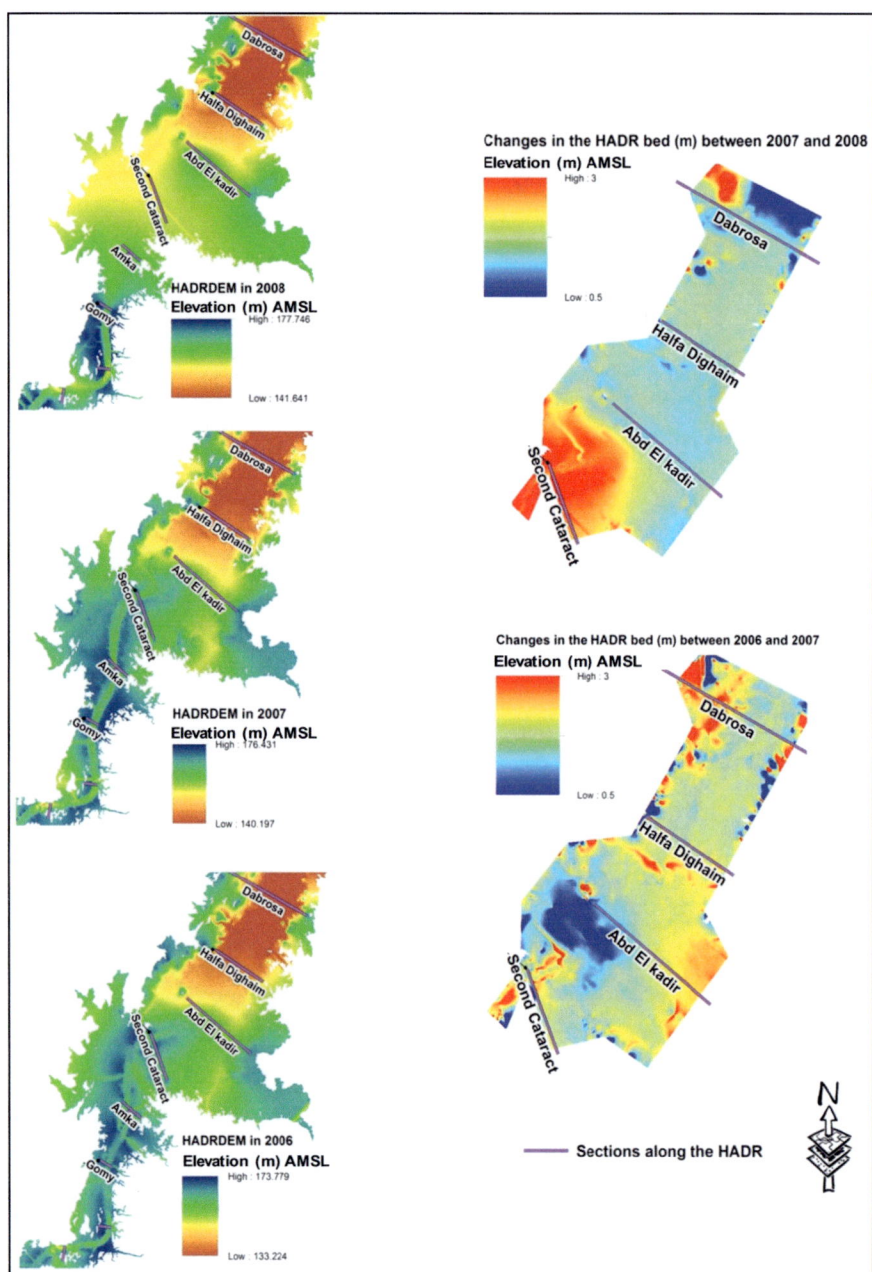

Figure 5.48: The HADR digital elevation models (HADRDEM) for 2006, 2007, 2008, and the corresponding changes of the lakebed

Results and Discussion 199

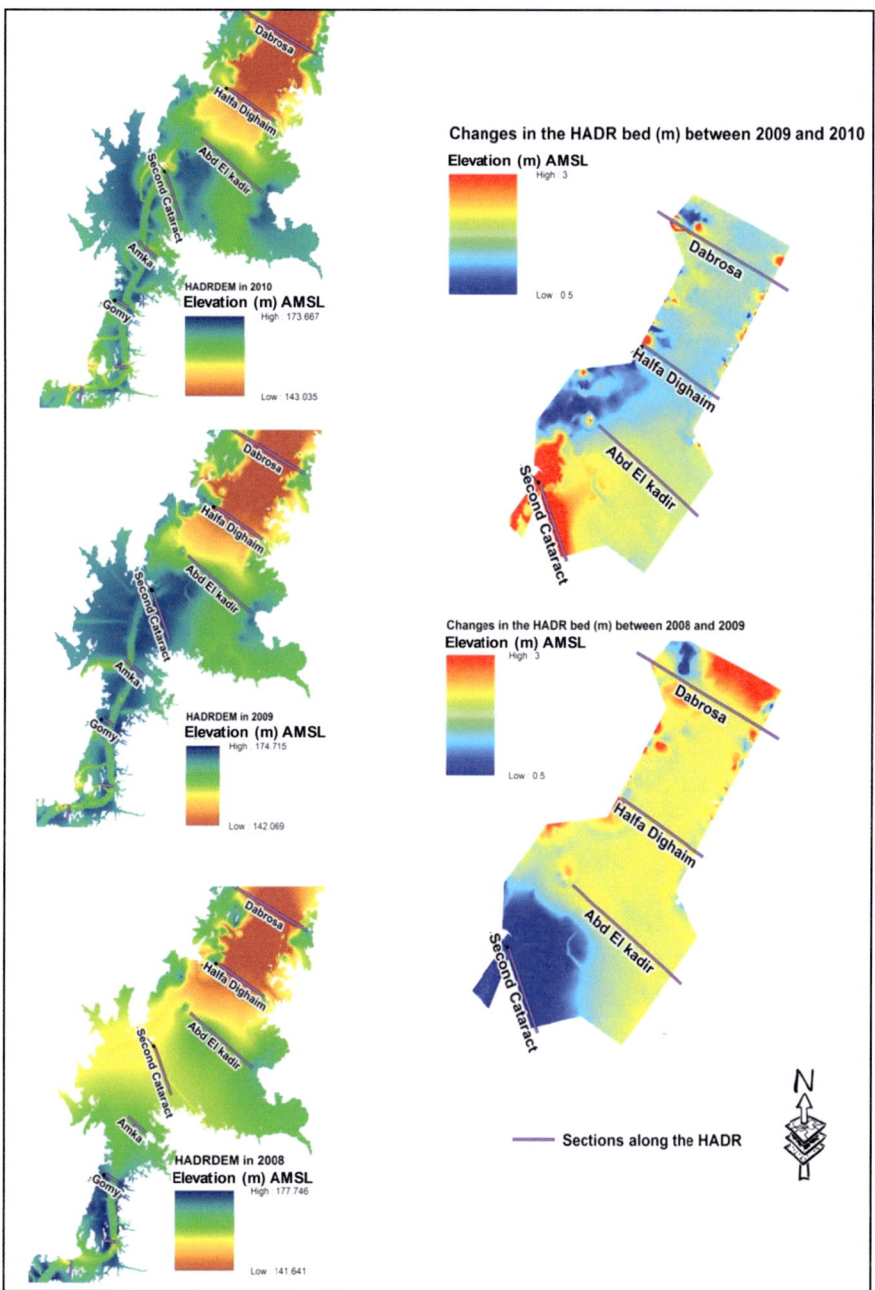

Figure 5.49: The HADR digital elevation models (HADRDEM) for 2008, 2009, 2010, and the corresponding changes of the lakebed

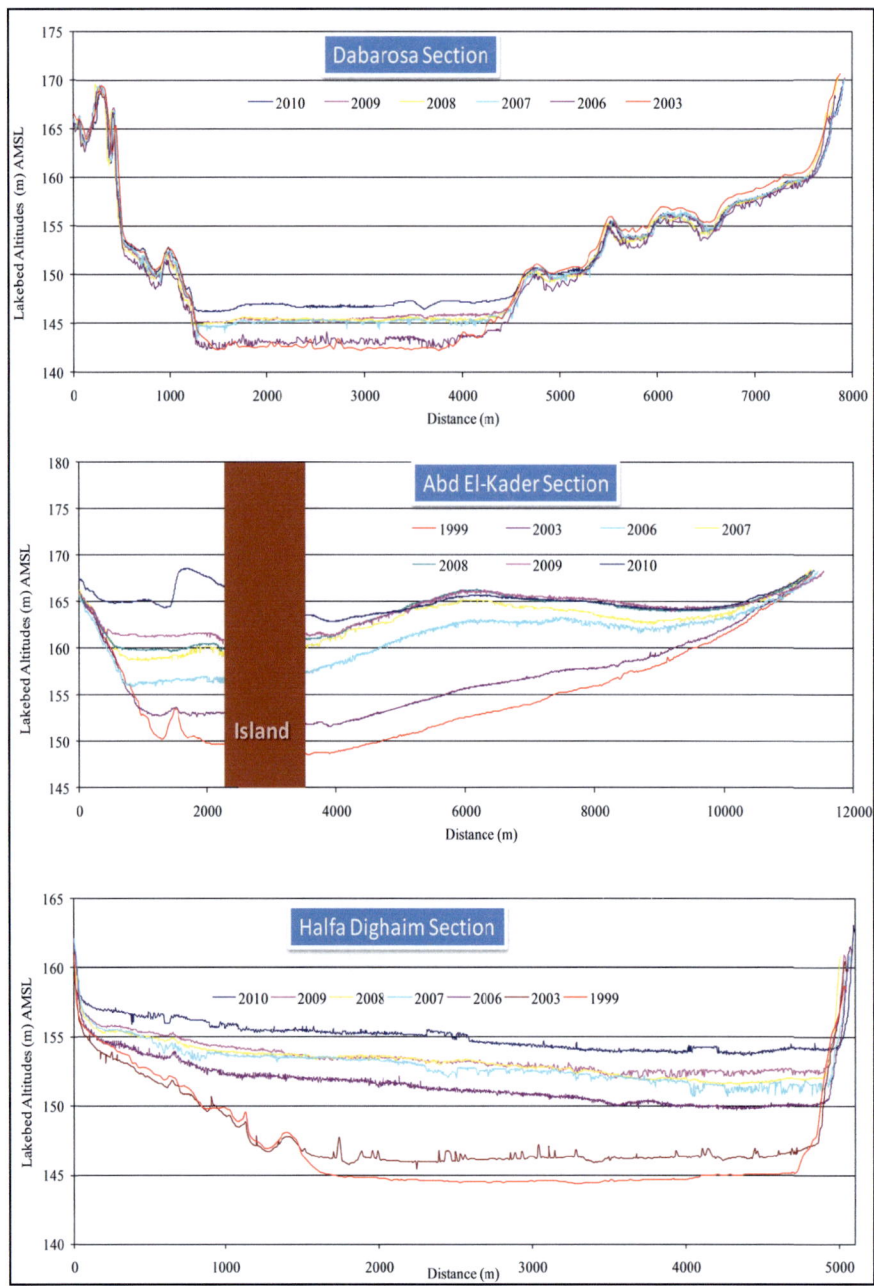

Figure 5.50: Cross sections through the HADR at Dabarosa, Abd Elkader and Halfa Dighaim sections

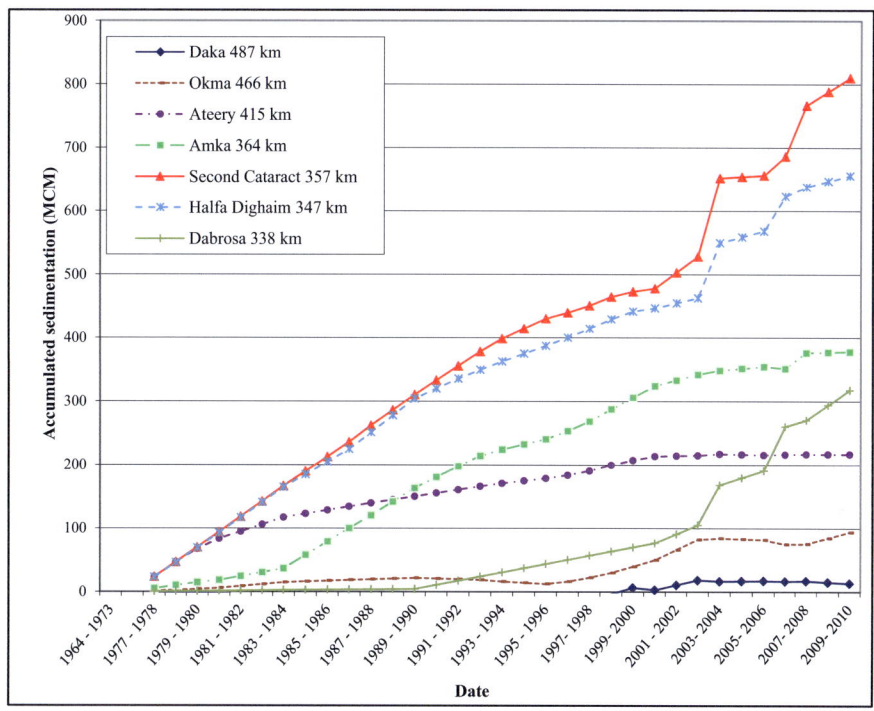

Figure 5.51: Sedimentation accumulated at major sections in Lake Nubia from 1977 to 2010

5.4 Scenarios

To address the problem of climate change effects on Egypt's natural resources, this study focuses on investigating the possibility of eliminating some khors in order to decrease the evaporation losses from the HADR. Different scenarios were studied to evaluate the significance of saving the evaporation losses by diminishing the lake surface area, and their influence on lake hydrological characteristics. Moreover, the possibility of controlling the HADR water levels by building a dam was investigated to analyze how building a new dam at the El-Madik section would affect water levels and possibly decrease evaporation losses. The investigated scenarios are explained below.

5.4.1 Eliminating Khors

As stated above in Chapter 5.3, the hydrological characteristics of 42 khors along the HADR were investigated. According to Table 5.16 and Figure 5.42, the khors with large surface areas such as the Khor Kalabsha on the western shore of the HADR and the Khor El-Alaky on the east shore were selected to be investigated. On the other hand, the Khor Toshka was not investigated because of its function as a spillway for the HAD. The Toshka spillway at the end of the Khor Toshka is considered the HAD's main protection from high floods (Abul-

Atta, 1978, Hereher, 2014, Warner, 2011). Furthermore, the intake of the Shiekh Zayed canal is located on this khor. This canal was built to irrigate new reclamation lands in the South of Egypt (Wahby, 2004). Some khors with intermediate surface areas such as the khor Genina on the western shore and the Khor Sara on the eastern shore were chosen to be investigated due to their high evaporation rates, as shown in Figure 5.17, which have reached approximately 9 mm per day. Each khor was isolated from the HADRDEM to investigate its hydrological characteristics and the position of its possible elimination. Details of the elimination options and their effects on the HADR's hydrology are described below.

5.4.1.1 Eliminating Khor Kalabsha

Khor Kalabsha is the largest khor in the HADR. It is situated 30 km upstream the HAD, in the Western desert. It has a surface area of over 600 km^2 at 181.6 m AMSL, approximately 10% of the lake's entire area. According to Figure 5.17, the Khor Kalabsha loses 2800 mm of water annually, accounting to 7.7 mm per day, due to evaporation losses. Thus, it is expected that eliminating it would lead to significant reduction of evaporation losses. As shown by the DEM of Khor Kalabsha (Figure 5.52), the khor depth can reach up to 32 m. The average water depth is 10 m. The khor can be eliminated by building a dam to close it or by filling it.

Scenario1: Closing the Khor using a dam

Six alternatives were analysed to select the optimal location for closing the khor by constructing a dam, as shown in Figure 5.52 and Appendix H. The alternatives were analyzed and evaluated based on the following parameters: the cross section area of the dam, the length of the dam, the reduced surface area downstream the dam, and the water volume downstream the dam. The results of this evaluation are presented in Figure 5.52 and Figure 5.53. Alternative 4 has the smallest cross section area and can save almost the same space and water volume as the other alternatives. Thus, Alternative 4 is considered the optimum location for constructing a dam to close the khor. This dam can be constructed using local material from sediment deposits of the lake to decrease construction costs.

Scenario 2: Filling the Khor

This scenario involves the possibility of filling Khor Kalabsha using local filling material from the surrounding area. Khor Kalabsha and the surrounding areas were separated from the DEM of the HADR (HADRDEM), using the 200 m contour line. Based on the clipped DEM, new equations were generated for the relationship between water level, surface area, and water volume, as stated in Chapter 4.2.3.2. The availability of filling material from areas around the khor was investigated at elevations 182, 190, 195, and 200 m AMSL, based on the new equations. The results show that, using the filling material available in the khor up to an elevation of 182 m AMSL, would be enough to fill the khor up to an elevation of 172.8 m AMSL. Cutting the areas around the khor up to an altitude of 200 m AMSL would be adequate to fill the khor to an altitude of 181.6 m AMSL, as shown in Figure 5.54. Moreover, the entire area for Khor Kalabsha (630 km^2) and the area around it up to altitude 200 m AMSL would comprise approximately 1200 km^2, as shown in Figure 5.55. This leveled area might also be used for new settlements.

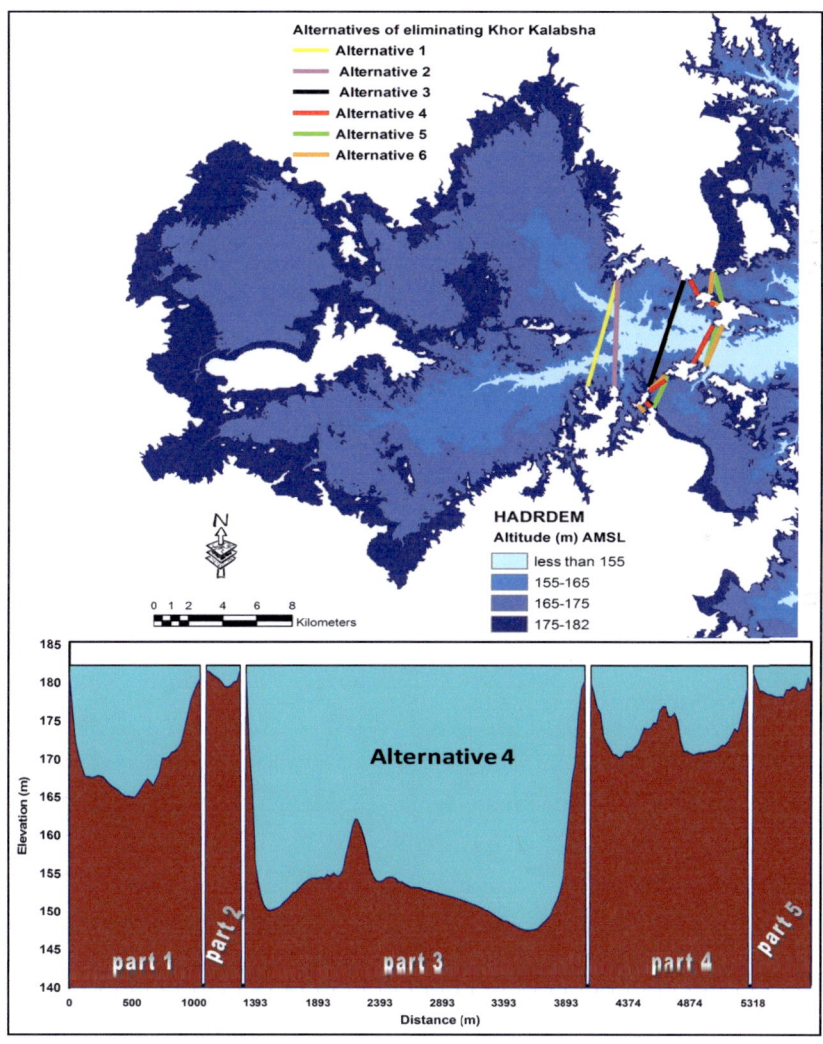

Figure 5.52: DEM of Khor Kalabsha, the alternatives suggested for eliminating Khor Kalabsha and the cross section at the optimal alternative

Results and Discussion

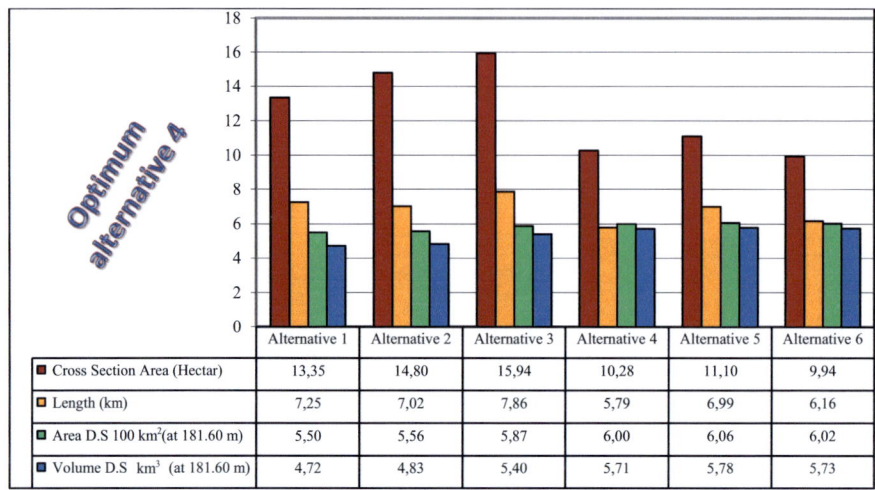

	Alternative 1	Alternative 2	Alternative 3	Alternative 4	Alternative 5	Alternative 6
■ Cross Section Area (Hectar)	13,35	14,80	15,94	10,28	11,10	9,94
■ Length (km)	7,25	7,02	7,86	5,79	6,99	6,16
■ Area D.S 100 km² (at 181.60 m)	5,50	5,56	5,87	6,00	6,06	6,02
■ Volume D.S km³ (at 181.60 m)	4,72	4,83	5,40	5,71	5,78	5,73

Figure 5.53: Chart for assessing alternatives to eliminate Khor Kalabsha

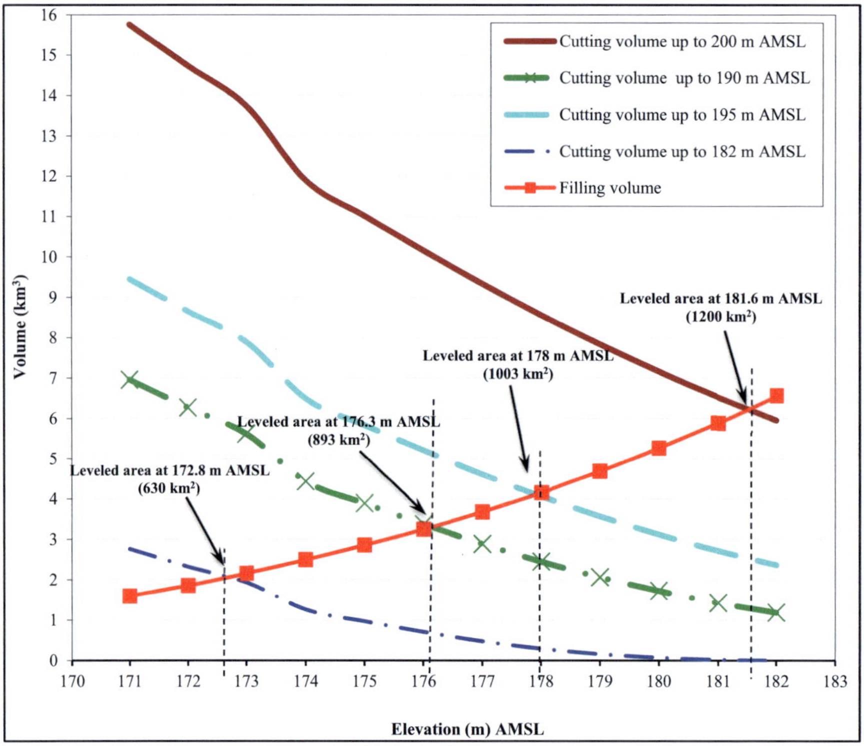

Figure 5.54: The alternatives of filling and cutting off of Khor Kalabsha

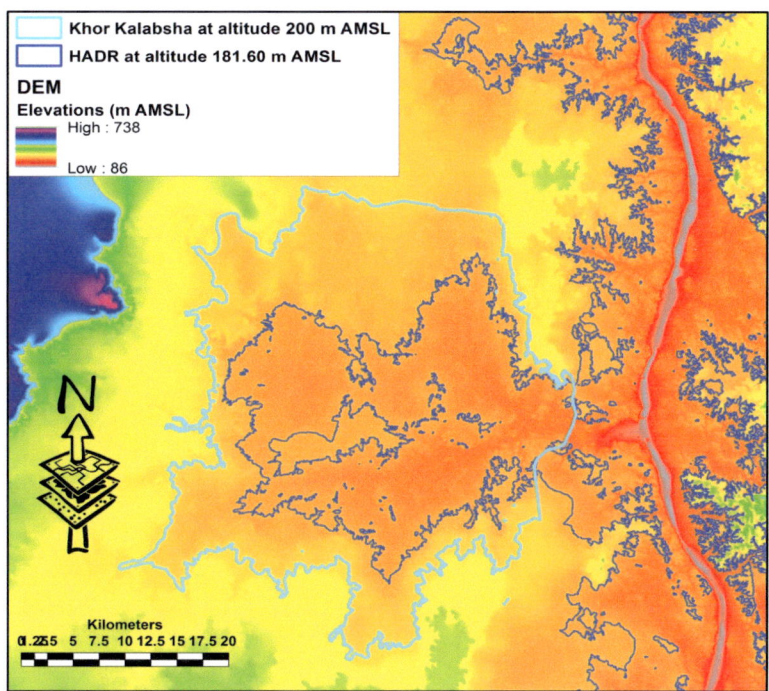

Figure 5.55: The area after filling Khor Kalabsha using the filling material from the surrounding area up to 200 m AMSL

Effects of eliminating Khor Kalabsha on the HADR hydrology

The HADRDEM, exclusive of Khor Kalabsha, was analyzed and new equations were generated for the relationship between water level, surface area, and water volume, as stated in Chapter 4.2.3.2. These equations were used to study the impact of the elimination changes on the hydrological characteristics of the HADR. Supposing that Khor Kalabsha had been eliminated two decades ago, the HADR's hydrological characteristics had changed considerably. The average water levels per month form August 1990 to July 2010 were calculated using the hydrological database HADRHYDB, and the corresponding surface areas were estimated using the mathematical models generated for the HADR before and after the potential removal of Khor Kalabsha. The change in the surface areas were computed, as the difference between the surface area of the actual HADR, and the surface area of the HADR without Khor Kalabsha. The reduced surface areas expected for different water levels, recorded form August 1990 to July 2010, are shown in Figure 5.56. The curves show that the higher the water levels the larger the reduced surface area would be. The lake surface area can be reduced by approximately 600 km^2 at most for water levels up to 181 m AMSL. The results of the elimination process are presented in Figure 5.57. The remaining lake surface area would be approximately 6000 km^2 at water level 181 m

AMSL, and the lake water volume would be reduced by approximately 5 km³ compared to the actual storage capacity of the HADR. Consequently, the lake water levels, would rise by utmost 90 cm after eliminating Khor Kalabsha.

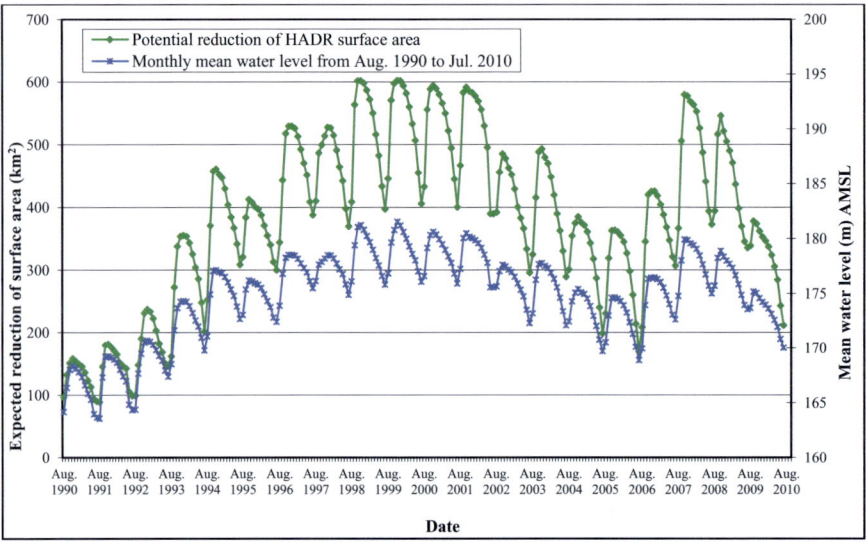

Figure 5.56: The mean monthly water level in m AMSL and the expected reduction of surface area in km² after a potential elimination of Khor Kalabsha two decades ago

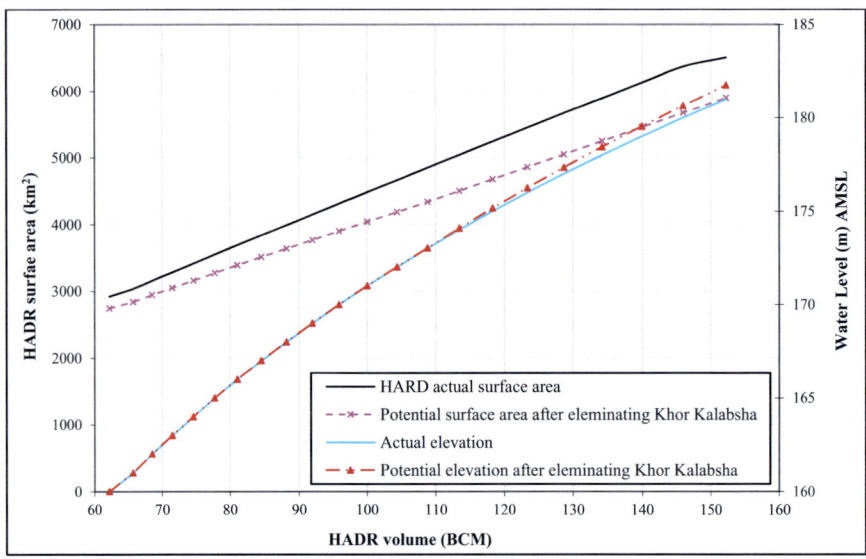

Figure 5.57: The effect of removing Khor Kalabsha on the HADR's hydrological characteristics

Effects of eliminating Khor Kalabsha on the HADR evaporation losses regarding climate change

The monthly evaporation losses were estimated before and after eliminating the khor, based on the monthly evaporation rates stored in the meteorological database HADRMTDB. The reductions of evaporation loss were calculated and are listed in Table 5.17. Approximately 1.1 km^3 of water, which represents approximately 2% of the total evaporation, could have been reduced if Khor Kalabsha had been eliminated two decades ago.

Table 5.17: Annual evaporation losses before and after a potential elimination of Khor Kalabsha for the past two decades and the expected reduction in evaporation losses in km^3

Year	Annual measured evaporation losses (km^3)	Annual expected evaporation losses (km^3)	Expected Annual reduction in evaporation losses (km^3)	Reduction in evaporation losses %
90/91	9.68	9.34	0.34	3
91/92	10.02	9.64	0.38	4
92/93	10.64	10.17	0.47	4
93/94	12.26	11.49	0.76	6
94/95	13.70	12.68	1.02	7
95/96	13.41	12.44	0.97	7
96/97	14.92	13.68	1.23	8
97/98	14.98	13.74	1.24	8
98/99	16.01	14.62	1.39	9
99/00	16.22	14.80	1.42	9
00/01	15.99	14.59	1.39	9
01/02	16.00	14.60	1.39	9
02/03	14.12	13.02	1.10	8
03/04	14.00	12.92	1.08	8
04/05	12.78	11.92	0.86	7
05/06	12.37	11.58	0.78	6
06/07	13.32	12.37	0.96	7
07/08	15.38	14.08	1.30	8
08/09	14.64	13.45	1.18	8
09/10	12.77	11.91	0.86	7
Mean	15.00	14.00	1.00	8

Table 5.18 shows the expected reductions in term of evaporation losses from the HADR due to the elimination of Khor Kalabsha under the current climatic conditions and according to the results of ECHAM5 and HadCM3 climate models computed by MWRI, as stated in Chapter 5.1.1.6 (LNFDC, 2008). According to HADRMTDB, the annual evaporation losses average up to 2800 mm in Khor Kalabsha under the climatic conditions of the past two decades. Thus, 1.60 km^3 could have been reduced under current climatic conditions. Based on the results of global climate change models and the climatic scenarios by MWRI using the

ECHAM5 model, it is predicted that evaporation losses would increase to 2900 mm by 2030, 2950 mm by 2050, and 3100 mm by 2100. For the 600 km^2 accounted for Khor Kalabsha, the reductions in terms of evaporation losses are expected to be two km^3 by the end of the twenty-first century. Using the HadCM3 model, it is predicted that evaporation losses would increase to 2890 mm by 2030, 2940 mm by 2050, and 3040 mm by 2100. For the 600 km^2 calculated for Khor Kalabsha, the reductions are expected to be approximately 1.8 km^3 by the end of the twenty-first century. Based on the HADRHYDB, the mode of the measured water levels during the past two decades is 178 m AMSL, as shown in Figure 5.19. The reductions at this level are estimated to be 1.4 km^3 under current climatic conditions, 1.6 km^3 based on the ECHAM5 model, and 1.5 km^3 based on the HadCM3 model, by the end of the twenty-first century.

Table 5.18: Expected reduced evaporation losses in km^3 from the HADR after a potential elimination of Khor Kalabsha under current climatic conditions and according to the ECHAM5 and HadCM3 climate models

Elevation	Current climatic conditions	ECHAM5			HadCM3		
		2030	2050	2100	2030	2050	2100
160	0.15	0.16	0.16	0.17	0.16	0.16	0.17
165	0.29	0.31	0.32	0.33	0.31	0.32	0.33
170	0.57	0.61	0.62	0.65	0.61	0.62	0.64
175	1.01	1.08	1.10	1.15	1.08	1.09	1.13
178	1.37	1.46	1.49	1.56	1.46	1.49	1.54
180	1.57	1.68	1.72	1.79	1.68	1.71	1.77
182	1.63	1.74	1.78	1.85	1.74	1.77	1.83

5.4.1.2 Khor El-Alaky

Khor El-Alaky is 100 km upstream the HAD and extends into the Eastern desert. It has a large surface area of 500 km^2 and undergoes 2500 mm evaporation losses annually, accounting to 6.9 mm/day as shown in Figure 5.17. Thus, it is expected that eliminating this khor would also help in decreasing evaporation losses. By separating Khor El-Alaky's DEM, the bed elevations of the khor vary from 150 m to 182 m AMSL with an average water depth of 18.5 m, as shown in Figure 5.58. Hence, this khor is considered a deep khor, and it is difficult to eliminate by filling it. However, the khor can be eliminated by building a dam to close it or by covering it.

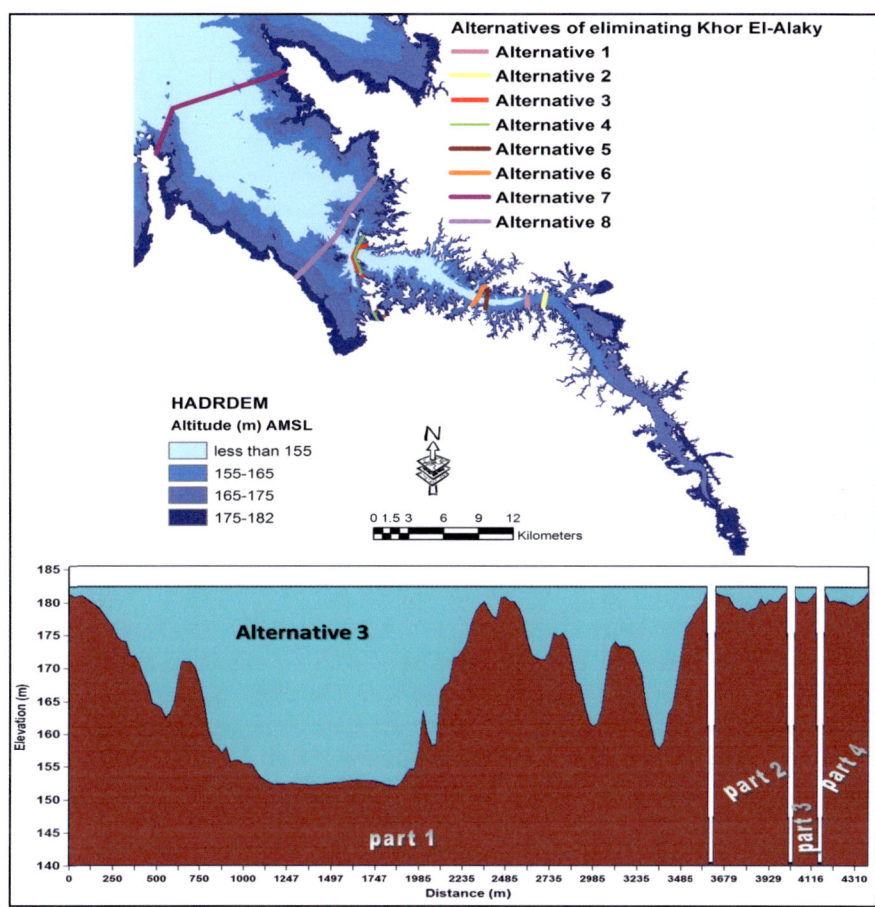

Figure 5.58: DEM of Khor El–Alaky, the alternatives suggested for eliminating Khor E–Alaky and the cross section at the optimal alternative

The different alternatives to close the khor are shown in Figure 5.58 and Appendix H. The alternatives were analyzed and evaluated with respect to four parameters: the cross section area of the dam, the length of the dam, the reduced surface area downstream the dam, and the water volume downstream the dam. The results of this evaluation are included in Figure 5.58 and Figure 5.59. As alternatives 7 and 8 have a huge cross section area, they should be abandoned for economic reasons. Alternatives 1, 2, 5, 6 would save around 100 km^2; therefore, it is not relevant to perform any measures there. Hence, alternatives 3 and 4 should be considered. Alternative 3 has has less cross section area than Alternative 4 and reduces quite the same surface area. Therefore, it is recommended to construct a dam at Alternative 3 and eliminate approximately 167 km^2.

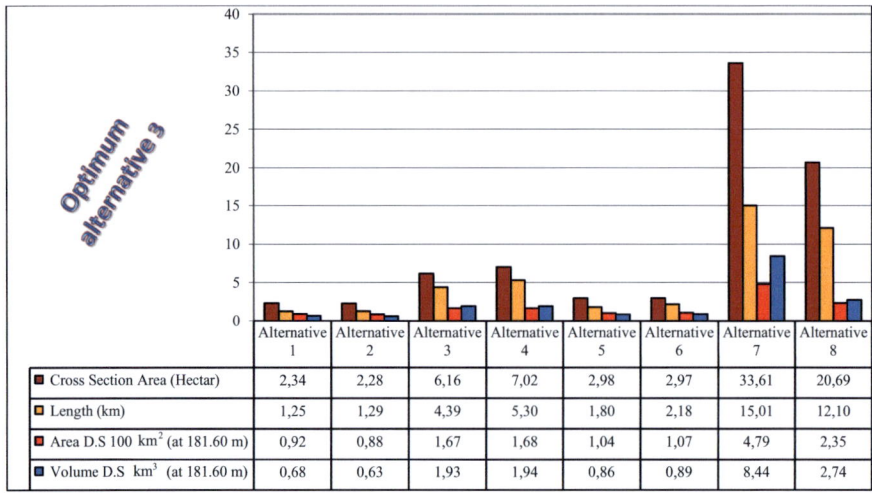

Figure 5.59: Chart for assessing alternatives for the elimination of Khor El-Alaky

Effects of eliminating Khor El-Alaky on the HADR hydrology

In case of covering the khor according to Alternative 3, the HADR's hydrology will not be affected. In case of building a dam to eliminate the khor, the HADR's storage capacity would be affected. Therefore, Khor El-Alaky was separated from the HADRDEM to study the impact of its elimination on the HADR's hydrological characteristics. Supposing that Khor El-Alaky would have been eliminated two decades ago, the HADR's hydrological characteristics would have changed. The average water levels per month from August 1990 to July 2010 were calculated using the HADRHYDB, and the corresponding surface areas were estimated using the mathematical models generated for the HADR before and after the potential removal of Khor El-Alaky. The change in the surface areas were computed, as the difference between the surface area of the actual HADR, and the surface area of the HADR without Khor El-Alaky. The reduced surface areas expected for different water levels, recorded form August 1990 to July 2010, are revealed in Figure 5.60. The curves show that the higher the water levels, the larger the reduced surface area would be. The lake surface area can be reduced by approximately 170 km^2 at most for water levels up to 181 m AMSL. The results of the elimination process are presented in Figure 5.61. The remaining lake surface area would be approximately 6,340 km^2 at water level 181 m AMSL, and the lake water volume would be reduced by approximately 1.9 km^3 compared to the actual storage capacity of the HADR. Consequently, the lake water levels would rise by at most 25 cm after the potential elimination of Khor El-Alaky.

Results and Discussion

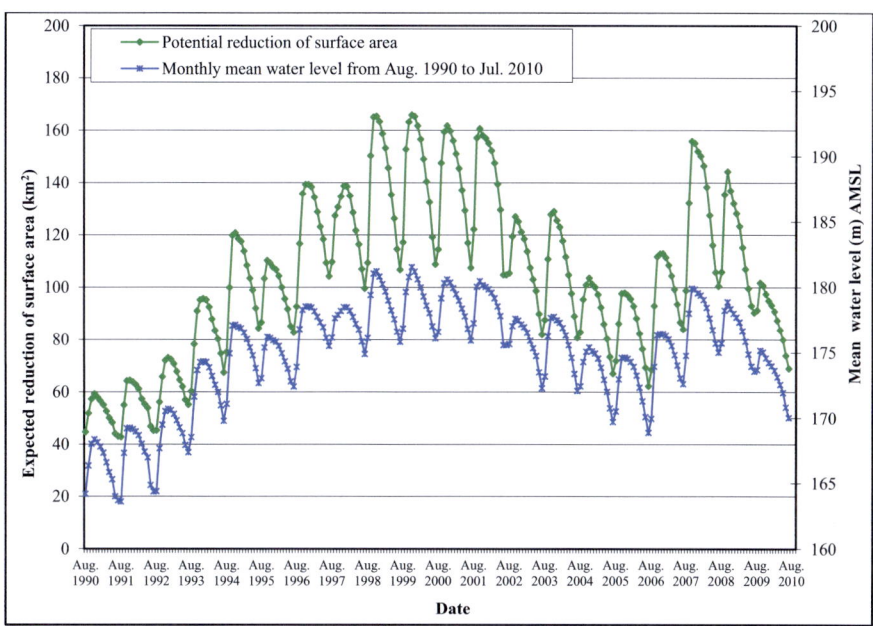

Figure 5.60: The monthly mean water level in m AMSL and the expected reduction of surface area in km² after a potential elimination of Khor El-Alaky, calculated in the past two decades

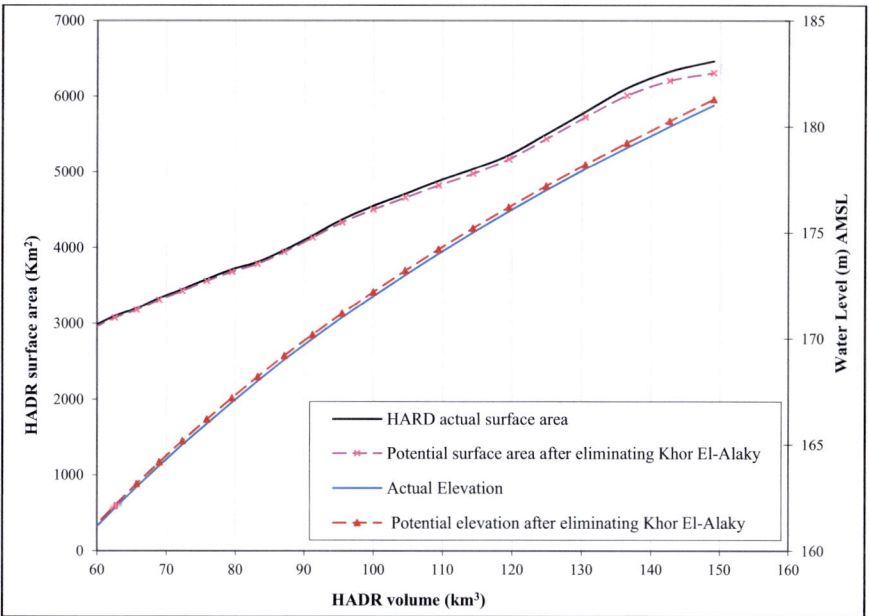

Figure 5.61: The effect of eliminating Khor El-Alaky on the hydrological characteristics of the HADR

Effects of eliminating Khor El-Alaky on the HADR evaporation losses regarding climate change

The monthly evaporation losses were estimated before and after the potential elimination of khor El-Alaky, based on the monthly evaporation rates stored in the HADRMTDB. The expected annual reduced evaporation losses were estimated and are listed in Table 5.19. The average expected reduced water loss in case of eliminating Khor El-Alaky two decades ago was approximately 0.3 km^3, representing some 2% of the total evaporation. Table 5.20 shows the expected reduced evaporation losses from the HADR due to the elimination of Khor El-Alaky under current climatic conditions and according to the ECHAM5 and HadCM3 climate models. According to the HADRMTDB, the annual evaporation losses averaged to 2500 mm in Khor El–Alaky under the climatic conditions of the recent past. Thus, the reduced evaporation loss is estimated to be 0.40 km^3 under current climatic conditions.

Table 5.19: Annual evaporation losses before and after a potential elimination of Khor El-Alaky for the past two decades and the expected reduction of evaporation losses in km^3

Year	Annual measured evaporation losses (km^3)	Annual expected evaporation losses (km^3)	Expected Annual reduction in evaporation losses (km^3)	Reduction in evaporation losses %
90/91	9.68	9.54	0.14	1.4
91/92	10.02	9.87	0.15	1.5
92/93	10.64	10.47	0.17	1.6
93/94	12.26	12.04	0.22	1.8
94/95	13.70	13.43	0.28	2.0
95/96	13.41	13.15	0.26	2.0
96/97	14.92	14.59	0.33	2.2
97/98	14.98	14.66	0.33	2.2
98/99	16.01	15.63	0.37	2.3
99/00	16.22	15.84	0.38	2.4
00/01	15.99	15.61	0.37	2.3
01/02	16.00	15.63	0.37	2.3
02/03	14.12	13.82	0.29	2.1
03/04	14.00	13.72	0.29	2.0
04/05	12.78	12.54	0.24	1.9
05/06	12.37	12.14	0.22	1.8
06/07	13.32	13.06	0.26	2.0
07/08	15.38	15.04	0.35	2.2
08/09	14.64	14.32	0.31	2.1
09/10	12.77	12.53	0.24	1.9
Mean	15.00	13.00	0.28	2.0

In view of the results of the global climate change model and the prepared climatic scenarios by MWRI, the ECHAM5 model predicted that evaporation losses would increase to 2600 mm by 2030, 2650 mm by 2050, and 2750 mm by 2100. For the reduced 160 km² after a potential elimination of Khor El-Alaky, the reduced evaporation loss is expected to be approximately 0.50 km³ by the end of twenty-first century. Using the HadCM3 model, it is predicted that evaporation losses would increase to 2570 mm by 2030, 2600 mm by 2050, and 2700 mm by 2100. For the 160 km² calculated for Khor EL-Alaky, the reductions are expected to be approximately 0.45 km³ by the end of the twenty-first century. As stated before, the mode of the measured water levels during the past two decades was 178 m AMSL. The reduced evaporation losses at this level are estimated to be 0.33 km³ under current climatic conditions and 0.36 km³ by the end of twenty-first century.

Table 5.20: Expected reduced evaporation losses in km³ from the HADR after a potential elimination of Khor El-Alaky under the current climatic conditions and according to the ECHAM5 and HadCM3 climate models

Elevation	Current climatic conditions	ECHAM5			HadCM3		
		2030	2050	2100	2030	2050	2100
160	0.08	0.08	0.09	0.09	0.08	0.09	0.09
165	0.12	0.12	0.12	0.13	0.12	0.12	0.13
170	0.17	0.18	0.18	0.19	0.18	0.18	0.19
175	0.25	0.26	0.26	0.28	0.26	0.26	0.27
178	**0.33**	**0.34**	**0.35**	**0.36**	**0.34**	**0.35**	**0.36**
180	0.39	0.40	0.41	0.43	0.40	0.41	0.43
182	0.42	0.43	0.44	0.46	0.43	0.44	0.45

5.4.1.3 Khor Genina

Khor Genina is found in the Eastern desert at approximately 180 km upstream HAD and covers approximately 103 km² at water level 182 m. It is steep and relatively narrow with a rocky bottom. According to Figure 5.17, Khor Genina loses 3100 mm of water annually account for 8.5 mm per day, due to evaporation. By isolating Khor Genina's DEM, the khor mean depth is estimated to be 14 m, which is also considered a deep khor. Since it would be difficult filling this khor in order to eliminate it, building a dam to close the khor was investigated. This khor has many fingers with tiny width average of 500 m as shown in Figure 5.62. Thus, a covering system for this khor can be another option to decrease the HADR's surface area. Different alternatives to close the khor are shown in Figure 5.62 and Appendix H. The alternatives were analyzed and evaluated with respect to four parameters: the cross section area of the dam, the length of the dam, the reduced surface area downstream the dam, and the water volume downstream the dam. The results of this evaluation are demonstrated in Figure 5.62 and Figure 5.63. Alternative 2 has a larger cross section area than Alternative 1, hence, it should be neglected for economic reasons, specially because there is slight difference in the reduced surface area between both options. Therefore, it is recommended to construct a dam at Alternative 1 to eliminate approximately 103 km^2 of surface area.

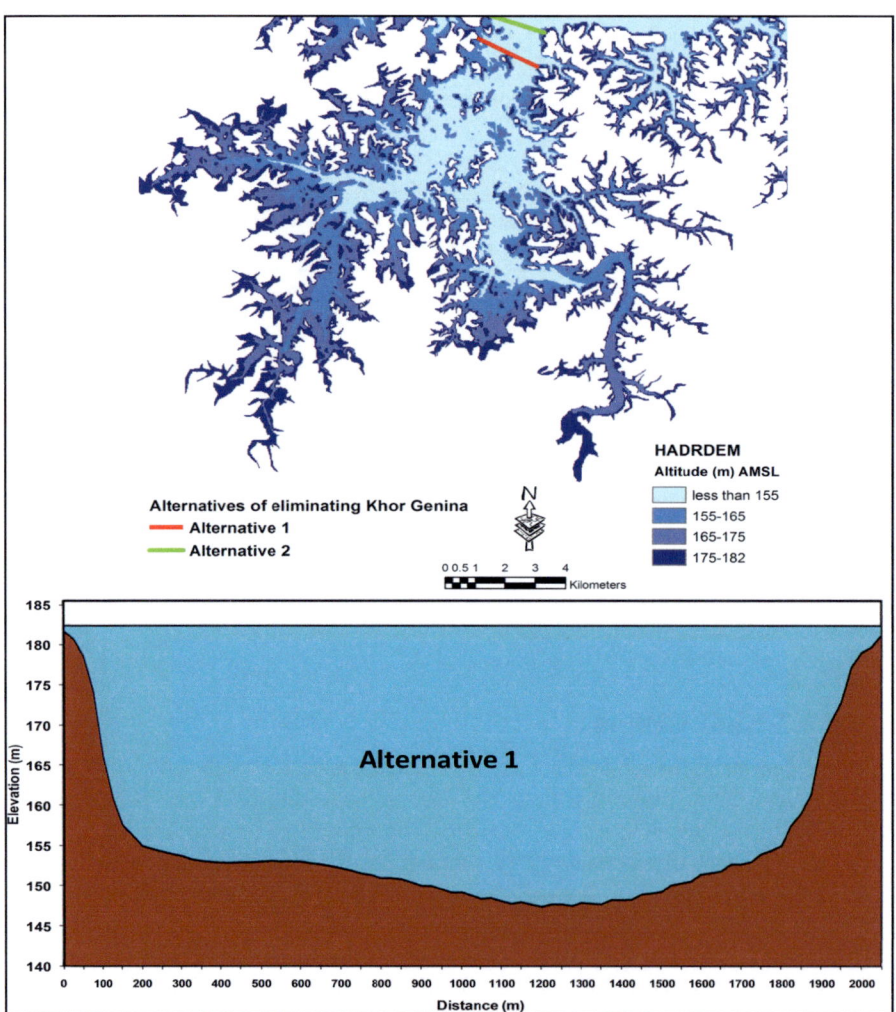

Figure 5.62: DEM of Khor Genina, the alternatives suggested for eliminating Khor Genina and the cross section at the optimal alternative

Results and Discussion   215

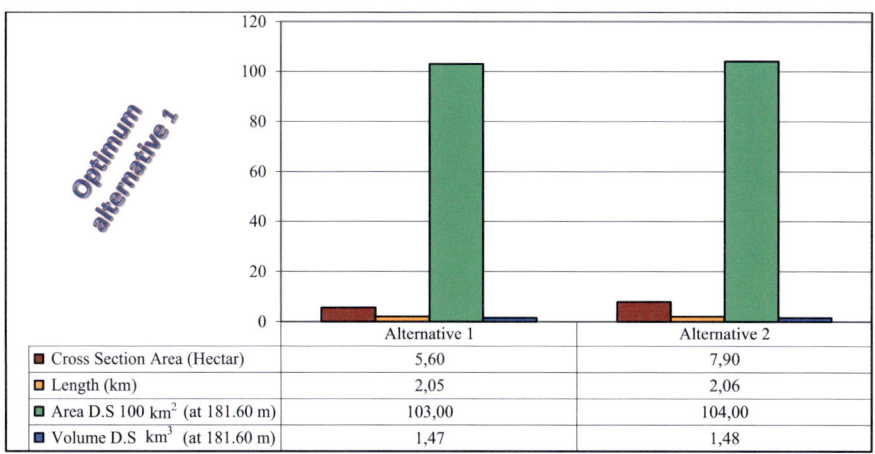

Figure 5.63: Chart for assessing alternatives for the elimination of Khor Genina

Effects of eliminating Khor Genina on the HADR hydrology

If Khor Genina is covered, the HADR's storage capacity would not be changed, and consequently, the HADR's hydrology would not be affected. Khor Genina was separated from HADRDEM to study the impact of its elimination, by building a dam, on the hydrological characteristics of the HADR. Supposing that Khor Genina was eliminated two decades ago, the HADR's hydrological characteristics would have changed. The average water levels per month from August 1990 to July 2010 were calculated using the HADRHYDB, and the corresponding surface areas were estimated using the mathematical models generated for the HADR before and after the potential removal of Khor Genina. The change in the surface areas were computed, as the difference between the surface area of the actual HADR, and the surface area of the HADR without Khor Genina. The reduced surface areas expected for different water levels, recorded form August 1990 to July 2010, are revealed in Figure 5.64. The curves show that the higher the water levels, the larger the reduced surface area would be. The lake surface area can be reduced by approximately 101 km^2 at most for water levels up to 181 m AMSL. The results of the elimination process are presented in Figure 5.65. The remaining lake surface area would be approximately 6400 km^2 at water level 181 m AMSL, and the lake water volume would be reduced by approximately 1,4 km^3 compared to the actual storage capacity of the HADR. Consequently, the lake water levels, would rise by at most 25 cm after a potential elimination of Khor Genina.

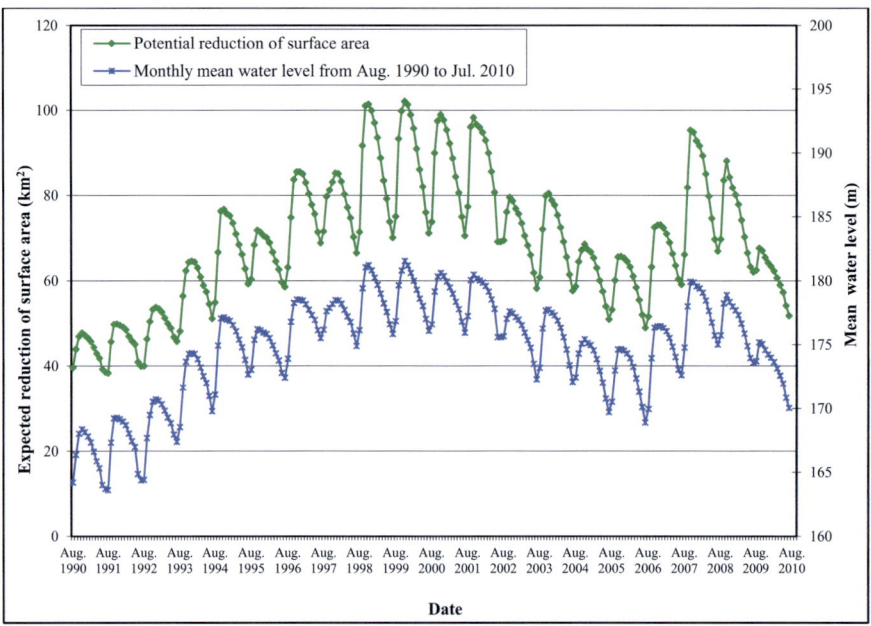

Figure 5.64: The mean monthly water level in m AMSL and the expected reduction of surface area in km² after a potential elimination of Khor Genina, from 1990 to 2010

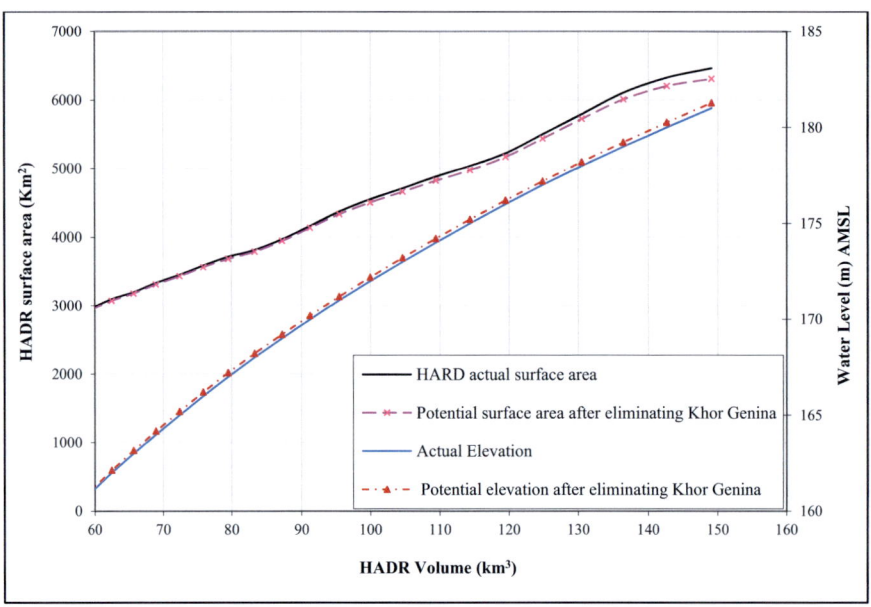

Figure 5.65: The effect of eliminating Khor Genina on HADR's hydrological features

Effects of eliminating Khor Genina on the HADR evaporation losses regarding climate change

The monthly evaporation losses were estimated before and after the potential elimination of Khor Genina, based on the monthly evaporation rates stored in the HADRMTDB. The expected annual reduced evaporation losses were estimated and are listed in Table 5.21. The average expected reduced evaporation losses after a potential elimination of Khor Genina two decades ago are approximately 0.23 km^3, some 1.7% of the total evaporation. Table 5.22 shows the expected reduced evaporation losses from HADR after the potential elimination of Khor Genina under the current climatic conditions and according to the results of ECHAM5 and HadCM3 climate models, computed by MWRI, as stated in Chapter 5.1.1.6 (LNFDC, 2008). According to HADRMTDB, the annual evaporation losses are averaged to 3100 mm in Khor Genina under climatic conditions of the recent past. Thus, 0.32 km^3 could have been reduced of the evaporation losses, under current climatic conditions.

Table 5.21: Annual evaporation losses before and after a potential elimination of Khor Genina for the past two decades and the expected reduced evaporation losses in km^3

Year	Annual measured evaporation losses (km^3)	Annual expected evaporation losses (km^3)	Expected Annual reduction in evaporation losses (km^3)	Reduction in evaporation losses %
90/91	9.68	9.53	0.15	1.5
91/92	10.02	9.86	0.15	1.5
92/93	10.64	10.48	0.16	1.5
93/94	12.26	12.06	0.20	1.6
94/95	13.70	13.48	0.23	1.7
95/96	13.41	13.19	0.22	1.6
96/97	14.92	14.66	0.26	1.7
97/98	14.98	14.72	0.26	1.7
98/99	16.01	15.72	0.29	1.8
99/00	16.22	15.92	0.30	1.8
00/01	15.99	15.70	0.29	1.8
01/02	16.00	15.71	0.29	1.8
02/03	14.12	13.88	0.24	1.7
03/04	14.00	13.77	0.24	1.7
04/05	12.78	12.57	0.21	1.6
05/06	12.37	12.17	0.20	1.6
06/07	13.32	13.10	0.22	1.6
07/08	15.38	15.11	0.27	1.8
08/09	14.64	14.39	0.25	1.7
09/10	12.77	12.56	0.21	1.6
Mean	15.00	13.43	0.23	1.7

In view of the global climate change model results, and the prepared climatic scenarios by MWRI, the ECHAM5 model predicted that evaporation losses would increase to 3190 mm by 2030, 3260 mm by 2050, and 3400 mm by 2100. For the reduced 100 km² accounted for by Khor Genina, the reduced evaporation losses are expected to be 0.35 km³ by the end of twenty-first century. Using the HadCM3 model, it is predicted that evaporation losses would increase to 3190 mm by 2030, 3240 mm by 2050, and 3360 mm by 2100. For the 100 km² calculated for Khor Genina, the reductions are expected to be approximately 0.34 km³ by the end of the twenty-first century. The reductions at 178 m AMSL are estimated to be 0.25 km³ under current climatic conditions, 0.28 km³ based on the ECHAM5 model, and 0.27 km³ based on the HadCM3 model, by the end of the twenty-first century.

Table 5.22: Expected reduced evaporation losses in km³ from the HADR after a potential elimination of Khor Genina under the current climatic conditions and according to the ECHAM5 and HadCM3 climate models

Elevation	Current climatic conditions	ECHAM5			HadCM3		
		2030	2050	2100	2030	2050	2100
160	0.09	0.10	0.10	0.10	0.10	0.10	0.10
165	0.13	0.13	0.13	0.14	0.13	0.13	0.14
170	0.16	0.16	0.17	0.18	0.17	0.17	0.17
175	0.21	0.21	0.22	0.23	0.21	0.22	0.22
178	**0.25**	**0.26**	**0.27**	**0.28**	**0.26**	**0.27**	**0.27**
180	0.30	0.31	0.31	0.33	0.31	0.31	0.32
182	0.32	0.33	0.34	0.35	0.33	0.33	0.34

5.4.1.4 Khor Sara

Khor Sara is located 325 km upstream the HAD reaching into the Western desert. It is one of the miniature khors in HADR and has a surface area of only 40 km². According to Figure 5.17, Khor Sara loses 3200 mm of water annually accounting for 8.7 mm per day, due to evaporation. Based on the DEM of Khor Sara, it is found that the khor's depths reach up to 32 m, as shown in Figure 5.66. The khor's mean depth is estimated to be 16.5 m, which is considered a deep khor. Since it will be difficult to eliminate this khor by filling it, building a dam to close the khor was investigated. This Khor has a small surface area, thus a covering system for this khor can be another option to decrease the HADR's surface area. Four alternatives were analyzed, as shown in Figure 5.66 and Appendix H.

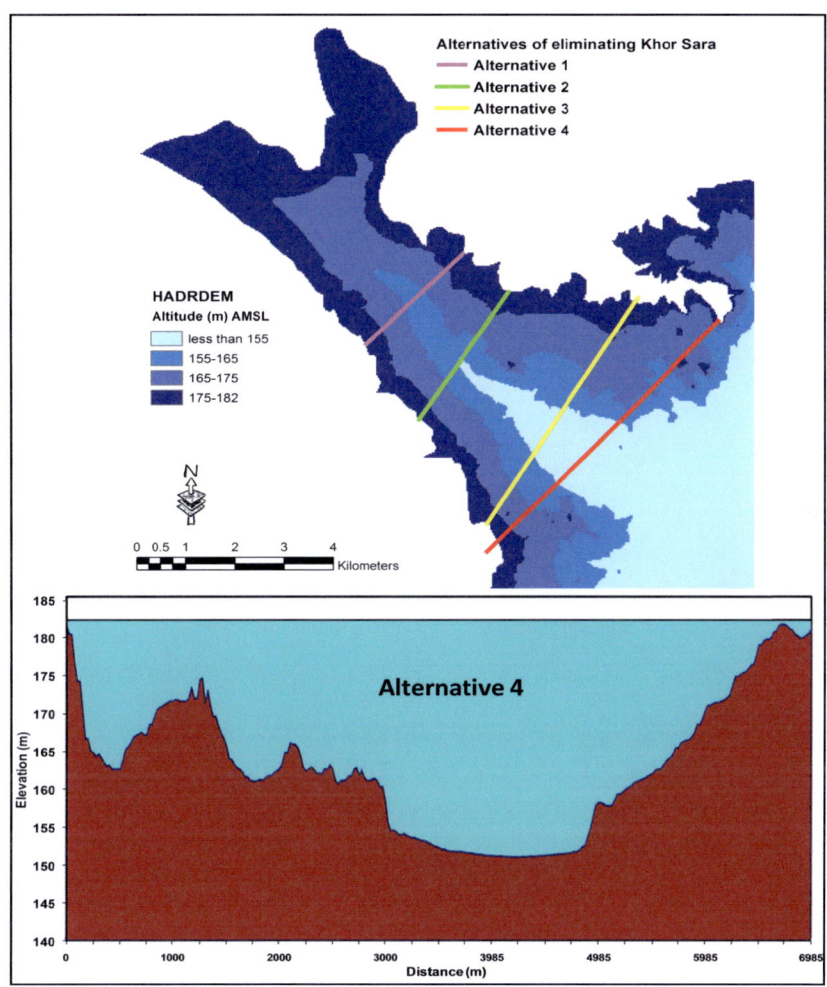

Figure 5.66: DEM of Khor Sara, the alternatives suggested for eliminating Khor Sara and the cross section at the optimal alternative

The alternatives were analyzed and evaluated with respect to four parameters; the cross section area of the dam, the length of the dam, the reduced surface area downstream the dam, and the water volume downstream the dam. The results of this evaluation are demonstrated in Figure 5.66 and Figure 5.67. Since Alternative 1 and Alternative 2 can save small surface area, they should be neglected for economic reasons. The results of Alternative 3 and Alternative 4 reflect that the reduced areas are quite small and the needed dams would be long with huge cross sections. On the other hand this area is undergoing a high annual evaporation rate of 3200 mm. Therefore, it is recommended to apply a covering measure to this khor up to Section 4, to decrease the HADR surface area by approximately 40 km².

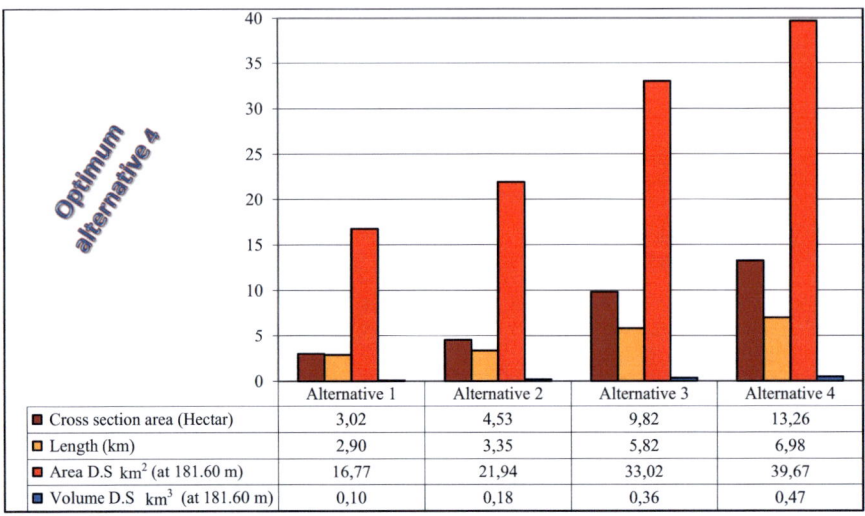

Figure 5.67: Chart for assessing alternatives for elimination of Khor Sara

Effects of eliminating Khor Sara on the HADR hydrology

If Khor Genina is covered, the HADR's storage capacity will not be changed, and consequently, the HADR's hydrology will not be affected. Supposing that Khor Sara was eliminated before two decades ago, the HADR hydrological characteristics wouldn't have changed. The average water levels per month were calculated using the HADRHYDB, and the corresponding surface areas were estimated for the HADR before and after covering Khor Sara. The expected reduced surface areas for different water levels recorded from 1990 to 2010 are revealed in Figure 5.68. The curves show that the higher the water levels the more the reduced surface area would be. The lake surface area can be reduced by approximately 40 km² at most for water levels up to 181 m AMSL.

Effects of eliminating Khor Sara on the Lake evaporation losses regarding climate change

Using the monthly evaporation rates stored in the HADRMTDB, the monthly evaporation losses were estimated before and after covering Khor Sara. The expected annual reduced evaporation losses were estimated and listed in Table 5.23. The average expected reduced evaporation losses in case of eliminating Khor Sara two decades ago are approximately 0.1 km³ some 0.6% of the total evaporation. Table 5.24 shows the expected reduced evaporation losses from HADR after the potential coverage of Khor Sara under the current climatic conditions and according to the ECHAM5 and HadCM3 climate models. According to HADRMTDB, the annual evaporation losses are averaged to 3200 mm in Khor Sara under climatic conditions of the recent past. Thus, the reduced evaporation losses are estimated to be 0.12 km^3 under current climatic conditions.

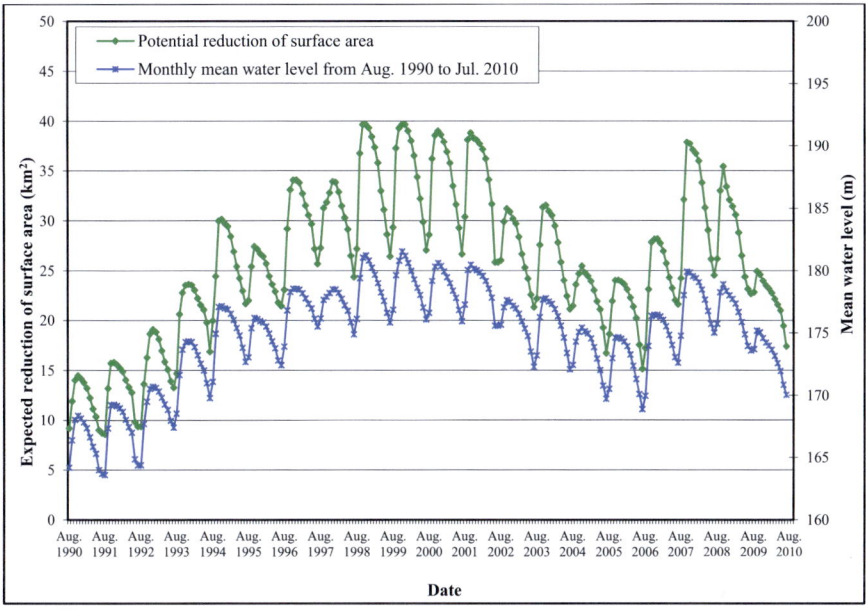

Figure 5.68: The monthly mean water level in m AMSL and the expected reduction of surface area in km^2 after a potential coverage of Khor Sara two decades ago

In view of the global climate change model results, and the prepared climatic scenarios by MWRI, the ECHAM5 model predicted that evaporation losses will increase to 3290 mm by 2030, 3360 mm by 2050, and 3500 mm by 2100. For the reduced 40 km^2 accounted for by Khor Sara, the reduced evaporation losses are expected to be 0.14 km^3 by 2100. Using the HadCM3 model, it is predicted that evaporation losses will increase to 3290 mm by 2030, 3340 mm by 2050, and 3460 mm by 2100. For the 40 km^2 calculated for Khor Sara, the reductions are expected to be approximately 0.14 km^3 by 2100. The reduced evaporation losses at 178 m AMSL are estimated to be 0.1 km^3 under current climatic conditions and 0.11 km^3 by the end of twenty-first century, based on the results of both ECHAM5 and HadCM3 climate models.

Table 5.23: Annual evaporation losses before and after a potential coverage of Khor Sara for the past two decades and the expected reduction of evaporation losses in km³

Year	Annual measured evaporation losses (km³)	Annual expected evaporation losses (km³)	Expected Annual reduction in evaporation losses (km³)	Reduction in evaporation losses %
90/91	9.68	9.64	0.04	0.4
91/92	10.02	9.97	0.04	0.4
92/93	10.64	10.59	0.05	0.5
93/94	12.26	12.19	0.07	0.6
94/95	13.70	13.62	0.09	0.6
95/96	13.41	13.33	0.08	0.6
96/97	14.92	14.82	0.10	0.7
97/98	14.98	14.88	0.10	0.7
98/99	16.01	15.89	0.11	0.7
99/00	16.22	16.10	0.12	0.7
00/01	15.99	15.87	0.11	0.7
01/02	16.00	15.88	0.11	0.7
02/03	14.12	14.02	0.09	0.6
03/04	14.00	13.91	0.09	0.6
04/05	12.78	12.70	0.07	0.6
05/06	12.37	12.30	0.07	0.6
06/07	13.32	13.24	0.08	0.6
07/08	15.38	15.28	0.11	0.7
08/09	14.64	14.54	0.10	0.7
09/10	12.77	12.70	0.07	0.6
Average	15.00	13.57	0.09	0.6

Table 5.24: Expected reduced evaporation losses in km³ from HADR after a potential coverage of Khor Sara under the current climatic conditions and according to the ECHAM5 and HadCM3 climate models

Elevation	Current climate conditions	ECHAM5			HadCM3		
		2030	2050	2100	2030	2050	2100
160	0.02	0.02	0.02	0.02	0.02	0.02	0.02
165	0.03	0.03	0.03	0.03	0.03	0.03	0.03
170	0.06	0.06	0.06	0.06	0.06	0.06	0.06
175	0.08	0.08	0.08	0.09	0.08	0.08	0.09
178	**0.10**	**0.11**	**0.11**	**0.11**	**0.11**	**0.11**	**0.11**
180	0.12	0.13	0.13	0.13	0.13	0.13	0.13
182	0.13	0.13	0.13	0.14	0.13	0.13	0.14

5.4.1.5 Other small Khors

As listed in Table 5.16, 42 khors in total were studied. Khors other than Kalabsha, El-Alaky, Genina, Sara, Toshka east and west, Abu Simbel east and west, Masmas, and Forkondi have a large surface area of approximately 1050 km^2 at water level 181.6 m AMSL accounting for over 15% of the entire surface area of the HADR and for 30% of the total khors' surface area. The small khors on the west bank of the HADR have an average surface area of less than 30 km^2 and a total surface area of over 420 km^2 with an average depth of 16 m, making them deep khors. The small khors on the east bank of the HADR have an average surface area of less than 35 km^2, and total surface area of over 630 km^2 with an average depth of 14 m, classifying them deep khors. Therefore, these khors can be covered to reduce evaporation losses.

Effects of eliminating small Khors on the HADR evaporation losses regarding climate change

Table 5.25: Expected reduced evaporation losses in km^3 from the HADR after a potential coverage of small Khors under the current climatic conditions and based on the results of ECHAM5 and HadCM3 climate models

Elevation	Current climate conditions	ECHAM5			HadCM3		
		2030	2050	2100	2030	2050	2100
160	0.4	0.4	0.4	0.4	0.4	0.4	0.4
165	0.7	0.7	0.7	0.8	0.7	0.7	0.8
170	1.2	1.2	1.3	1.3	1.2	1.3	1.3
175	1.7	1.8	1.8	1.9	1.8	1.8	1.9
178	**2.2**	**2.3**	**2.4**	**2.5**	**2.3**	**2.4**	**2.4**
180	2.8	2.9	2.9	3.1	2.9	2.9	3.0
182	3.1	3.2	3.2	3.4	3.2	3.2	3.3

Using monthly evaporation rates stored in the HADRMTDB, the monthly evaporation losses were estimated before and after eliminating these khors. Table 5.25 shows the expected reduced evaporation losses from HADR after a potential coverage of the small Khors under current climatic conditions and according to the results of ECHAM5 and HadCM3 climate models. According to HADRMTDB, the mean annual evaporation losses are approximately 2800 mm in the HADR under climatic conditions of the recent past, and thus, the reduced evaporation losses are estimated to be 3.1 km^3. In view of the results of the ECHAM5 model, it is predicted that the evaporation losses will increase to 2860 mm by 2030, 2940 mm by 2050, and 3100 mm by 2100. For the covered 1050 km^2 accounted for the Khors, the reduced evaporation losses are expected to be 3.4 km^3 by 2100. Using the HadCM3 model, the reductions are expected to be approximately 3.3 km^3 by 2100. The reduced evaporation losses at 178 m AMSL are estimated to be 2.2 km^3 under current climatic conditioned, and 2.5 km^3 by 2100.

5.4.2 Control of water levels

The HADR's surface area can be reduced by controlling water levels in the reservoir. This could be achieved by constructing a new dam upstream of the HAD in order to control water level and storage. There were originally two sites considered for the HAD and that one of these sites was located at El-Madik reach (MWRI, 1990). El-Madik, which means "a section of narrow width" in Arabic, has the narrowest width in the HADR, as shown in Figure 5.43. The width of El-Madik sections varies from 1300 m to 1600 m, about half of the width of the HAD. Moreover, as shown in Figures 5.44 and 5.45, the lakebed at this site has not changed significantly over the course of the past five decades. Figure 5.69 shows the cross sections at El-Madik based on the HADR bathometric surveys in 2000, 2004, 2007, and 2010 (HDA-MWRI, 2010, NWS, 2012). Only slight changes were observed in the lakebed at this location, which means that there was only minor erosion or sediment accumulation during the past decade. A new dam at the El-Madik section, that is, about 130 km away from HAD, is proposed to control water levels in the HADR. The main objectives of this dam are not limited to the reduction of evaporation losses, which can be achieved by reducing the lake surface area. This dam would also allow for the generation of hydroelectric power through the elevation difference between the upstream and downstream parts of the new dam.

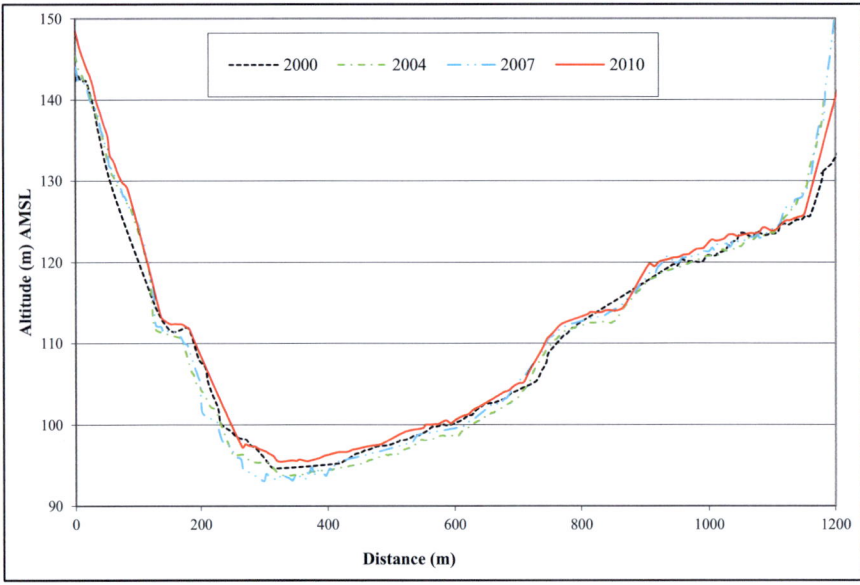

Figure 5.69: Cross sections at El-Madik section in 2000, 2004, 2007, and 2010.

5.4.2.1 Establishing a new dam at the El-Madik reach

Five alternatives at the El-Madik section were assessed to find the optimal position for the dam, as shown in Appendix H and Figure 5.70. The parameters considered in this evaluation were as follows: dam width, dam depth, and the area of dam cross section. The evaluation of these alternatives is shown in Figure 5.71.

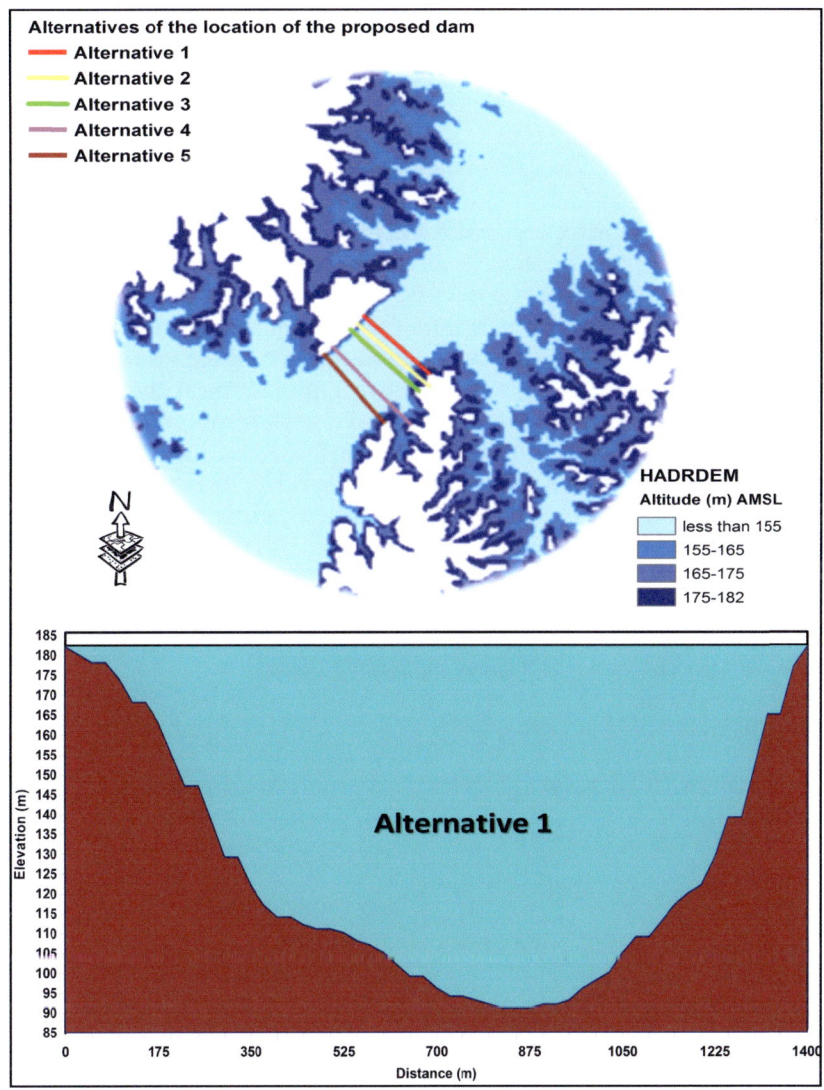

Figure 5.70: Alternatives for choosing the optimial location of the proposed new dam at El-Madik section, and the cross section of the optimal alternative.

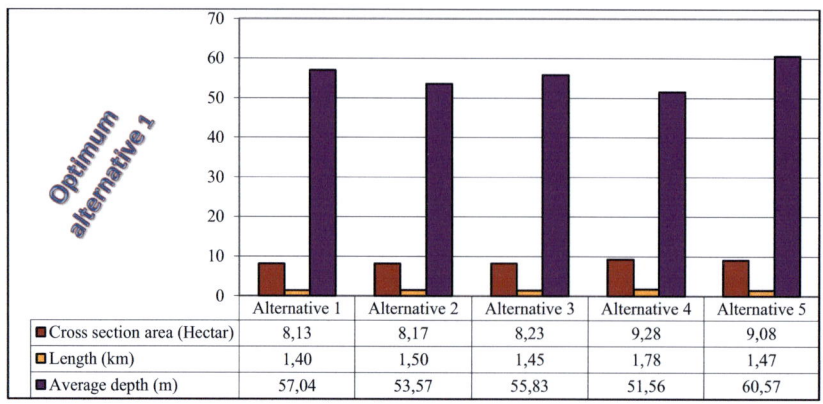

Figure 5.71: Evaluation of the different alternatives for the dam location at El-Madik section

Alternative 1 has the narrowest width and the smallest cross section area, and it is therefore the optimal location to divide the lake into two parts. Part One is upstream of the new dam and covers the area from the El-Daka section to the new dam, while Part Two extends from downstream the new dam to the HAD. The results of the analysis of the hydrological features of each part are shown in Figure 5.72. The reservoir is divided into two almost identical divisions in terms of volume and surface area. The hydrological characteristics of the two parts of the HADR were analyzed with SPSS, and mathematical models in a form of equations were developed to describe the relationships between water level in meter, surface area in km², and volume in km³. Equations from 6.1 to 6.4 show the mathematical relationship for Part One, whereas the equations from 6.5 to 6.8 those for Part Two. A_1 and A_2 refers to the surface area of Part One and Part Two, respectively, of the HADR in km², WL_1 and WL_2 to the water level of Part One and Part Two of the HADR in meters AMSL, and V_1 and V_2 to the water volume of Part One and Part Two of the HADR in km³.

$A_1 = 2.6283 \times 10^{-12} \times (WL_1)^{6.6851}$ \hfill Equation 6.1

$V_1 = 3.9197 \times 10^{-16} \times (WL_1)^{7.6611}$ \hfill Equation 6.2

$WL_1 = 102.5789 \times (V_1)^{0.1305}$ \hfill Equation 6.3

$A_1 = 73.04828 \times (V_1)^{0.8726}$ \hfill Equation 6.4

$A_2 = 3.4684 \times 10^{-14} \times (WL_2)^{7.5204}$ \hfill Equation 6.5

$V_2 = 1.7691 \times 10^{-17} \times (WL_2)^{8.241}$ \hfill Equation 6.6

$WL_2 = 107.8466 \times (V_2)^{0.1213}$ \hfill Equation 6.7

$A_2 = 67.2924 \times (V_2)^{0.9124}$ \hfill Equation 6.8

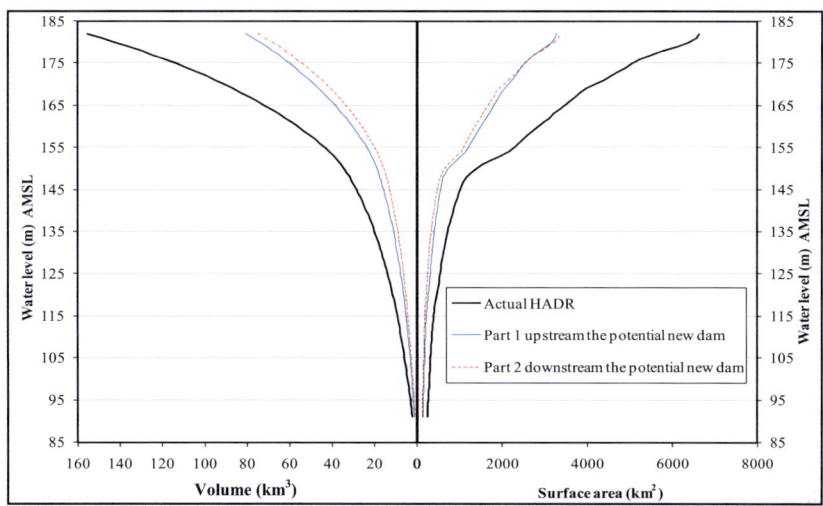

Figure 5.72: The HADR hydrological characteristics, and its two parts after the construction of the potential new dam

5.4.2.2 Scenarios for investigating the optimal operational water levels upstream and downstream of the proposed new dam

Ten scenarios were investigated to find the optimal water levels upstream and downstream of the new dam in order to identify the smallest surface area for each altitude. These scenarios were developed as follows: water levels downstream of the new dam would be decreased in one-meter intervals from one to ten meters. Assuming that the storage capacity does not change in terms of altitude before and after the construction of the new dam, the water level upstream of the new dam would rise. The minimum water level downstream of the new dam must not be below an altitude of 155 m AMSL, as lower levels would affect the operation of the HAD's hydropower turbines (MWRI, 2010; INTEA, 2010). As stated in Chapter 3.1.4.1, the upstream water levels must not exceed 182 m AMSL to protect the lands around the lake in the Sudan section from flooding events, as stipulated in an agreement between Egypt and Sudan made in 1959 (FAO, 1997b; IWLP, 2008). For each water level (WL), the HADR volume (V) was calculated based on Equation 5.2, and the surface area (A) based on Equation 5.1 in Chapter 5.2.2. For each scenario, the drop (1 m to 10 m) in water levels was subtracted from the HADR water level (WL) to calculate the water level in Part Two (WL_2). The water volume in Part Two (V_2) was calculated based on Equation 6.6, and the surface area (A_2) based on Equation 6.5. The water volume in Part One (V_1) is the difference between V and V_2. The corresponding water level (WL_1) was calculated using Equation 6.3, and the surface area of Part One (A_1) using Equation 6.1. The potential change in the surface area after the construction of the new dam is the difference between A and the sum of A_1 and A_2.

The results of these ten scenarios are given in Appendix H. The optimal drop of water levels downstream of the new dam corresponds to the biggest reduction in surface area for each altitude are listed in Table 5.26, and are shown in Figure 5.73. Over 257 km² can be eliminated, for example, from the reservoir's surface area at an altitude of 179 m AMSL if the water level downstream of the new dam decreases three meters to 176 m AMSL and the upstream water level increases to 181.52 m AMSL. The reduced surface area is equivalent to approximately 43% of the surface area of Khor Kalabsha (600 km²), and about 106% of the surface area of Khor El-Alaky (235 km²). The surface areas that would be eliminated due to the construction of the new dam are scattered around the lake shores downstream of the new dam. The optimal drop of water levels downstream of the new dam that should be considered at altitude 178 m AMSL is 5 m, which would raise the water level by about 4 m upstream of the new dam and would decrease the surface area by about 200 km².

Table 5.26: The optimum drop of the water level downstream the new dam at different water levels and the potential reduction in the surface area

Drop of water level in part 2 (m)	Change in Surface area (km²)										Max reduced surface area (km²)	Optimal reduced water level (m)
Water Level (m) AMSL	1m	2m	3m	4m	5m	6m	7m	8m	9m	10m		
156	6.36										6.36	1
157	2.78	7.49									7.49	2
158	0.64	-2.98	3.21								3.21	3
159	0.49	2.28	1.95	9.32							9.32	4
160	2.89	6.08	12.76	20.95	30.49						30.49	5
161	-7.67	6.80	17.24	23.66	34.53	48.74					48.74	6
162	9.88	14.66	28.28	43.82	52.29	63.75	83.56				83.56	7
163	-23.95	-5.75	5.20	19.68	38.36	58.06	73.68	88.40			88.40	8
164	4.68	-2.51	16.52	36.68	61.12	75.42	83.24	88.22	97.70		97.70	9
165	-11.27	6.48	12.94	33.99	45.48	54.17	63.01	64.38	69.88	84.97	84.97	10
166	3.87	20.94	35.01	26.26	37.02	39.37	48.06	62.99	77.72	94.64	94.64	10
167	9.63	19.16	16.65	21.51	7.24	22.83	39.60	62.23	81.70	93.93	93.93	10
168	-30.16	-30.94	-32.15	-34.41	-19.63	-14.90	7.60	22.66	42.69	65.97	65.97	10
169	-0.34	-31.21	-28.23	-12.03	1.83	18.35	19.15	44.54	59.15	72.28	72.28	10
170	20.30	35.36	27.70	41.27	55.66	68.93	87.80	81.86	86.63	83.17	87.80	7
171	40.24	94.95	116.61	106.70	123.47	132.62	126.60	122.22	100.59	102.15	132.62	6
172	34.70	84.16	136.48	158.41	139.16	123.60	115.11	101.38	91.46	61.38	158.41	4
173	15.31	50.97	98.26	128.05	121.92	84.58	66.21	48.08	39.87	49.17	128.05	4
174	21.87	34.40	37.17	58.78	79.14	61.82	25.31	25.79	44.48	60.64	79.14	5
175	0.81	-18.01	-30.50	-37.12	-26.65	8.28	28.14	25.35	69.00	124.46	124.46	10
176	-18.58	-47.46	-76.68	-90.73	-74.69	-13.86	59.78	127.03			127.03	8
177	41.77	1.24	-21.52	-13.45	17.41	95.94					95.94	6
178	**42.02**	**78.69**	**88.36**	**122.30**	**191.82**						**191.82**	**5**
179	66.25	150.15	257.22								257.22	3
180	49.54	178.57									178.57	2
181	83.44										83.44	1

Effect of building the new dam on the HADR's hydrology

The HADRHYDB was used to obtain data on water levels for the two decades from August 1990 to July 2010. Supposing that the new dam had been built two decades ago, the HADR's hydrological characteristics would have changed upstream and downstream of the proposed new dam. The HADR water levels were simulated based on the optimal scenarios for each altitude, as presented in Figure 5.73. The optimal drops downstream of the new dam were used to calculate the downstream water levels in Part Two. The upstream water levels in Part One were calculated as stated above, as shown in Figure 5.74. The surface areas of the two parts of HADR, and the reduction of the surface area after the construction of the dam were calculated. The potential reductions in surface area at different water levels between 1990 to 2010 are given in Figure 5.75.

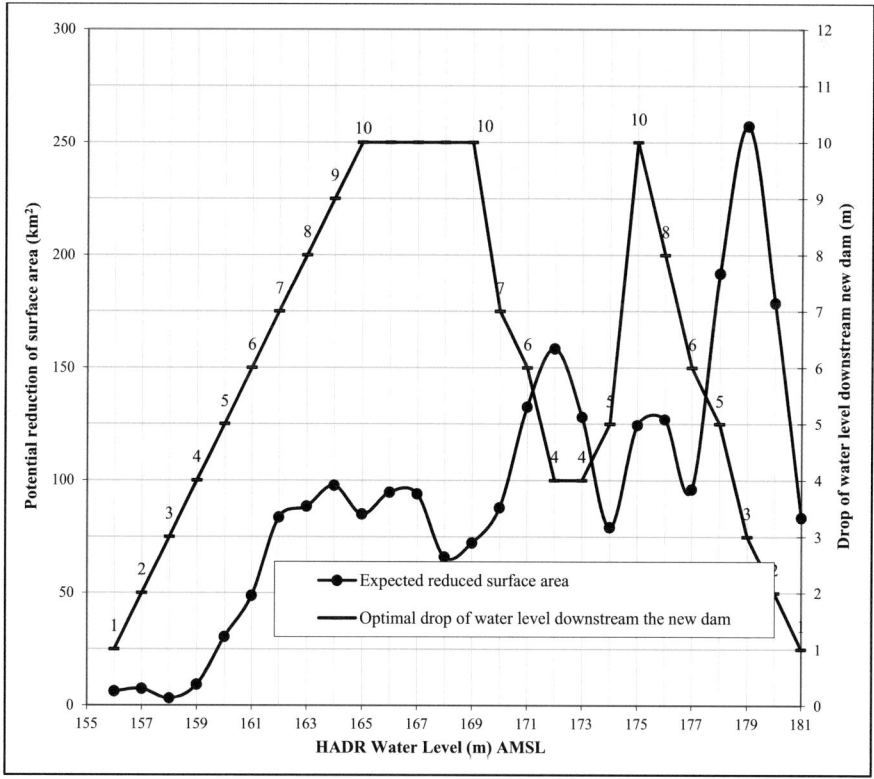

Figure 5.73: The optimal drop of the water level downstream the new dam at different water levels and the resulting reduction of the surface area

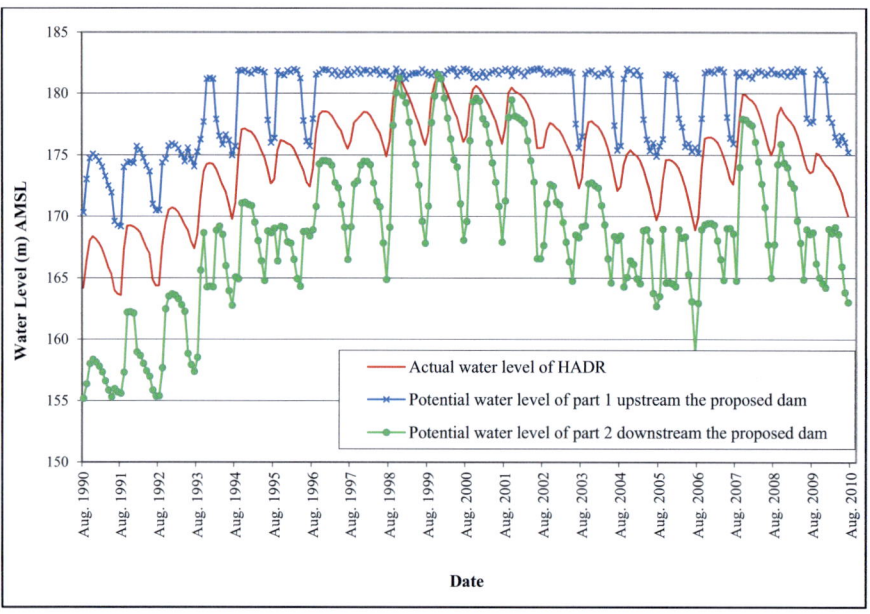

Figure 5.74: Variations of the HADR water level after a potential constration of the new dam from 1990 to 2010

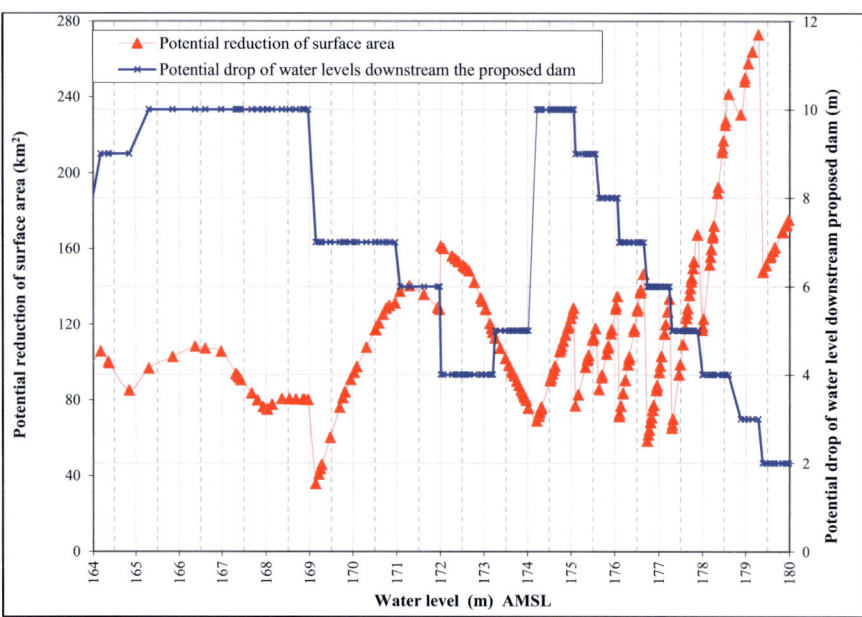

Figure 5.75: The potential drops of water levels downstream of the new dam and the reduction of surface area exptected for different water levels after a potiential construction of the new dam from 1990 to 2010.

Effect of building the new dam on the HADR evaporation losses regarding climate change

The monthly evaporation losses were calculated before and after the construction of the new dam based on the monthly evaporation rates stored in the HADRMTDB. The expected reduction of annual evaporation losses were calculated based on the assumption that the new dam had been constructed two decades ago. The results are listed in Table 5.27. On average, total evaporation before the construction of the dam would have been reduced by 2% to 3%. The expected maximum reduction of evaporation losses was 0.413 km^3 for the water year 2001/2002. The maximum reduction in surface area at altitude 178 m AMSL would be about 200 km^2, and the water savings are expected to be approximately 0.54 km^3 for current climatic conditions. By the end of the twenty-first century, the reduction of annual evaporation losses, as calculated based on the evaporation rates predicted by ECHAM5 and HadCM3 models, would lead to savings of approximately 0.6 km^3.

Table 5.27: The mean water level measured from 1990 to 2010, calculated annual evaporation losses before and after a potential construction of the new dam, and the potential reductions of evaporation losses

Year	Annual Evaporation Losses (km^3)		Potential reduction of evaporation losses	
	Computed before constructing the proposed Dam	Expected after constructing the proposed Dam	km^3	%
90/91	9.680	9.420	0.260	2.70
91/92	10.020	9.800	0.220	2.20
92/93	10.640	10.400	0.240	2.26
93/94	12.259	11.981	0.278	2.27
94/95	13.705	13.399	0.306	2.23
95/96	13.411	13.115	0.296	2.21
96/97	14.916	14.575	0.341	2.28
97/98	14.984	14.630	0.354	2.36
98/99	16.005	15.710	0.295	1.84
99/2000	16.220	15.914	0.306	1.89
2000/01	15.986	15.619	0.367	2.30
01/02	15.998	15.585	0.413	2.58
02/03	14.115	13.822	0.294	2.08
03/04	14.002	13.699	0.303	2.17
04/05	12.778	12.474	0.304	2.38
05/06	12.366	12.076	0.290	2.35
06/07	13.323	13.029	0.294	2.20
07/08	15.381	15.013	0.368	2.39
08/09	14.637	14.283	0.354	2.42
09/10	12.770	12.484	0.286	2.24
Mean	15.00	13.352	0.308	2.27

5.4.3 Lowering the lakebed level

As stated in Chapter 5.3.3, the HADR suffered from high sediment loads during the past 50 years, as shown in Figures 5.44 and 5.51. The total sediments were approximately 1.1 km^3 in Lake Nasser and approximately 5.5 km^3 in Lake Nubia, resulting in a total of 6.6 km^3. These sediments raised lakebed levels and decreased the storage capacity of the HADR by 6.6 km^3. As stated in Chapter 3.4.6, the sediment deposited during river flood season forms a delta, which has increased every year. Continuing accumulation of sediment reduces water supply volume in the reservoir, which shortens the life span time of the HAD in Egypt (HDA-MWRI, 2009, NWS, 2012, Moussa, 2013). Therefore, it is important to develop a plan to mitigate these serious consequences, to restore this crucial waterway, and make use of the sediments. Furthermore, lowering the lakebed leads to more storage capacity and lower surface water levels and smaller surface areas. Hence, it is expected that lowering the lakebed will decrease the surface area and evaporation losses and prevent blockages in Egypt's vital waterway, the Nile River.

Effect of removing sediments on the HADR evaporation losses regarding climate change.

Two alternatives were investigated to study the effects of removing sediment deposits from the HADR and lowering the lakebed. Alternative 1 investigated the effects of removing sediment deposits from Lake Nasser, which are estimated to be 1.1 km^3. Alternative 2 investigated the effect of removing sediment deposits from both Lake Nasser and Lake Nubia, which are estimated to be 6.6 km^3. Table 5.28 shows the results of Alternative 1 and the effect of removing 1.1 km^3 of sediment deposits from Lake Nasser on evaporation losses. The maximum reduced surface area is approximately 73 km^2 at altitude 178 m AMSL while the minimum reduced surface area is approximately 23 km^2. The average reduced surface area is approximately 45 km^2, which accounts for over 0.7% of the HADR surface area. Based on a mean annual evaporation rate of 2700 mm, the maximum water savings can reach up to 0.20 km^3 under current climatic conditions. By 2100, the maximum reduction with regard to annual evaporation losses could reach approximately 0.22 km^3, based on the results of ECHAM5 and HadCM3 models.

Table 5.29 shows the results of Alternative 2 and the effect of removing 6.6 km^3 of sediment deposits from the entire HADR on evaporation losses. The maximum reduced surface area is approximately 355 km^2 at altitude 178 m AMSL while the minimum reduced surface area is approximately 180 km^2. The average reduced surface area is approximately 270 km^2, which accounts for over 4% of the HADR surface area. Based on a mean annual evaporation rate of 2700 mm, the maximum water savings can reach up to approximately one km^3 under current climatic conditions. By 2100, the maximum reduction with regard to annual evaporation losses could reach approximately 1.1 km^3, based on the results of ECHAM5 and HadCM3 models.

Table 5.28: The effect of a potential removal of the sediment deposits from the Lake Nasser on the evaporation losses

HADR Current conditions			HADR conditions after a potential removal of the sediment deposits from the Lake Nasser			Effect of a potential removal of the sediment deposits		
Elevation (m) AMSL	Volume (km^3)	Surface area (km^2)	Elevation (m) AMSL	Volume (km^3)	Surface area (km^2)	change in Surface area (km^2)	Area saved %	Annual reduction of Evaporation losses (km^3)
155	43.28	2272.97	154.42	43.28				
156	45.61	2379.94	155.46	45.61	2321.65	58.29	2.45	0.16
157	48.05	2492.20	156.49	48.05	2435.41	56.79	2.28	0.15
158	50.59	2601.61	157.53	50.59	2550.45	51.16	1.97	0.14
159	53.25	2720.40	158.57	53.25	2669.27	51.13	1.88	0.14
160	56.04	2844.82	159.61	56.04	2795.92	48.90	1.72	0.13
161	58.94	2964.46	160.64	58.94	2921.74	42.72	1.44	0.12
162	61.97	3098.85	161.68	61.97	3055.55	43.30	1.40	0.12
163	65.12	3199.62	162.71	65.12	3170.10	29.52	0.92	0.08
164	68.39	3334.65	163.73	68.39	3298.18	36.46	1.09	0.10
165	71.78	3452.35	164.75	71.78	3422.71	29.63	0.86	0.08
166	75.30	3587.20	165.76	75.30	3554.97	32.23	0.90	0.09
167	78.95	3715.42	166.77	78.95	3685.78	29.63	0.80	0.08
168	82.72	3813.85	167.77	82.72	3790.85	23.00	0.60	0.06
169	86.61	3969.21	168.76	86.61	3931.41	37.80	0.95	0.10
170	90.68	4162.59	169.75	90.68	4114.39	48.20	1.16	0.13
171	94.94	4373.53	170.75	94.94	4321.49	52.04	1.19	0.14
172	99.41	4552.90	171.76	99.41	4509.94	42.96	0.94	0.12
173	104.04	4711.98	172.76	104.04	4674.59	37.39	0.79	0.10
174	108.84	4882.72	173.76	108.84	4842.58	40.14	0.82	0.11
175	113.80	5040.57	174.76	113.80	5002.61	37.97	0.75	0.10
176	118.93	5225.60	175.75	118.93	5179.23	46.37	0.89	0.13
177	124.30	5502.41	176.74	124.30	5431.75	70.65	1.28	0.19
178	129.95	5799.07	177.75	129.95	5725.98	73.09	1.26	0.20
179	135.90	6116.49	178.78	135.90	6045.61	70.88	1.16	0.19
180	142.14	6357.38	179.81	142.14	6310.80	46.58	0.73	0.13
181	148.60	6551.88	180.83	148.60	6519.18	32.71	0.50	0.09
182	155.1829	6623.478	181.84	155.1829	6611.94	11.54	0.17	0.03
Min						23.00	0.50	0.06
Max						73.09	2.45	0.20
Mean						44.98	1.18	0.12

Table 5.29: The effect of a potential removal of the sediment deposits from the HADR on the evaporation losses

HADR Current conditions			HADR conditions after a potential removal of the sediment deposits from the HADR			Effect of a potential removal of the sediment deposits		
Elevation (m) AMSL	Volume (km^3)	Surface area (km^2)	Elevation (m) AMSL	Volume (km^3)	Surface area (km^2)	change in Surface area (km^2)	Area saved %	Annual reduction of Evaporation losses (km^3)
155	43.28	2272.97	152.12	43.28				
156	45.61	2379.94	153.21	45.61				
157	48.05	2492.20	154.30	48.05				
158	50.59	2601.61	155.40	50.59	2315.31	286.29	11.00	0.77
159	53.25	2720.40	156.49	53.25	2434.82	285.58	10.50	0.77
160	56.04	2844.82	157.58	56.04	2555.93	288.89	10.15	0.78
161	58.94	2964.46	158.68	58.94	2681.82	282.64	9.53	0.76
162	61.97	3098.85	159.77	61.97	2815.85	283.00	9.13	0.76
163	65.12	3199.62	160.85	65.12	2946.96	252.65	7.90	0.68
164	68.39	3334.65	161.93	68.39	3089.99	244.65	7.34	0.66
165	71.78	3452.35	163.01	71.78	3200.96	251.38	7.28	0.68
166	75.30	3587.20	164.08	75.30	3344.12	243.08	6.78	0.66
167	78.95	3715.42	165.15	78.95	3472.07	243.35	6.55	0.66
168	82.72	3813.85	166.20	82.72	3613.03	200.82	5.27	0.54
169	86.61	3969.21	167.25	86.61	3739.98	229.23	5.78	0.62
170	90.68	4162.59	168.30	90.68	3860.78	301.81	7.25	0.81
171	94.94	4373.53	169.36	94.94	4039.58	333.96	7.64	0.90
172	99.41	4552.90	170.43	99.41	4253.52	299.38	6.58	0.81
173	104.04	4711.98	171.50	104.04	4462.47	249.51	5.30	0.67
174	108.84	4882.72	172.56	108.84	4641.39	241.33	4.94	0.65
175	113.80	5040.57	173.61	113.80	4816.36	224.21	4.45	0.61
176	118.93	5225.60	174.66	118.93	4987.20	238.41	4.56	0.64
177	124.30	5502.41	175.72	124.30	5173.54	328.87	5.98	0.89
178	129.95	5799.07	176.79	129.95	5444.31	354.75	6.12	0.96
179	135.90	6116.49	177.88	135.90	5762.61	353.88	5.79	0.96
180	142.14	6357.38	178.97	142.14	6107.55	249.83	3.93	0.67
181	148.60	6551.88	180.06	148.60	6372.31	179.58	2.74	0.48
182	155.18	6623.47	181.13	155.18	6577.78	45.69	0.69	0.12
Min						179.58	2.74	0.48
Max						354.75	11.00	0.96
Mean						268.63	6.77	0.73

5.5 Combining scenarios

Some of the scenarios presented in Chapter 5.4 can be combined to maximize the benefits from them and to increase water savings at the HADR. The various combinations are explained below.

5.5.1 Building a new dam at El-Madik section and filling Khor Kalabsha

This scenario assessed the consequences of building the new dam discussed above in addition to filling Khor Kalabsha up to 182 m in order to reduce surface area. Khor Kalabsha is located downstream of the proposed dam in Part Two. Khor Kalabsha was removed from the DEM of Part Two downstream of the new dam. The DEM of Part Two without Khor Kalabsha was reanalyzed, and new equations were developed for this part. Equation 6.9 calculates the new volume of Part Two (V_{2ek}), and Equation 6.10 calculates the new surface area of Part Two (A_{2ek}). For Part one, Equations 6.1, 6.2, 6.3, and 6.4 were used.

$$V_{2ek} = 1.9785 \times 10^{-16} \times (WL_2)^{7.762} \qquad \text{Equation 6.9}$$

$$A_{2ek} = 1.2868 \times 10^{-11} \times (WL_2)^{6.345} \qquad \text{Equation 6.10}$$

The ten scenarios created in this manner were again evaluated to find the optimal water levels upstream and downstream of the new dam under the same conditions as stated above. For both parts of the HADR, the surface areas were calculated, and then their sum (A_1+A_{2ek}) was subtracted from the actual surface area of HADR to calculate the potential reduction of surface area. The optimal drops of water levels downstream of the new dam were selected and are listed in Table 5.30 and shown in Figure 5.76. Over 700 km² could be eliminated from the HADR surface area at an altitude of 179 m AMSL if the water level in Part Two would be decreased to 175 m AMSL and the water level in Part One would be raised to 181.75 m AMSL. The optimal drop of water levels downstream of the new dam that should be considered at altitude 178 m AMSL is 5 m, which would raise the water level by approximately 3.45 m in Part One and decreases the surface area by approximately 530 km².

Similar steps were applied to water levels for the past two decades stored in HADRHYDB, based on the results presented in Figure 5.76. The expected reduction of surface area and the drops in water level in Part Two for each altitude are shown in Figure 5.77. The surface area could be reduced approximately 700 km² by lowering the water level by approximately 3 m downstream of the new dam. This would reduce evaporation losses by 2.0 km³ under current climate conditions, and would turn approximately 600 km² in Khor Kalabsha into a settlement area for the climate migrants arriving from the northern shores. The remainder of the reduced surface area (100 km²) are scattered along the lake shores and might be unsuitable for settlement purposes. In addition, another

600 km² around Khor Kalabsha would be added to the settlement area by leveling this area from altitude 200 m to 182 m AMSL. As shown in Figure 5.76, the maximum reduction of surface area at altitude 178 m AMSL is approximately 530 km². The water savings are expected to be around 1.5 km³ under current climatic conditions and approximately 1.7 km³ by the year 2100, based on the results of the ECHAM5 and HadCM3 models.

Table 5.30: The optimal drop of the water level downstream the suggested new dam at different water levels and the resulting reduction of surface area after a potential fill of Khor Kalabsha

Water Level (m) AMSL	Drop of water level in part 2 (m)	Change in Surface area (km²)										Max reduced surface area (km²)	Optimal reduced water level (m)
		1m	2m	3m	4m	5m	6m	7m	8m	9m	10m		
	156	36										36	1
	157	36	38									38	2
	158	36	32	39								39	3
	159	41	40	39	41							41	4
	160	50	49	51	59	66						66	5
	161	46	56	62	64	73	85					85	6
	162	71	71	80	91	95	104	119				119	7
	163	48	58	64	73	87	101	117	129			129	8
	164	82	73	83	97	116	128	134	137	146		146	9
	165	78	87	91	105	112	116	122	120	121	132	132	10
	166	107	115	120	110	113	112	113	123	130	144	144	10
	167	129	126	115	113	96	102	110	127	145	153	153	10
	168	106	94	80	69	74	74	91	101	113	130	130	10
	169	142	111	103	103	108	118	113	129	145	153	153	10
	170	185	186	175	178	178	180	195	187	194	187	195	7
	171	246	268	275	261	266	265	259	248	225	219	275	3
	172	282	299	317	327	307	288	267	246	236	206	327	4
	173	302	307	324	329	310	272	249	221	191	189	329	4
	174	345	333	313	304	294	271	224	207	199	202	345	1
	175	355	323	288	261	241	230	226	213	220	248	355	1
	176	369	326	288	240	224	237	261	293	319		369	1
	177	466	413	365	346	345	360	423				466	1
	178	525	517	495	494	526						526	5
	179	604	636	672	713							713	4
	180	630	696									696	2
	181	678										678	1

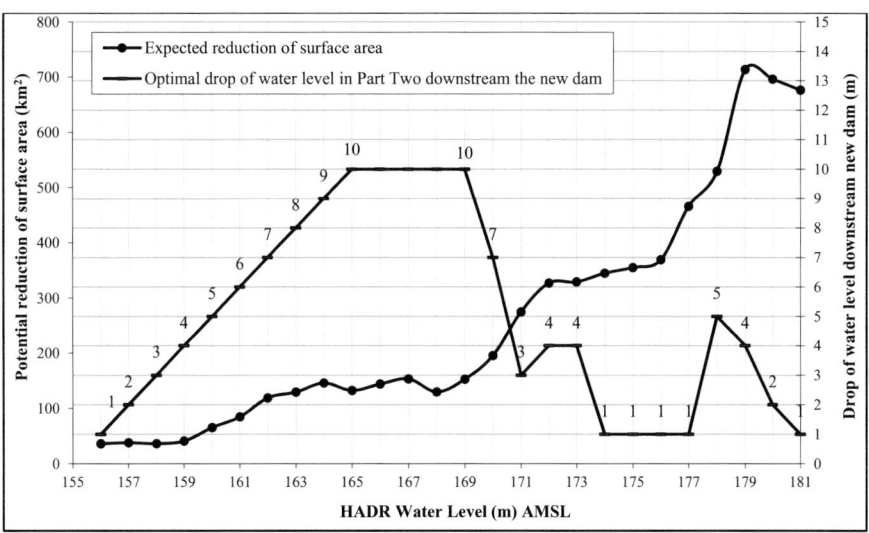

Figure 5.76: The optimal drop of water level in Part Two downstream of the proposed dam and the reduction of surface area after a potential fill of Khor Kalabsha

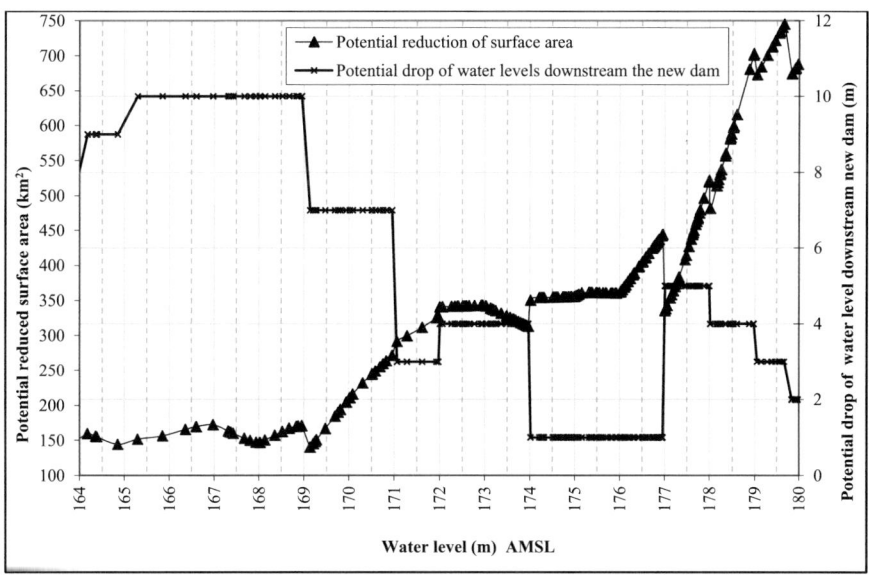

Figure 5.77: The optimal drop of water levels in Part Two downstream of the new dam and the reduction of surface area after a potential fill of Khor Kalabasha two decades ago

5.5.2 Building new dam at El-Madik section and removing the sediments from the HADR

This scenario assesses the consequences of building the new dam in addition to removing the sediments from the HADR in order to lower the lakebed level and to reduce the surface area. As indicated by Figure 5.44, the sediment loads are usually deposited upstream of the El-Madik reach. Thus, the volume of Part One upstream of the new dam was assumed to increase by the amount of sediment deposits and was then reanalyzed to create new equations for this part. Two alternatives were studied with respect to the amount and location of sediments. Alternative 1 evaluated the case for removing the sediments from Lake Nasser (1.1 km^3), and Alternative 2 the case of removing the sediments from the entire HADR (6.6 km^3). The volume of Part One at each water level is increased by 1.1 km^3 and 6.6 km^3 for Alternative 1 and Alternative 2, respectively. For Alternative 1, the water level of Part One was calculated based on the water volume after the removal of 1.1 km^3 sediment ($V_{1r1.1}$) using Equation 6.11. For Alternative 2, the water level of Part One was calculated based on the water volume after the removal of 6.6 km^3 of sediment ($V_{1r6.6}$) using Equation 6.12. The surface area (A_1) of Part One was calculated using Equation 6.1.

$WL_1 = 100.7625 \times (V_{1r1.1})^{0.1343}$ Equation 6.11

$WL_1 = 92.7725 \times (V_{1r6.6})^{0.1512}$ Equation 6.12

Ten scenarios based on both alternatives were again assessed to find the optimal water levels upstream and downstream of the dam under the conditions stated above. For both parts of the HADR, the surface areas were calculated, and the sum of both parts were subtracted from the actual surface area of the HADR to calculate the potential reduction of surface area. The optimal drop in the water level in Part Two downstream of the new dam were identified and are listed in Table 5.31 for Alternative 1 and in Table 5.32 for Alternative 2. Alternative 1 would reduce the surface area by approximately 285 km^2, accounts for 4% of the entire surface area of the HADR for a drop in water levels of 3 m downstream of the new dam. The maximum reduction of surface area at altitude 178 m AMSL would be approximately 160 km^2, which accounts for 3% of the entire surface area of the HADR for a drop in water levels of 4 m downstream of the new dam. The water savings are expected to be around 0.5 km^3 under current climatic conditions, and approximately 0.6 km^3 by 2100, based on the results of the ECHAM5 and HadCM3 models. Alternative 2 would reduce the surface area by approximately 492 km^2, which accounts for 8% of the entire surface area of the HADR for a drop in water levels of 3 m in Part Two. At 178 m AMSL, the maximum reduction of surface area for this water level would be approximately 443 km^2, which accounts for 7% of the entire surface area of the HADR for a drop in water levels of 4 m in Part Two. The water savings are expected to be around 1.2 km^3 under current climatic conditions and approximately 1.4 km^3 by 2100, based on the results of the ECHAM5 and HadCM3 models.

Table 5.31: The optimal drop of the water levels in Part Two downstream the new dam at different water levels, and the reduction of surface area after a potential removal of 1.1 km^3 of sediment from Lake Nasser

Water Level (m) AMSL	Drop of water level in part 2 (m)	Change in Surface area (km^2)									Max reduced surface area (km^2)	Optimal reduced water level (m)
		1m	2m	3m	4m	5m	6m	7m	8m	9m	10m		
	156	48.00										48.00	1
	157	44.88	50.14									50.14	2
	158	47.83	44.81	51.16								51.16	3
	159	51.44	53.38	53.71	60.61							60.61	4
	160	53.18	57.02	63.22	72.19	81.83						81.83	5
	161	50.25	64.21	75.43	81.45	92.35	106.96					106.96	6
	162	55.01	59.91	73.11	89.06	97.51	108.67	127.94				127.94	7
	163	35.34	53.59	64.94	79.13	97.73	116.87	133.39	147.96			147.96	8
	164	56.81	49.61	68.33	87.92	113.27	127.40	135.45	140.22	149.89		149.89	9
	165	46.71	64.37	70.25	92.15	103.87	112.32	121.32	122.39	127.20	142.01	142.01	10
	166	57.02	73.98	88.85	80.28	91.19	93.20	101.15	115.77	129.86	146.52	146.52	10
	167	50.31	59.60	57.22	62.16	47.10	62.34	78.41	100.75	120.16	132.03	132.03	10
	168	28.12	27.00	25.81	22.74	36.75	41.15	63.56	78.22	97.79	120.73	120.73	10
	169	65.06	33.71	35.85	51.28	65.03	81.10	81.38	106.37	120.51	133.14	133.14	10
	170	84.76	98.93	90.48	103.86	117.67	130.48	148.85	142.34	146.05	141.86	148.85	7
	171	91.00	144.89	166.29	155.76	172.02	180.53	173.40	168.18	145.70	146.46	180.53	6
	172	76.70	125.83	177.49	198.78	178.43	161.92	152.47	137.83	126.07	94.98	198.78	4
	173	56.93	91.89	138.44	166.94	159.73	121.35	101.04	81.76	74.41	82.89	166.94	4
	174	56.27	67.95	69.26	89.69	108.05	89.43	53.57	53.12	72.98	88.50	108.05	5
	175	35.32	14.88	1.09	-7.71	3.11	36.99	57.84	54.33	102.12	157.36	157.36	10
	176	23.28	-7.01	-38.60	-52.29	-37.40	24.50	102.06	169.07			169.07	8
	177	77.88	34.81	12.37	21.20	51.16	133.88					133.88	6
	178	74.55	111.49	121.91	159.60							159.60	4
	179	89.96	174.58	285.54								285.54	3
	180	69.77	202.79									202.79	2
	181	91.13										91.13	1

Table 5.32: The optimal drop of the water levels in Part Two downstream the new dam at different water levels, and the reduction of surface area after a potential removal of 6.6 km³ of sediments from the entire HADR

Water Level (m) AMSL	Drop of water level in part 2 (m)	Change in Surface area (km²)										Max reduced surface area (km²)	Optimal reduced water level (m)
		1m	2m	3m	4m	5m	6m	7m	8m	9m	10m		
	156	147.98										147.98	1
	157	163.49	171.18									171.18	2
	158	192.17	191.40	199.72								199.72	3
	159	200.58	204.51	206.96	214.20							214.20	4
	160	234.49	240.61	247.01	258.36	268.93						268.93	5
	161	208.50	222.60	236.30	242.04	253.54	269.67					269.67	6
	162	238.23	246.04	258.89	276.47	285.35	296.07	314.74				314.74	7
	163	230.77	248.69	261.66	275.52	294.31	312.02	331.18	346.19			346.19	8
	164	269.00	263.24	281.56	299.80	324.78	342.39	351.64	356.01	366.79		366.79	9
	165	256.24	273.66	278.13	303.21	314.79	324.16	334.21	334.59	337.12	351.24	351.24	10
	166	228.72	244.63	262.60	255.11	265.84	268.32	273.89	287.71	299.48	315.42	315.42	10
	167	242.99	254.46	253.04	258.85	242.31	255.66	269.28	290.80	310.59	321.41	321.41	10
	168	277.08	276.64	276.16	270.50	282.85	285.40	308.09	321.55	339.45	361.30	361.30	10
	169	340.14	309.52	308.98	321.75	334.91	350.50	348.93	372.68	386.06	397.09	397.09	10
	170	344.71	356.08	344.89	358.30	370.39	381.47	398.97	390.63	392.08	385.44	398.97	7
	171	307.28	358.33	379.50	366.81	382.21	388.66	379.12	371.11	345.81	343.84	388.66	6
	172	272.72	321.39	370.76	390.78	368.07	348.41	335.77	318.02	300.84	266.21	390.78	4
	173	248.79	281.30	326.19	351.68	340.92	299.03	272.98	250.33	243.78	249.36	351.68	4
	174	237.02	246.70	244.41	260.90	273.38	250.21	215.11	213.92	234.01	247.22	273.38	5
	175	248.61	224.07	205.94	190.56	201.09	231.23	254.28	256.63	308.10		308.10	9
	176	298.89	263.85	225.03	211.04	228.63	286.14	375.12				375.12	7
	177	356.04	305.12	282.27	292.42	335.07	414.20					414.20	6
	178	346.72	383.03	394.72	443.17							443.17	4
	179	285.81	370.20	492.45								492.45	3
	180	226.67	365.94									365.94	2
	181	149.24										149.24	1

5.5.3 Establishing new dam at El-Madik section simultaneously with filling Khor Kalabsha and removing the sediments from the HADR

This scenario assessed the impact of building the new dam, filling the khor, and removing sediments from the HADR to lower the lakebed level and to decrease surface area. The volume (V_{2ek}) and surface area (A_{2ek}) of Part Two were calculated using Equation 6.9 and Equation 6.10. The elevation (WL_1) of Part One was calculated using Equation 6.12. The surface area (A_1) of Part One was calculated using Equation 6.1. The ten scenarios were then assessed for this alternative to find the optimal water levels upstream and downstream of the new dam under the conditions stated above. For both parts of the HADR, the surface areas were calculated, and then the sum of both parts was subtracted from the actual surface area of the HADR and used to calculate the potential reduction of surface area.

Table 5.33: The optimal drop of the water level in part 2 downstream the suggested dam and the reduced surface area after a potential removal of 6.6 km³ sediment from the HADR and a potential fill of Khor Kalabsha

Water Level (m) AMSL	Drop of water level in part 2 (m)	Change in Surface area (km²)									Max reduced surface area (km²)	Optimal reduced water level (m)
		1m	2m	3m	4m	5m	6m	7m	8m	9m	10m		
	156	177										177	1
	157	196	201									201	2
	158	228	226	231								231	3
	159	241	242	243	248							248	4
	160	281	283	285	295	304						304	5
	161	262	272	281	283	292	306					306	6
	162	299	302	310	323	328	336	352				352	7
	163	302	312	321	329	343	355	371	387			387	8
	164	347	338	348	360	379	395	401	405	413		413	9
	165	346	355	356	374	380	387	392	391	390	399	399	10
	166	331	338	348	339	341	341	341	348	354	365	365	10
	167	365	361	352	350	332	335	342	356	374	382	382	10
	168	414	402	389	374	378	375	391	400	410	426	426	1
	169	484	452	440	437	441	450	444	458	473	479	484	1
	170	509	507	493	495	494	494	507	497	503	492	509	1
	171	513	531	538	522	524	522	514	499	473	463	538	3
	172	520	536	552	560	537	515	490	465	451	415	560	4
	173	536	538	552	554	531	488	462	427	396	392	554	4
	174	560	546	522	508	494	463	411	396	388	391	560	1
	175	603	567	526	491	465	456	448	439	450	487	603	1
	176	687	639	592	542	524	540	561	609	0		687	1
	177	781	719	671	650	649	680					781	1
	178	836	824	803	804	847						847	5
	179	824	858	908								908	3
	180	800	881									881	2
	181	736										736	1

The optimal drops of the water level downstream of the new dam, which resulted in the maximum reduction of surface area for each water level, were selected and are listed in Table 5.33. This maximum reduction in the surface area would be approximately 908 km^2, which accounts for 14% of the surface area of the HADR for a drop in water levels of 3 m downstream of the new dam. The maximum reduction of surface area at altitude 178 m AMSL would be approximately 847 km^2, which accounts for 13% of the surface area for a drop in water levels of 5 m downstream of the new dam. The water savings are expected to be around 2.3 km^3 under current climatic conditions, and approximately 2.6 km^3 by 2100, based on the results of the ECHAM5 and HadCM3 models.

5.5.4 Evaluation of the proposed measures to reduce the evaporation losses

Table 5.34 compares the different combinations based on the projected reduction in evaporation if these different scenarios were to become reality. Figure 5.78 shows the reduction of surface area for different scenarios, namely a) filling Khor Kalabsha, b) removing sediment deposits from Lake Nasser, c) removing sediment deposits from the HADR, d) building a new dam at the El-Madik section as well as the following combinations of scenarios: d) and a), d) and b), d) and c), as well as d), a), and c).

For water levels less than 172 m AMSL, removing the sediment deposits from the HADR could already effectively reduce the surface area, and there would be no need for further measures. Removing the sediments from the HADR would decrease its surface area by approximately 300 km^2, and it would reduce evaporation losses by 6.6% at an altitude of 172 m AMSL. For water levels above 172 m AMSL, filling Khor Kalabsha could efficiently reduce the surface area, and there would be no need to use other means. Filling Khor Kalabsha would decrease the surface area of the HADR by approximately 600 km^2, and it would reduce evaporation losses by 9.2% at an altitude of 181 m AMSL. In general, building a dam at the El-Madik section and simultaneously using other means would reduce surface area even further. For water levels less than 173 m, building the dam and removing the sediment deposits from the HADR could effectively reduce the surface area of the HADR by approximately 350 km^2, and would reduce evaporation losses by 7.5% at an altitude of 173 m AMSL. For water levels above 173 m AMSL, building the proposed dam and filling Khor Kalabsha would again efficiently reduce the surface area by approximately 700 km^2, and would reduce evaporation losses by 11.7% at an altitude of 179 m AMSL.

Results and Discussion

Table 5.34: Comparison between the different combinations based on the expected reduction in evaporation losses (%) after a potential execution of different measures

		% Expected reduction in the evaporation losses due to						
		Remove sediments from			Establish New dam at El-Madik combined with			
Elevation (m) AMSL	Filling Khor Kalabsha	Lake Nasser (1.1 km³)	Entire HADR (6.6 km³)	-----	Removing 1.1 km³ sediments	Removing 6.6 km³ sediments	Filling Khor Kalabsha	Removing 6.6 km³ sediments and filling Khor Kalabsha
156	1.2	2.4	0.0	0.3	2.0	6.2	1.5	7.4
157	1.3	2.3	0.0	0.3	2.0	6.9	1.5	8.1
158	1.5	2.0	11.0	0.1	2.0	7.7	1.4	8.9
159	1.7	1.9	10.5	0.3	2.2	7.9	1.5	9.1
160	1.9	1.7	10.2	1.1	2.9	9.5	2.3	10.7
161	2.1	1.5	9.5	1.7	3.6	9.1	2.9	10.3
162	2.4	1.4	9.1	2.7	4.1	10.2	3.8	11.4
163	2.6	0.9	7.9	2.8	4.6	10.8	4.0	12.1
164	2.8	1.1	7.3	2.9	4.5	11.0	4.4	12.4
165	3.1	0.9	7.3	2.5	4.1	10.2	3.9	11.6
166	3.5	0.9	6.8	2.6	4.1	8.8	4.0	10.2
167	3.8	0.8	6.5	2.5	3.6	8.6	4.1	10.3
168	4.0	0.6	5.3	1.7	3.2	9.5	3.4	11.2
169	4.3	1.0	5.8	1.8	3.4	10.0	3.9	12.2
170	5.0	1.2	7.3	2.1	3.6	9.6	4.7	12.2
171	5.7	1.2	7.6	3.0	4.1	8.9	6.3	12.3
172	6.3	0.9	6.6	3.5	4.4	8.6	7.2	12.3
173	6.8	0.8	5.3	2.7	3.5	7.5	7.0	11.8
174	7.1	0.8	4.9	1.6	2.2	5.6	7.1	11.5
175	7.4	0.8	4.4	2.5	3.1	6.1	7.0	12.0
176	7.7	0.9	4.6	2.4	3.2	7.2	7.1	13.1
177	8.3	1.3	6.0	1.7	2.4	7.5	8.5	14.2
178	8.7	1.3	6.1	3.3	2.8	7.6	9.1	14.6
179	9.0	1.2	5.8	4.2	4.7	8.0	11.7	14.8
180	9.2	0.7	3.9	2.8	3.2	5.8	10.9	13.9
181	9.2	0.5	2.7	1.3	1.4	2.3	10.3	11.2
	29	58		6	48	148	36	177
Min.	1.2	0.6	0.0	0.1	2.0	5.6	1.4	7.4
Max.	9.2	2.4	11.0	4.2	4.7	11.0	11.7	14.8
Mean	4.9	1.2	6.2	2.1	3.3	8.1	5.4	11.5

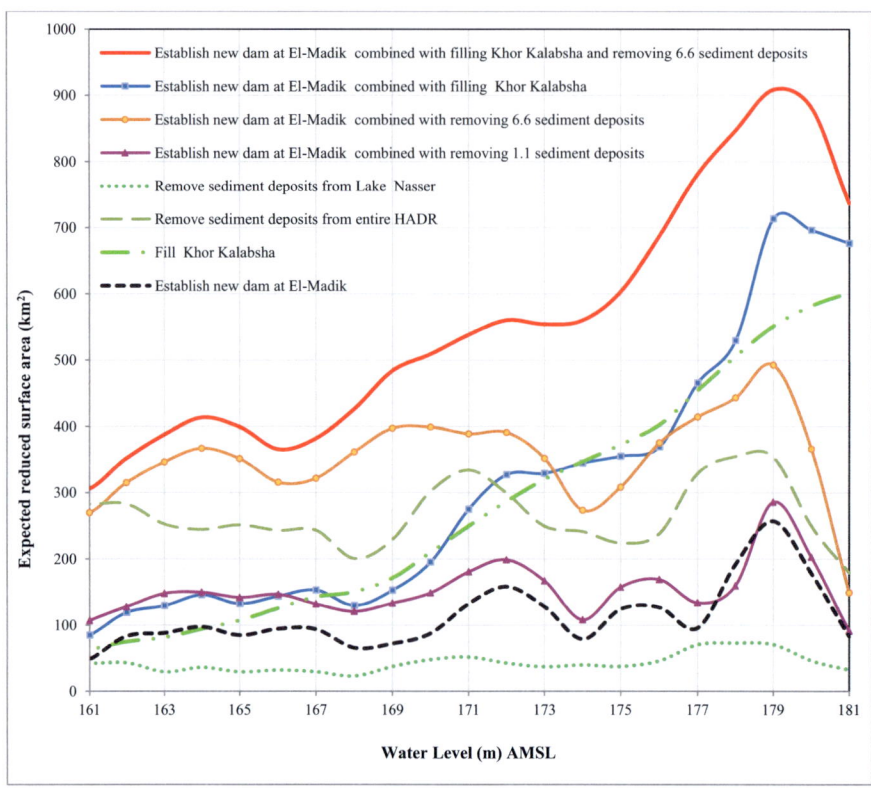

Figure 5.78: The expected reduced surface area after a potential execution of the different investigated measures

Water levels in the HADR have fluctuated between 171 and 180 m AMSL over the course of the past two decades. Due to this water fluctuation, filling Khor Kalabsha could efficiently reduce the surface area, and there would be no need to use other measures. Building the dam at the El–Madik section would, however, actually reduce surface area by a further reduction of 150 km^2 on average. Removing the sediment deposits from the HADR would strongly increase water savings if the new dam would be built and Khor Kalabsha to be filled. As a result, the surface area would be reduced by a further reduction of 300 km^2 on average. At altitude 178 m AMSL, the best alternative would be to build a dam at the El-Madik section, to fill Khor Kalabsha, and to remove sediment deposits from the HADR. All of these measures would reduce the surface area by approximately 850 km^2 and thereby reduce evaporation losses by approximately 2.3 km^3 (14.6%) annually, based on a mean evaporation rate of 7.5 mm/day under current climate conditions and approximately 2.6 km^3 by end of the twenty-first century, based on the results of the ECHAM5 and HadCM3 models.

5.6 Summary

The HADRMTDB, used for meteorological data, shows that with a mean air temperature of about 35°C, the period from June to September is the hottest one. The mean air temperature increased by about 1°C per decade. The period from June to September had the lowest mean relative humidity of about 34%. The variations in wind speed can be ignored since it varied only from 3.1 m s^{-1} to 3.8 m s^{-1}. The period from June to September was characterized by the maximum water temperature of about 28°C. The period from June till September had the highest mean evaporation rate of about 10 mm/day due to the high air temperatures and the low relative humidity in this period. The mean evaporation rate was increased by approximately 1 mm/day per decade equivalent to an increase of approximately 12%.

The HADRHYDB, developed for the hydrological data, shows that water levels usually started to rise from September to January. The most frequent surface water level was 178 m AMSL. The difference in the monthly mean surface water level between Aswan and Wadi Halfa gauge stations was about 11 cm due to the back water curve upstream of the HAD. Thus, the HADR surface water level can be considered the same along the entire reservoir. The HADRSPDB, used for seepage data, shows that the mean annual seepage losses were about 55 hm^3, which presents 0.1% of Egypt's share of the HADR and can be neglected. The HADRBSDB, used for bed soil sediment data, shows that the HADR usually had high silty and clay sediments in Lake Nasser, whereas sandy sediments increased in Lake Nubia and reached about 100% at end of the HADR.

The resulting HADRDEM shows that water capacity in the HADR had decreased by about seven km^3 due to sediment events during the past decades. The evaluation of the mathematical models developed to simulate the hydrological characteristics of the HADR shows that the resulting mathematical models are accurate enough to compute different parameters.

3000 km^2 is approximately the total area of the khors of the HADR, an area that represents approximately 46% of the surface area of the HADR. 44 km^3 of water are stored in khors, which represents approximately 28% of the total storage capacity of the HADR. 14 m is the average depth of the khors. 1350 m is the width of the narrowest cross section of the HADR, located at the El-Madik section, 130 km upstream of the HAD.

6.6 km^3 were the total sediment deposits in the HADR during the past five decades, with 5.5 km^3 deposited in Lake Nubia and 1.1 km^3 in Lake Nasser. 2.8 km^3 of sediment were deposited in the reach between the Amka to Dabrosa sections, which are located approximately 26 km apart. This amount represents approximately 60% of the total sediment load in Lake Nubia, which, in turn, represents 50% of the total sediment deposits in the HADR during the period from 1977 to 2010. On the other hand, in the reach between Korosko and El-Madik in Lake Nasser, there were only minor changes in bed altitudes during the past two decades. Erosion in the bed of HADR is very limited. 126 hm^3 was the total volume caused by erosion during the past five decades.

Three main scenarios were investigated to reduce evaporation losses. Scenario 1 investigated the possibility of reducing evaporation losses by eliminating some of the khors. Two options were investigated for Khor Kalabsha, including building a dam or using material from the surrounding area to fill it. This scenario suggests that eliminating Khor Kalabsha would reduce the surface area by over 600 km^2 and reduce the evaporation losses by approximately 2 km^3 by 2100. It also shows that in the case of filling Khor Kalabsha with material from the surrounding area up to an altitude of 200 m AMSL, the khor and its surroundings would be leveled at an altitude 182 m AMSL to create a total area of 1200 km^2. Two options were investigated for Khor El-Alaky, including building a dam or covering it. The scenario indicates that eliminating Khor El-Alaky would reduce the surface area of the HADR by approximately 170 km^2 and reduce the evaporation losses by approximately 0.5 km^3 by 2100. Two options were investigated for Khor Genina, including building a dam or covering it. The scenario suggests that eliminating Khor Genina would reduce the surface area by approximately 100 km^2 and reduce evaporation losses by approximately 0.3 km^3 by 2100. Two options were also investigated for Khor Sara, including building a dam or covering it. The scenario suggests that eliminating Khor Sara would reduce the surface area of the HADR by approximately 40 km^2 and reduce evaporation losses by approximately 0.1 km^3 by 2100. Other small khors covering 1000 km^2 of the surface area of the HADR were also assessed. This scenario suggests that covering these khors would reduce the surface area by approximately 1000 km^2 and reduce the evaporation losses by approximately 3.0 km^3 by end of the twenty-first century.

Scenario 2 investigated the possibility of controlling water levels by building a dam at the El–Madik section. The goal was to assess the impact of lowering water levels downstream of the proposed dam and rising water levels upstream. Ten scenarios were assessed to study the effect of lowering the water level from one to ten meters downstream of the proposed dam. These scenarios show that reducing the water level by 5 m at 178 m AMSL, the mode of the water levels for the past two decades, downstream of the proposed dam would reduce the surface area by approximately 200 km^2 and reduce evaporation losses by approximately 0.8 km^3 by end of the twenty-first century.

Scenario 3 investigated the possibility of lowering the lakebed level by removing sediment deposits from the bed. Two possibilities were assessed: In the first, sediment deposits were removed only from Lake Nasser, and in the second, deposits were removed from the entire HADR. These scenarios indicate that removing sediments from Lake Nasser would reduce the surface area by approximately 73 km^2 and reduce the evaporation losses by approximately 0.23 km^3 in 2100, whereas removing the sediments from the entire HADR would reduce the surface area by approximately 355 km² and reduce the evaporation losses by approximately 1.1 km³ in 2100.

Finally, combinations of several distinct scenarios were assessed to identify the measures that could be applied to maximize reduction of surface area. The optimal combination was building a dam at the El-Madik section, filling Khor Kalabsha, and removing sediments to reduce the surface area by approximately 900 km² and reduce evaporation losses by approximately 2.7 km³ by end of the twenty-first century.

Chapter 6

Prospective projects

In this chapter, prospective projects, which contribute to overcoming the problems due to the effect of climate change on land and water resources in Egypt, are suggested and discussed. These projects include the identification of new measures for potential evaporation loss reduction by: establishing new dam at El-Madik section, filling Khor Kalabsha, covering El-Alaky Khor and other small khors, and lowering lakebed. The projects include also the identification of new settlement areas for potential migrants from the Mediterranean coast by defining the potential areas to be colonized and the forms of land use, and presenting the necessary infrastructure to ensure sustainable development.

6.1 Identification of new measures to reduce evaporation losses

As mentioned in Chapter 5, there are three alternatives for reducing the water surface area and decreasing evaporation losses. These alternatives are controlling the water levels of the HADR, eliminating some khors, and lowering the lakebed. Each alternative has advantages and disadvantages, which must be considered before they are applied. Each alternative can be applied by itself or in combinations with other alternatives. Different alternatives and their respective advantages and disadvantages are described below.

6.1.1 Building a new dam at the El-Madik section

As stated in Chapter 5.4.2, the lake surface area could be reduced by approximately 250 km^2 by controlling the water levels in the lake after the construction of the new dam at the El-Madik section upstream of HAD. This area of 250 km^2 is scattered along the shorelines of the HADR and is not permanent, due to fluctuations of water levels. By reducing the HADR surface area, the new dam would not only reduce evaporation losses by approximately 0.8 km^3, but it would also allow for the generation of hydroelectric power due to the difference in elevation between the two bodies of water upstream and downstream of the new dam. The advantages and benefits of this new dam can be summarized as follows:

- reduction of evaporation losses from the HADR by controlling and managing water levels upstream and downstream of the new dam,
- protection of the HAD in case of unexpected structural problems, which could be caused by natural, industrial, or human disasters,

- reduction of the cost of pumping water from the HADR to the South Valley project (Toshka project) due to higher water levels upstream of the new dam,
- increase in energy production (hydroelectric power),
- establishment of new development projects, to provide social and economic benefits to Egypt. It is very likely, for instance, that this project will improve food security,

Building a dam usually has many disadvantages. In general any new dam destroys private property or cultural heritage in the reservoir zone, and it leads to the loss of agricultural or forest lands or other important areas such as wetlands. Even if the measures recommended above, including the construction of a new dam, are taken, they are unlikely to have a negative impact. For example, since the maximum water level upstream of the new dam will not exceed 182 m, the lands around the HADR in Sudan would not be affected and still be protected from flooding events. Furthermore, the negative impact of building a dam, for instance damages to flora and fauna in the surrounding area, do not have to be considered since the reservoir has already been there for the past 50 years. Thus, there will only be a minor impact on the local ecosystems if the new dam is built. Moreover, there will not be any major changes in water characteristics such as temperature, salinity, or oxygen content. Sediment load patterns may, however, change due to the change in current velocity after the construction of the dam. The proposed dam is unlikely to have a greater negative impact than the HAD when it was first constructed, but the new dam will improve water and energy resources, which, in turn, provide numerous social and economic benefits to Egypt.

6.1.2 Filling Khor Kalabsha

As stated in Chapter 5.4.1.1, the HADR surface area can be reduced by approximately 600 km^2 if Khor Kalabsha is eliminated. Building dams to close the khor is usually more expensive than filling it. Furthermore, filling the khor provides a level area that can be used for other purposes such as agriculture and new communities, providing additional economical and social befits. The khor can be filled by material from the surrounding area, which will be leveled from an altitude of 200 m AMSL to an altitude of 181.6 m AMSL. The area created in this manner will be approximately 1200 km^2. Filling Khor Kalabsha will not only reduce evaporation losses by approximately 2 km^3, but the new lands will also be used for agricultural purposes. The urban planners and architects of the new communities will have to take into account that two kilometers will have to remain as a buffer zone between farms or settlements and the shoreline of the HADR, as required by Decree 203/2002 (see Chapter 3.4.5). Moreover, organic fertilizers and organic pesticides must be applied, in a similar way as they have been used at SEKEM farms in Egypt (SEKEM, 2010), to protect the groundwater reservoir around the lake and to prevent other forms of pollution. This approach is likely to result in economic stability and sustainable lives for riparian people around the khor since they usually cultivate the lands around it, as discussed in Chapter 1.2 (Near East Foundation, 2010, OWARA, 2008). These cultivated lands have not been permanent and have varied in terms of size due to water level fluctuations. Although the filling of the khor will have a negative short-term impact on non-human habitants such as birds, fish, or other animals, the ecosystem will remain essentially unchanged, and the other khors will still available for these inhabitants.

6.1.3 Covering El-Alaky Khor and other small khors

Khor El-Alaky is sometimes conflated with Wadi El-Alaky, but they are not the same. Since 1993, Wadi El-Alaky has been a UNESCO Biosphere, as stated on the UNESCO website. The Wadi El-Alaky Biosphere is located between latitudes 20°20' and 22°10'N, and between longitudes 32°40' and 33°40'E, as shown in Figure 6.1 (UNESCO, 1993). Khor El-Alaky is located near to Wadi El-Alaky, but it is not part of the area protected by the UNESCO. For this reason, this study also considers whether it is feasible to use physical covering for Khor El-Alaky in order to reduce evaporation losses and to protect the local ecosystem. The khor can be covered up to Section 4, which covers an area of approximately 170 km^2 (see Figure 5.58). As stated in Chapter 2.3.2.2, the floating covers, such as an E-VapCap, effectively reduce evaporation by 95%, whereas floating objects, such as Aquacap, do so only by 70%. Compared to floating objects, floating covers require, however, complex and costly installation procedures. Moreover, floating covers can have a severe impact on the environment such as loss of habitat and disruption of existing ecosystems. They may cause aerobic conditions in water bodies, which, in turn, have, by increasing iron, manganese, and ammonia levels, a negative impact on water quality including also a positive impact because they reduce algal growth (Craig et al., 2005, Jennison, 2003). In contrast to floating covers, floating objects have a less severe impact on the environment and water quality because part of the surface is left uncovered. They are considered the most economic and environmentally friendly alternative to reduce evaporation losses (AQUA, 2006). For this reason, Khor El-Alaky should not be covered with floating covers but with floating objects, which may help to reduce evaporation losses by approximately 0.35 km^3 by 2100.

With a total surface area of approximately 1200 km^2, Khor Sara, Khor Genina, and other small khors, are likewise important areas where floating objects could be used to reduce evaporation losses. This measure will reduce evaporation losses by approximately 2.5 km^3 by the end of this century. Thus, total water savings will be 2.85 km^3 by 2100 if Khor El-Alaky and the other smaller khors have been covered. To maintain the balance of the ecosystem, to protect the environment, and to achieve higher economic benefits, the floating objects could be manufactured using natural materials such as palm fronds. This approach is likely to reduce evaporation losses by up to 60%, as stated in Chapter 2.3.1.3 (Alam and Al Shaikh, 2013, Al–Hassoun et al. 2011). Covering Khor El-Alaky and the other smaller khors, which cover a total surface area of approximately 1370 km^2, with palm fronds could reduce evaporation losses by approximately 2.45 km^3 by end of this century. Additional research must be conducted to evaluate the effectiveness of other natural materials available in the region, for example papyrus, water hyacinth, or any other plants that could be used to manufacture floating objects.

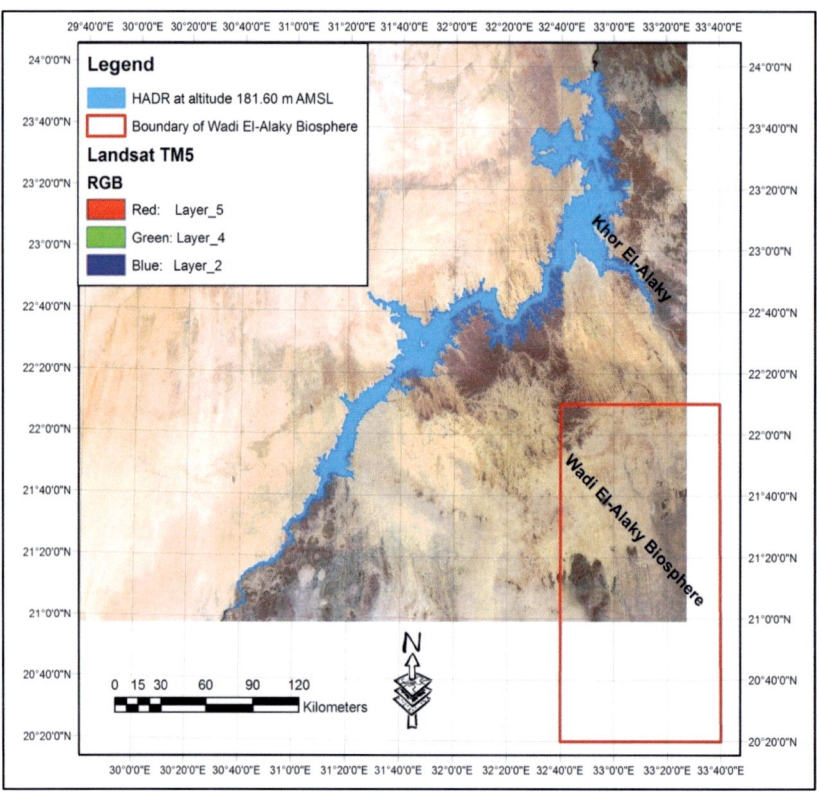

Figure 6.1: The boundaries of Wadi El-Alaky Biosphere, HADR, and Khor El-Alaky, overlaid in a mosaic of false color Landsat TM 5 images taken in 1986, composites of bands 5, 4, and 2

6.1.4 Lowering lakebed altitudes

As stated in Chapter 5.4.3, removing 1.1 km^3 of sediment deposits from Lake Nasser will reduce the surface area of the HADR by approximately 45 km^2 and consequently reduce evaporation losses by 0.14 km^3 by 2100. Removing 6.6 km^3 of sediment deposits from the HADR will reduce the surface area by approximately 260 km^2 and evaporation losses by 0.78 km^3 by 2100. Removing these deposits could be done by dredging (IADC, 2005, Sciortino, 2010,

Helmke, 2010, Halcrow, 2001, Allen and Dunbar, 2005, Allen et al., 2005). Removing the sediment and lowering the lakebed would not only reduce evaporation losses, but also prevent the blockage of the Nile River at the growing delta in Lake Nubia. The sediments could be used to reclaim, create, or develop arable land in Egypt. Furthermore, the sediments could be used to protect the Nile delta from flooding by covering the areas that will be affected by sea level rise crisis.

6.1.5 Measures to reduce evaporation losses

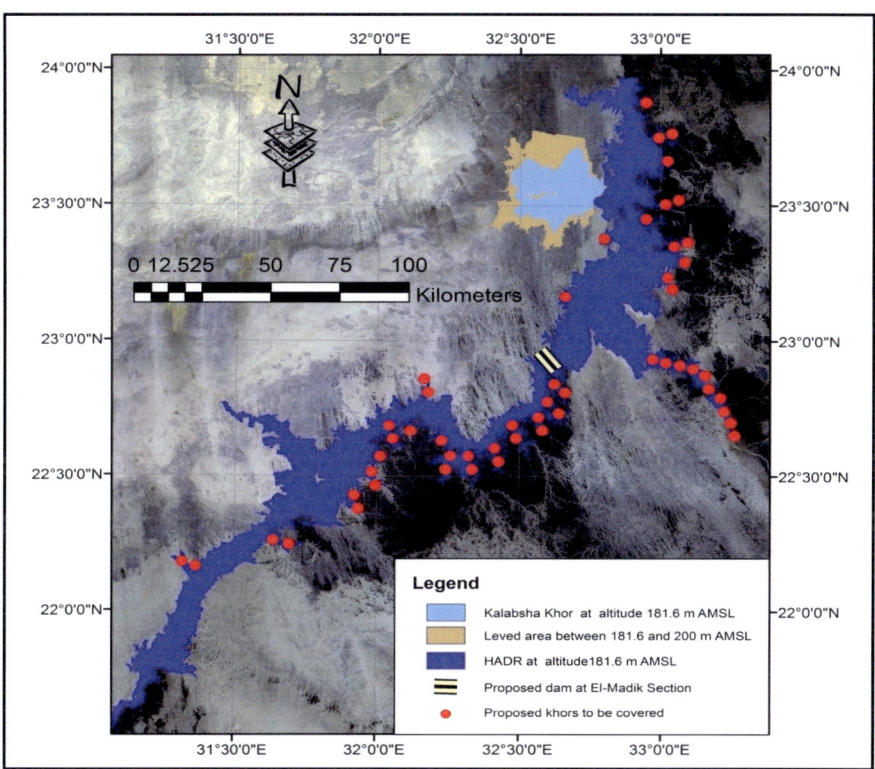

Figure 6.2: Proposed measures for reducing evaporation losses from the HADR

The optimal combination of the measures discussed above, which is shown in Figure 6.2, includes the removal of sediments from the HADR, the construction of a new dam at the El−Madik section, the filling of Khor Kalabsha, as well as the covering of Khor El-Alaky and the smaller khors. As stated in Chapter 5.5.3, building the proposed dam at the El-Madik section, filling Khor Kalabsha, and removing the sediments from the HADR could reduce the surface area by approximately 900 km^2, as shown in Figure 5.78. Evaporation losses will be reduced by approximately 2.7 km^3 by the year 2100. Moreover, as mentioned in

Chapter 6.1.3, covering Khor El-Alaky and the small khors could reduce evaporation losses by approximately 2.85 km^3. Thus, applying all of these measures could reduce the annual evaporation losses by approximately 5.0 km^3 under current climatic conditions and by approximately 5.55 km^3 by end of this century. This amount is equivalent to 50% of the evaporation losses, which is stated in 1959 agreement (Appendix A), as well as 10% of Egypt annual share from the Nile.

6.2 Identification of new settlement areas for potential migrants from the Mediterranean coast

As stated in Chapter 1.1.2, Egypt's coastal zones, with their relatively low elevations, are vulnerable to climate change hazards because of the expected sea level rise. If global sea levels rise by 50 cm, the coastal zones' water resources, agricultural resources, tourism, and human settlements will be severely affected. These coastal lands are primarily arable lands, which are used to produce approximately 50% of the national agricultural output and which are occupied by over 30% of the Egyptian population (Dasgupta et al., 2007). The Egyptian Coastal Research Institute suggests that a loss of approximately 760 km^2 of land, which represents 3% of the total fertile land in the Nile Delta, will also lead to the relocation of 6 to 7 million people currently living in this area (CoRI, 2009). These people have to adapt to the changing conditions, and the government has to offer arable lands and settlements to encourage people to migrate from the lands that will be lost to new lands around the valley. In 2008, over 800 farmers from the north resettled in a new settlement area near the village of Kalabsha, which has a simple infrastructure (OWARA, 2008). This study recommends that climate migrants currently living in areas need to move to settlements in the south. The different possibilities for new settlement areas are discussed below.

6.2.1 Potential areas to be settled and forms of land use

As discussed in Chapter 5.4.1.1 and Chapter 6.1.2, Khor Kalabsha could be filled up to an altitude of 181.6 m AMSL by leveling the areas surrounding the khor from an altitude of 200 m AMSL. The total area, including the khor, would be approximately 1200 km^2 (280000 Faddans), which could be used for new settlements for the climate migrant from the northern coastal zone, as shown in Figure 6.3. As stipulated by to Decree 203/2002, a buffer zone of two km must be left between the shores of the HADR and the cultivated lands, farms, and settlements. Organic fertilizers and pesticides must be used to protect the groundwater reservoir around the HADR.

Currently, some areas around the khor are cultivated, as shown in Figure 6.4. When the water level was 175 m AMSL, the cultivated areas were located directly on shore. These areas, which are well equipped with irrigation infrastructure, will be flooded when the water level rises to 180 m AMSL. These areas are difficult to map, and the riparian people usually move from one location to another depending on water levels. Riparian people living in these areas tend to ignore regulations requiring a buffer zone of two km from the shore to protect the environment. Moreover, no one controls the types of fertilisers and pesticides used by these farmers, and runoff from the fields may cause damage to the local ecosystem. Therefore, filling this khor and using it for agricultural purposes will lead to stability, continuity, and sustainability around the khor.

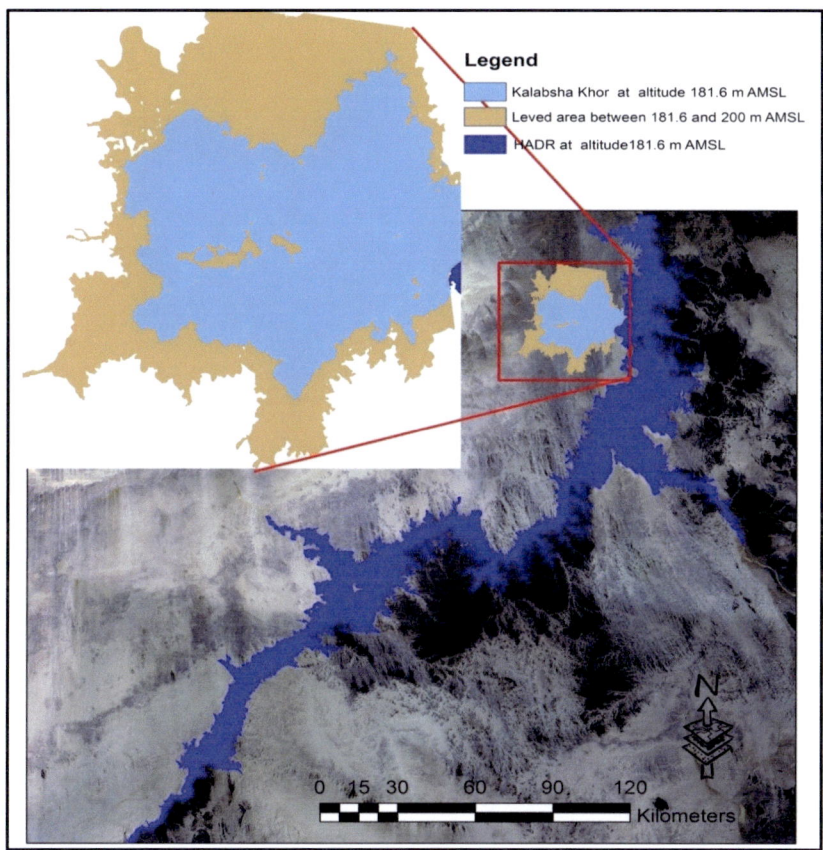

Figure 6.3: Proposed new settlements and cultivated lands on Khor Kalabsha.

Building only the dam will reduce the surface area and evaporation losses, but the reduced areas will not be used to settle the people settling in the region, as these areas are scattered along the HADR. Building the dam will help the government to control water levels by lowering the water levels downstream of the dam. Khor Kalabsha is located downstream of the new dam, and therefore, this dam could protect the new settlements at Khor Kalabsha from flooding hazards and provide several economic and social benefits, such as stability and sustainability to the riparian. These conditions need to be advertised, and people need to be encouraged to migrate from the north to the south with the help of media programs that provide detailed information on climate change hazards, adaptation methods, and measures that mitigate the impact of global warming.

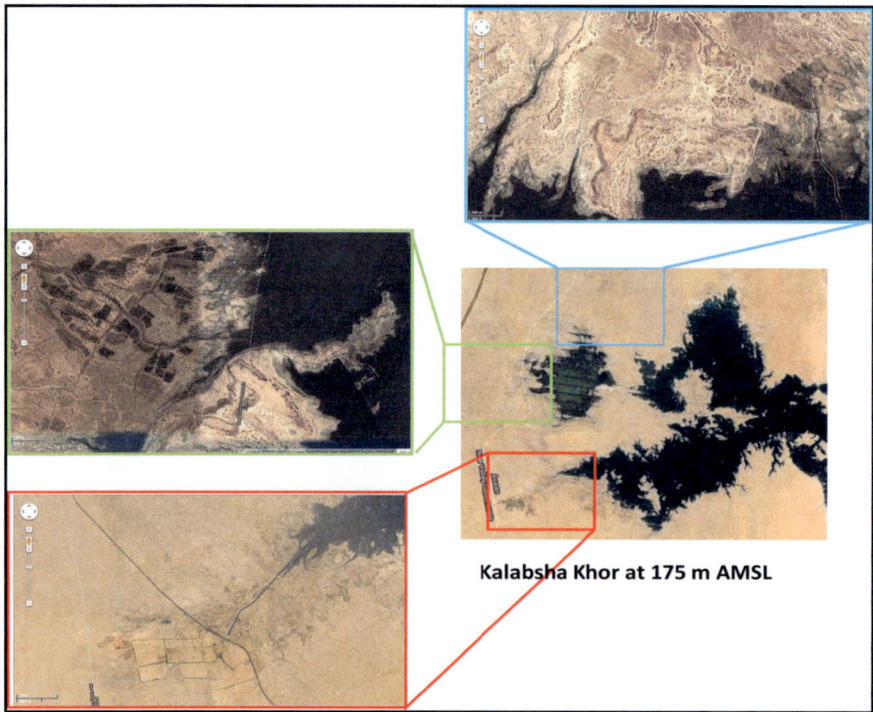

Figure 6.4: Satellite image provided by Google Earth of Khor Kalabsha and cultivated areas around it in 2012.

6.2.2 The required infrastructure for sustainable development

In order to establish communities for the populations from the northern coastal region, it is necessary to identify suitable areas and to develop these new areas in a sustainable manner. The needs of present generation have to be considered, as well as the ability of future generations to lead healthy and productive lives must not be compromised, to encourage more people to move to these settlements. Sustainable development can be achieved if economic, environmental, and social aspects are all taken into consideration, and water, energy, and land resources are used frugally. Khor Kalabsha and the surrounding areas could be an ideal location for the people from the Nile Delta because they would, as this land could be primarily used for cultivation, be moving from one agricultural area to another. The water could be pumped into the main canal by a pump station at the HADR, to irrigate this area, as shown in Figure 6.5. The irrigation system must use modern irrigation techniques to maximize the efficiency of irrigation and minimize water losses in a tropical climate. A comprehensive study is needed to design the optimal irrigation system there. Due to the shortage of irrigation water in Egypt, drainage systems have to be designed to drain the water back into the HADR to be reused. Consequently, only organic fertilizers and biological pesticides should be allowed in this area to protect the water quality in the HADR and the surrounding areas.

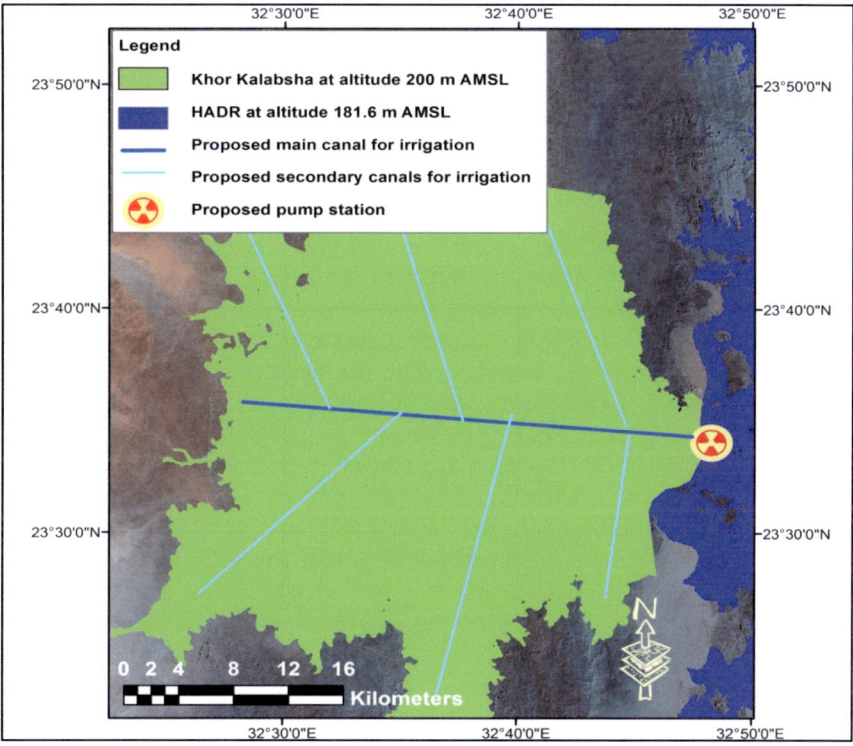

Figure 6.5: Proposed locations for the main and secondary irrigation canals and the main pump station at Khor Kalabsha.

Providing enough energy resources is also essential to develop the area in a sustainable manner, in particular electricity. A secure and reliable supply of electricity is important because it affects three levels of sustainable development, namely environmental, economical, and social conditions. Sufficient energy can lead to economic growth but conventional sources of energy such as coal or oil can be major sources of environmental pollution. Thus, clean energy is required to protect the environment and promote sustainable economic development. The availability of electricity also encourages social progress with regard to communication, healthcare, and education. In this area, electricity could be provided through the new dam's hydropower station. Moreover, clean energy such as solar or wind power can be easily generated in this area. Hence, it would be economically feasible to build new factories, which could provide job opportunities for the growing population.

Access to safe drinking-water is essential to achieve social, economic, and political stability as well as sustainable development at the local level. New technologies and new clean energy resources such as solar energy could be used to power portable water stations. Sanitation systems will need to use advanced technologies for water treatment so that this water can be reused for irrigation and some industrial purposes. Of course, schools and hospitals would have to be built to provide essential services to the people. Adequate road coverage is also required to connect this area with existing network of roads to facilitate trading and communication.

6.3 Summary

This chapter summarized the projects to mitigate the impact of climate change in the region around the HADR, and described the advantages and disadvantages of each plan. These plans are meant to achieve two major goals: the identification of new measures to reduce evaporation losses and the identification of new settlement areas for people relocated from the Mediterranean coast. The new measures to reduce evaporation losses included building a new dam at the El–Madik section, filling Khor Kalabsha, removing 6.6 km^3 of sediment from the HADR, and covering Khor El-Alaky and other small khors with floating objects to reduce annual evaporation losses by approximately 5.0 km^3 under the current climate conditions and approximately 5.55 km^3 by the end of this century. Khor Kalabsha and the surrounding areas represent an ideal location for new settlements for the potential climate migrants affected by the sea level rise in particular. Finally, the infrastructure required for a sustainable development of the region was also proposed.

Chapter 7

Conclusions and Recommendations

In this chapter, an executive summary of the content of this thesis and its findings is provided. The chapter also includes recommendations and a discussion of future projects based on the present study.

7.1 Summary of findings

Egypt is vulnerable to climate change, especially with regard to its water and land resources. Climate change will increase the evaporation losses from the High Aswan Dam Reservoir (HADR) and lead to loss of lands on the northern coast due to sea level rise. In order to address this problem, this project assesses several adaptation tools that could be used to mitigate the impact of climate change. For example, Egypt could protect its water resources by decreasing evaporation losses from the HADR, or it could maximize land resources by identifying new settlement sites for climate migrants, especially from the northern costal zones. To evaluate these kinds of tools and to fulfill related objectives, this thesis draws on a broad variety of sources such as meteorological data; hydrological data, including reservoir water levels as well as discharge and seepage data; topographic data, including old topographic maps, the HADR bathymetric survey, satellite images, and aerial photos; and bed soil sedimentary data. The data was used to develop the following databases:

- HADR Meteorological Database (HADRMTDB)
- HADR Hydrological Database (HADRHYDB)
- HADR Seepage Database (HADRSPDB)
- HADR Bed Soil Sedimentary Categories Database (HADRBSDB)

Satellite images of the HADR were processed to extract the shorelines for the water levels between 154.5 m AMSL and 181.6 m AMSL, and they were then converted into contour lines. For the altitudes below 154 m, the formal Nile morphology before the construction of the HAD was extracted from old aerial photos taken in 1955 to simulate the contour line of 110 m AMSL. The spot heights measured by bathymetric survey of the HADR in 2010 were used to identify the altitude of the HADR bed. The contour lines and spot heights of the HADR were used to generate the HADR Digital Elevation Model (HADR-DEM). The HADRDEM was used to calculate the HADR's hydrological characteristics: surface area, water volume, and water level. These characteristics were used to create a model of the HADR in a form of equations representing the relationship between the hydrological characteristics mentioned above. The results generated with the databases are as follows:.

- The HADRMTDB shows that the mean air temperature increased by about 1°C per decade between 1970 and 2010. In addition, the mean evaporation rate increased by about 1.0 mm/day per decade for the same period, This represents an increase of 10%.

- The HADRHYDB shows that the water level usually starts to rise in September until January. The most frequent estimated surface water level was 178 m AMSL. The difference in the monthly mean surface water level between the Aswan and Wadi Halfa gauge stations was about 11 cm due to the back water curve upstream of the HAD. Therefore, the surface water level of HADR is considered identical on the entire reservoir.

- The HADRSPDB shows that the mean annual seepage losses were about 55 hm^3, which presents 0.1% of Egypt's share of the HADR. Because this is a relatively small amount of water, the issue of seepage loss can be neglected.

- The HADRBSDB shows that there were primarily silt and clay sediments in Lake Nasser, whereas sand sediments increased in Lake Nubia and reached about 100% at the southern end of the HADR.

- The HADRDEM shows that water surface area estimated by the HADRDEM and the surface area estimated by the Nile Water Sector (NWS) are highly correlated, and the relationship between them can be presented with a line inclined at 45°. Both water volumes have parallel curves with a difference of about 7 km^3 due to sediment events over the past five decades.

- The evaluation of the mathematical models developed to simulate the hydrological characteristics of the HADR shows that these models are accurate enough to compute the different parameters.

The total area of the khors in the HADR is about 3000 km^2, which represents 46% of the HADR's surface area. With a mean depth of 14 m, the khors hold 44 km^3 of water, which accounts for 28% of the HADR's total water volume. The cross section with narrowest width is located at the El-Madik section, approximately 130 km downstream of the HAD with a width of 1350 m at a water level of 181.6 m AMSL. A closer examination of sediment conditions in the HADR shows that the total erosion volume was only about 126 hm^3 while the total sediment volume reached over 6.6 km^3 during the past five decades. Of these 6.6 km^3, 5.5 km^3 were deposited in Lake Nubia and 1.1 km^3 in Lake Nasser. In the reach between Korosko, located 182 km upstream of the HAD, and the El-Madik section, there were no changes to the altitude of the HADR's bed during the past decade.

To reduce evaporation losses, three main scenarios were investigated. Scenario 1 investigated the possibility of reducing evaporation losses by eliminating several khors. Two options were investigated for Khor Kalabsha, namely either building

Conclusions and Recommendations 261

a dam to close it or filling it with materials from the surrounding area. This scenario shows that eliminating Khor Kalabsha will reduce the surface area by over 600 km^2 and reduce evaporation losses by approximately 2 km^3 by 2100. Two options were also investigated for Khor El-Alaky, namely either building a dam to close it or covering it. The study shows that eliminating Khor El-Alaky will reduce the surface area of the HADR by approximately 170 km^2 and reduce evaporation losses by approximately 0.5 km^3 by 2100. Two options were investigated for Khor Genina, namely either building a dam to close it or covering it. This scenario indicates that eliminating Khor Genina will reduce the surface area by approximately 100 km^2 and reduce evaporation losses by approximately 0.4 km^3 by 2100. Two options were also investigated for Khor Sara, namely either building a dam to close it or covering it. The scenario shows this approach will reduce the surface area by approximately 40 km^2 and reduce evaporation losses by approximately 0.1 km^3 by 2100. The possibility of covering other small khors, which represent 1050 km^2 of the total surface area of the HADR, was also investigated. This scenario shows that this approach will reduce evaporation losses by approximately 3.0 km^3 by end of the twenty-first century.

Scenario 2 investigated the possibility of controlling the water level in the HADR by building a dam in the El-Madik reach. This scenario assessed the effect of lowering the water level downstream of the proposed dam and raising the water level upstream. Ten scenarios were investigated to study the effect of lowering the water level from one to ten meters in one-meter intervals downstream of the proposed dam. The study shows that reducing the water level by 5 m at an altitude of 178 m AMSL downstream of the proposed dam will reduce the surface area by approximately 200 km^2 and reduce evaporation losses by approximately 0.6 km^3 by the end of the twenty-first century.

Scenario 3 investigated the possibility of lowering the lakebed level by removing sediment loads from the bed. Two options were considered: the first assessed the impact of removing sediments only from Lake Nasser and the second of removing sediment from the entire HADR. The study shows that removing the sediments from Lake Nasser will reduce the surface area by about 73 km^2 at an altitude of 178 m AMSL and reduce evaporation losses by approximately 0.22 km^3 in 2100, whereas removing the sediments from the entire HADR will reduce the surface area by approximately 355 km^2 at an altitude of 178 m AMSL and reduce evaporation losses by approximately 1.1 km^3 in 2100. Finally, combinations of some of these scenarios were assessed to identify optimal measures that could be applied to maximize reductions in terms of water surface area. The optimal combination was building a dam at the El-Madik section, filling Khor Kalabsha, and removing the sediments from the HADR. This combined approach is likely to reduce evaporation losses by approximately 2.6 km^3 by end of the twenty-first century.

The tools assessed above will reduce evaporation losses and thereby help, on the one hand, to protect Egypt's limited water resources and, on the other hand, to optimize land use, and mitigate some problems resulting from climate change. Prospective project will identify new measures to reduce evaporation losses. The project includes establishing a new dam at the El-Madik section, filling Khor Kalabsha, removing 6.6 km³ of sediment from the HADR, as well as covering El−Alaky Khor and other small khors with floating objects. This project will help to reduce the evaporation losses by approximately 5.0 km³ under the current climate conditions and 5.55 km³ by the end of this century. This amount is equivalent to 50% of the evaporation losses stated in 1959 agreement, as well as 10% of annual share of Egypt from the Nile. Another prospective project, building on the insights of this thesis, will identify and assess new settlement areas for climate change migrants from the Mediterranean coast. The total area of 1200 km² that will be created by filling Khor Kalabsha and leveling the surrounding areas, can be used by the populations relocated from the northern coastal zones. The study proposed infrastructure that will be needed to make settlements in these new areas feasible, sustainable, and socially acceptable.

7.2 Recommendations and future study

This study assessed possibilities for reducing evaporation losses from the HADR and for maximizing land resources by identifying new settlement sites for climate change migrants from the northern costal zones. Based on its results, the study recommends the following measures:

- filling Khor Kalabsha to reduce evaporation losses by 2 km³ and to create a leveled area of approximately 1200 km², which could be used for agriculture and new communities. As required by Decree 203/2002, a buffer zone of two km must be established between farms and the shoreline of the HADR in order to protect water supplies. BIO organic fertilizers and pesticides must be used to protect the groundwater reservoir and the environment around the HADR.

- establishing new dam at the El-Madik section to provide many benefits to Egypt by, for example, reducing evaporation losses and increasing energy resources. In this manner, the social and economic condition of southern Egypt could be improved, and the issue of food security could be addressed as well. The control of the water level downstream of the proposed dam could provide stability and sustainability to riparian people around Khor Kalabsha by obtaining permanent lands with minimal fluctuations in the water level and by preventing the new settlement areas from flooding hazards.

- covering part of Khor El-Alaky and other small khors with palm fronds, which are available in the region to reduce evaporation losses by 2.45 km³ by 2100. This measure could improve the balance of the local ecosystem, protect the environment, and lead to higher economic benefits for inhabitants of southern Egypt, as, for instance, farmers would be able to sell palm fronds to the government.

- lowering the altitude of the lakebed by removing 6.6 km^3 of sediment from the HADR to reduce evaporation losses by approximately 0.78 km^3 by 2100.
- combining the measures described above is likely to reduce evaporation losses by approximately 5.0 km^3 under current climatic conditions and by approximately 5.55 km^3 by the end of this century.

To apply these measures and to encourage migration to the area proposed above, it is recommended that:

- People need to be encouraged to migrate from the north to the south. This could be achieved with media programs, which provide information about climate change hazards, adaptation methods, and measures for mitigating the impact of global warming. These programs may convince them to leave the delta for the south valley and to settle in new communities established in this region.
- Advanced irrigation systems need to be established, which will provide the water to the new area created after filling Khor Kalabsha.
- Micro irrigation techniques such as micro sprinklers and drip irrigation need to be applied in newly reclaimed lands.
- Pipelines have to be used to transfer water in the new areas, especially in locations characterized by soils with high porosity and high evaporation losses.
- New technologies have to be introduced for canal maintenance and weed control.
- Volumetric water pricing has to be provided by metering at the tertiary canal level.
- New crop varieties have to be cultivated. These new varieties have been genetically engineered to produce new seeds, and they are characterized by higher productivity, high resistance to diseases, and reduced water consumption.
- An indicative cropping pattern has to be designed for this region. This new pattern matches climatological conditions, soil characteristics, and the availability of water resources in terms of both quantity and quality.
- Farmers have to be encouraged to follow the indicative cropping pattern to consume less water. The farmers who fail to comply with guidelines in this respect will be required to pay for excess water.
- A drainage system has to be built that drains water back to the HADR so that it can be used again. To protect the environment and inhabitants, only BIO fertilizers and pesticides should be allowed in this area.
- A supply of safe drinking water has to be established, to provide a healthy

environment for the new settlement communities and that will decrease health care costs.
- A sanitation system has to be built to collect domestic sewage. A basic treatment of water before it is discharged, is needed to protect water supplies and public health.
- Energy resources such as electricity from the new hydropower station and clean solar and wind energy need to be provided.
- New factories and other businesses need to be established to provide job opportunities for climate migrants.
- Industrial wastewater needs to be separated from domestic sewage to reduce the costs of treating domestic sewage.
- Industrial waste water needs to be used in cooling systems of factories and other businesses.
- Polluting materials used by firms need to be taxed and/or non-polluting materials need to be subsidized so that they can compete with polluting ones.
- Schools and hospitals need to be established to provide crucial public services.
- An adequate infrastructure, for example roads, needs to be built in order to connect this area with the rest of the country and to encourage trade and communication.
- Environmental education needs to be promoted by public institutions, the media, NGOs, and other stakeholders.
- The public's awareness of major environmental concerns such as water scarcity and related problems such as the rationalization of water, needs to be increased.
- The government and the private sector need to be held accountable for pollution.

This thesis offers new mathematical models to describe the current hydrological characteristics of the HADR in a form of equations. As highlighted in Chapter 1, several limitations are, however, beyond of scope of this project, and further data is required to conduct detailed feasibility studies, which would be required before the measures recommended above could be implemented. More specifically, the following projects could be pursued:

- The potential impact of climate change needs to be assessed with the help of recent climate change models.
- The effect of new development projects in the upper Nile basin countries, which affect the natural flow to the HADR, needs to be evaluated.
- Methods for optimizing water use downstream of the HAD have to be investigated.

- Environmental impact assessments and feasibility studies for measures that reduce evaporation losses have to be conducted.
- Environmental impact assessments and feasibility studies for new resettlement proposed above also need to be conducted.
- A feasibility study of the optimal methods for removing sediment deposits from HADR needs to be conducted as well.
- A feasibility study that evaluates the possibilities for using sediment from the HADR has to be conducted, and a pilot project needs to be completed to evaluate the characteristics of sediment for its potential use in agricultural land reclamation and to identify how sediments affect crop yields.
- Detailed feasibility studies of the design and the costs of the new dam and the other hydraulic structures need to be conducted.
- Detailed feasibility studies of the design and capital costs of the new hydropower station and subsequent electricity production also need to be conducted.
- Natural materials available in southern Egypt, for example papyrus, the stems of maize or sugar cane, or banana leafs, which could all be used to manufacture floating objects that will cover khors of the HADR, need to be investigated.
- The entire HADR and its khors need to be surveyed as accurately as Lake Nubia in order to update the HADRDEM.
- A detailed geological survey of the HADR including its khors needs to be conducted.
- Additional meteorological stations on the HADR need to be established to collect more meteorological data, which, in turn, could then be used in other follow-up studies.

References

Asmal, K., Lakshmi, C. J., Henderson, J., Lindahl, G., Scudder, T., Cariño, J., Blackmore, D., Patkar, M., Goldemberg, J., Moore, D., Veltrop, J. and Steiner, A. (2000). Dams and development, a new framework for decision-making: the report of the World Commission on Dams. World Commission on Dams. Earthscan, London, Sterling, VA.

Awulachew, S. B., McCartney, M., Steenhuis, T., Ahmed, A. and Bashar, K. (2009). Improved water and land management in the Ethiopian highlands and implications for the downstream Blue Nile. In: Intermediate Results Dissemination Workshop. Addis Ababa, Ethiopia, 5-6 February 2009. International Livestock Research Institute (ILRI). International Water Management Institute (IWMI). http://publications.iwmi.org/pdf/H041758.pdf.

Abdel-Aziz, T.M. (1997), Prediction of bed profile in the longitudinal and transverse direction in Aswan high dam reservoir, Ph.D. Thesis, Cairo University, Giza, Egypt.

Abdel Moneim, A. A. (2005). Overview of the geomorphological and hydrogeological characteristics of Eastern Desert of Egypt. Hydrogeol J, 13 (2), 416–425.

Abdin, A.E., and Gaafar, I. (2009). Rational water use in Egypt. In: Technological perspectives for rational use of water resources in the Mediterranean region. Options Méditerranéennes: Série A. Séminaires Méditerranéens; n. 88. Bari, Italy. CIHEAM. Italy, pp. 11–27. http://ressources.ciheam.org/om/pdf/a88/00801177.pdf.

Abul-Atta, A. (1978). Egypt and the Nile After the Construction of the High Aswan Dam, Report, Ministry of water reserouces and irrigation, EL-Qanater, Egypt.

Abulnaga, B. E. and El-Sammany, M. S. (2004). De-Silting Lake Nasser with Slurry Pipelines. In: Critical Transitions in Water and Environmental Resources Management. American Society of Civil Engineers, pp. 1–13.

Abu-Zeid, M. A. and El-Shibini, F. Z. (1997). Egypt's High Aswan Dam. International Journal of Water Resources Development, 13 (2), 209–217.

Ahmed, A. A. and Ismail, U. H. A. E. (2008). Sediment in the Nile River system. Report, UNESCO, International hydrological programme, international sediment initiative, Khartoum, Sudan. http://www.irtces.org/isi/isi_document/sediment%20in%20the%20nile%20river%20system.pdf

Alam, S. and Al Shaikh, A. A. (2013). Use of palm fronds Randall shaded cover for evaporation reduction to improve water storage efficiency. Journal of King Saud University - Engineering Sciences, 25 (1), 55–58.

Al-Hassoun, S. A., Al-Shaikh, A. A., Al-Rehaili, A. M. and Misbahuddin, M. (2011). Effectiveness of using palm fronds in reducing water evaporation. Canadian Journal of Civil Engineering, 38 (10), 1170–1174.

Allam, M. N. and Allam, G. I. (2007). Water resources in Egypt: Future challeges and opportunities. International Water Resources Association, Water International Journal, 32, 205–218.

Allen, R. G. (1998). Crop evapotranspiration Guidelines for computing crop water requirements. FAO irrigation and drainage paper, 56, Rome, Italy.

Allen, D., Averett, D. and Barth, F. (2005). Contaminated Sediment Remediation Guidance for Hazardous Waste Sites. Report, USA Environmental Protection Agency, Office of Solid Waste and Emergency Response, USA.

Allen, P. M. and Dunbar, J. A. (2005). Dredging Versus New Reservoir. Report, Texas Water Development Board, Texas, USA.

Amini, A. and Heller, P. (2014). Comprehensive numerical simulations of sediment transport and flushing of a Peruvian reservoir. In: Reservoir sedimentation: Special session on reservoir sedimentation of the Seventh International Conference on Fluvial Hydraulics (River Flow 2014), IAHR Committee on Fluvial Hydraulics, EPFL Lausanne, Switzerland, 3-5 September 2014, pp. 211–219: Leiden, The Netherlands.

Aras, T. (2009). Cost analysis of sediment removal techniques from reservoir. Master of Science, Middle East Technical University, Ankara, Turkey.

AQUA (2006). Introducing Aqua Guardian Group: AquaArmour. Online Technical documents, Aqua Guardian Group Limited, Australia. www.aquaarmour.com.au/storage/cms-files/AquaGuardianCorpProfile05.pdf.

Attia, B.B. (2008). Assessment of vulnerability and adaptation of water resources to climate change in Egypt. In: 13th IWRA world water congerce, International Water Reseources Association, 1-4 September 2008, France.

Awulachew, S. B. (2012): The Nile River basin. Water, agriculture, governance and livelihoods. Abingdon, Oxon, New York, USA: Routledge.

AWTT (2012). Hexprotect - Wind resistant: Technical data Wind resistant. Online Technical documents, Advanced Water Treatment Technologies Inc. (AWTT), USA. www.awtti.com/hexprotect_cover.php.

Bates, B. C., Kundzewicz, Z. W., Palutikof, J. and Shaohong, W. (2008). Climate Change and Water Technical Paper of the Intergovernmental Panel on Climate Change. IPCC Technical paper VI, IPCC Secretariat, Geneva.

Baumann, P.R. (2010). Remote sensing: Landsat program. From the beginning. http://www.oneonta.edu/faculty/baumanpr/geosat2/RS%20Landsat/RS-Landsat.htm.

Bernacsek, G. M. (1984). Guidelines for dam design and operation to optimize fish production in impounded river basins. Based on a review of the ecological effects of large dams in Africa. Food And Agriculture Organization Of The United Nations (FAO). CIFA Tech. Pap., 11, p. 98. http://www.fao.org/docrep/005/ac675e/AC675E00.htm#TOC.

Biswas, A. K. and Tortajada, C. (2004). Hydropolitics and impacts of the High Aswan Dam, Advisory panel project on water management. Report, Ministry of Water Resources and Irrigation, Egypt.

Boroujeni, H. S. (2012). Sediment Management in Hydropower Dam: Case Study – Dez Dam Project. Hydropower - Practice and Application. InTech (978-953-51-0164-2), 115–142.

Brown, P. (2001). Basics of Evaporation and Evapotranspiration. University of Arizona Horticulture Publications. Arizona, USA. http://ag.arizona.edu/pubs/water/az1194.pdf.

Brutsaert, W. (1982). Evaporation into the atmosphere: Theory, history, and applications. Environmental fluid mechanics (1), Springer, ISBN-13: 978-9027712479

Burston, I. A. (2008). Conservation of water from open storages by minimising evaporation. Doctor of philosophy, PhD Thesis, RMIT University, Australia.

Camberlin, P. (2009). Nile Basin Climates. In: The Nile. Origin, environments, limnology and human use. Springer.89, pp. 307–333, Monographiae biologicae. https://hal.archivesouvertes.fr/file/index/docid/391068/filename/NileBasinClimates.pdf.

Carvalho, N. O., Júnior, N. P. F., Santos, P. M. C. and Lima, J. E. W. (2000). Reservoir sedimentation assessment guideline, Brazilian electricity regulatory agency - Hydrological Studies and Information Department - SIH, Brasilia.

Caruso, J. (2005). Controlling evaporation losses from farm dams: Fact Sheet. http://npsi.gov.au/files/products/national-program-sustainable-irrigation/pf050873/pf050873.pdf.

Cedergren, H. R. (1998). Seepage, drainage and flow nets, 3rd edn., Wiley, New York, USA.

CEQA (2011). Klamath settlement process - sediment management in the reservoirs. Report, California Environmental Quality Act (CEQA), California, USA.

Chao, B. F.; Wu, Y. H.; Li, Y. S. (2008). Impact of Artificial Reservoir Water Impoundment on Global Sea Level. Science 320 (5873), pp. 212–214. DOI: 10.1126/science.1154580.

Clarke, D. (2009). Reducing evaporation losses from farm dams. Online Report. http://ramblingsdc.net/Australia/EvapReduc.html.

CoRI (2009). Second national communication report. The coastal research institute, Ministry of Water Resources and Irrigation (MWRI), Alexandria, Egypt., MWRI.

Craig, I., Green, A., Scobie, M. and Schmidt, E. (2005). Controlling evaporation loss from water storages: Rural water use efficiency initiative. NCEA Publication No 1000580/1, Queensland department of natural resources and mines, National Centre for Engineering in Agriculture, University of Southern Queensland, Toowoomba, Australia.

Das, M. M. and Saikia, M. D. (2009). Hydrology, PHI Learning Private Limited, New Delhi.

Dasgupta, S., Laplante, B., Meisner, C. and Yan, J. (2007). The impact of sea level rise on developing countries: A comparative analysis: World Bank policy research working Paper 4136.
http://www.wds.worldbank.org/external/default/WDSContentServer/IW3P/IB/2007/02/09/000016406_20070209161430/Rendered/PDF/wps4136.pdf.

Digout, D. (2001). Major river basins of the world. Atlas of Population and Environment. Source: United Nations Environmental Program (UNEP), World conservation monitoring center (WCMC), World Resources Institute (WRI), American Association of Advancement of Science (AAAS). UNEP/GRID-Arendal.
http://www.grida.no/graphicslib/detail/major-river-basins-of-the-world_b6d2

Dougherty, J. P. (1975). Evaporation Data in Texas. Report 192, Texas water development board.

Ebaid, H. M. I. and Ismail, S. S. (2010). Lake Nasser evaporation reduction study. Journal of Advanced Research, Elsevier, 1 (4), 315–322.

EEAA (2001). The National Environmental Action Plan (NEAP). Environmental Policy Report, Egyptian Environmental Affairs Agency, Cairo, Egypt. http://www.eeaa.gov.eg/english/main/policies3.asp

EEAA (2005). Study report on comprehensive support strategies for environment and development in the early 21st century: Arab Republic of Egypt. Report, Egyptian environmental affairs agency (EEAA). Japanese's Overseas environmental cooperation center, Egypt.
http://www.env.go.jp/earth/coop/coop/document/c_report/egypt_h16/english/pdf/016.pdf, checked on 3/20/2013.

Eid, H., M., El Marsafawy, S., M. and Ouda, S.,A. (2007). Assessing the economic impact of climate change on agriculture in Egypt: A Ricardian approach, Centre for Environmental Economics and Policy in Africa (CEEPA), discussion paper no. 16, Special Series on Climate Change and Agriculture in Africa, World Bank.
http://elibrary.worldbank.org/doi/pdf/10.1596/1813-9450-4293

Elewa, A. S. (1985). Effect of flood water on the salt content of Aswan High Dam Reservoir. Hydrobiologia, 128 (3), 249-254.

El-Fellaly, S. H., and Saleh, E. M. A. (2004). Egypt's experience with regard to water demand management in agriculture. In: Eighth International Water Technology Conference, IWTC8, 26-28 March 2004, International Water Technology Association (IWTA), Alexandria, Egypt. pp. 5–28.
http://www.iwtc.info/2004_pdf/01-1.pdf.

El-Moattassem, M., Abdel-Aziz, T. M., EL-Sersawy, H.E M. (2010). Modelling of sedimentation process in Aswan High Dam Reservoir. UNESCO Publications. Flanders-UNESCO Science Trust Fund cooperation, Nile Research Institute (NRI), National Water Research Center (NWRC), Cairo, Egypt.
http://www.unesco.org/fileadmin/MULTIMEDIA/FIELD/Cairo/pdf/MODELLING_OF_SEDIMENTATION_PROCESS.pdf.

El-Moattassem, M., Abdel-Aziz, T. M., El-Sersawy, H. (2013). Effect of upstream structures on delta deposit progress in Aswan High Dam Reservoir. UNESCO Publications. Flanders-UNESCO Science Trust Fund cooperation, Nile Research Institute (NRI), National Water Research Center (NWRC), Cairo, Egypt.
http://www.unesco.org/fileadmin/MULTIMEDIA/FIELD/Cairo/pdf/EFFECT_OF_UPSTREAM_STRUCTURES_ON_DELTA_DEPOSIT_PROGRESS_IN_ASWAN_HIGH_DAM_RESERVOIR.pdf.

El-Shabrawy, G. M. (2009). Lake Nasser–Nubia. Springer Science - Business Media B.V.2009, 125–155.

Elshamy, M. E., Sayed, M. A. and Badawy, B. (2009). Impacts of Climate Change on the Nile Flows at Dongola Using Statistical Downscaled GCM Scenarios. Nile Basin Water Engineering Scientific Magazine, 2, 1–14.

Elsharkawy, H., Rashed H. and Rached, I. (2009). Climate Change: The Impacts of Sea Level Rise on Egypt. In: 45th ISOCARP Annual World Congress, 18-22 October 2009, International Society of City and Regional Planners (ISOCARP), Porto, Portugal. http://www.isocarp.net/data/case_studies/1456.pdf.

El-Tahlawi, M. R., Farrag, A. A. and Ahmed, S. S. (2008). Groundwater of Egypt: an environmental overview. Environmental Geology, 55 (3), 639–652.

ENVI EX (2013). Quick find menu. ENVI EX online User's Guide - Exelis Visual Information Solutions.
www.exelisvis.com/portals/0/pdfs/enviex/ENVI_EX_User_Guide.pdf.

ESRI (2011). ArcGIS Help 10.1: Fundamentals of Surfaces. Technical Support. ARCGIS online User's Guide.
http://resources.arcgis.com/en/help/main/10.1/index.html#//00q80000005z000000.

Fahmy, A. (2006). The Identification of Hydrological Unites in the Nile Basin. Data Availability. UNESCO Publications. FRIEND/Nile (FN) UNESCO-Flanders Science Trust Fund Project, Cairo, Egypt http://www.unesco.org/fileadmin/MULTIMEDIA/FIELD/Cairo/pdf/hydrology_study.pdf.

FAO (1986). Irrigation water management, Training manuals Chapter 3: Crop water needs. Series of training manuals on irrigation, Food and Agriculture Organization Of The United Nations (FAO) - Natural Resources Management and Environment Department, Rome, Italy.

FAO (1997a). Irrigation potential in Africa, a basin approach. FAO land and water bulletin, no. 4, FAO, Rome.
http://www.fao.org/docrep/W4347E/w4347e00.htm

FAO (1997b). Treaties concerning the non-navigational uses of international watercourses. Africa = Traités concernant l'utilisation des cours d'eau internationaux à des fins autres que la navigation. Rome (FAO legislative study, 61). ISBN 92-5-004079-2. http://www.fao.org/docrep/W7414B/w7414b13.htm

FAO (2001). Manual: Installation, Operation and Maintenance of Buoy Operated Automatic Meteorological Stations Established in Lake Nasser, Food and Agriculture Organization of The United Nations (FAO), Uganda.

FAO (2005). Fertilizer use by crop in Egypt Land and plant nutrition management service land and water development division. FAO Publications. Food and Agriculture Organization of The United Nations (FAO), Rome, Italy.
ftp://ftp.fao.org/docrep/fao/008/y5863e/y5863e00.pdf

FAO (2011). Information Products for Nile Basin Water Resources Management Population prospects in the Nile Basin. FAO Publications. Food and Agriculture Organization of The United Nations (FAO), Rome, Italy.
http://www.fao.org/nr/water/faonile/products/Docs/Poster_Maps/POPULATIONBIG.pdf

FAO (2014). FAOSTAT - The latest statistics for FAO. FAO Publications. Food and Agriculture Organization of The United Nations (FAO), Rome, Italy.
http://faostat.fao.org/site/535/default.aspx#ancor.

Ferrari, R. L. (2006). Reclamation Managing water in the west - Reconnaissance Technique for Reservoir Surveys. U.S. Department of the Interior Bureau of Reclamation Technical Service Center, Denver, Colorado.

Finch, J. and Calver, A. (2008). Methods for the quantification of evaporation from lakes, World Meteorological Organization's Commission for Hydrology, CEH Wallingford Wallingford Oxfordshire OX10 8BB UK.

Flögl, W. (2010). The History of the World Register of Dams. ICOLD publications. International Commission on Large Dams (ICOLD), Paris, France.
http://www.icold-cigb.org/userfiles/files/CIGB/History_of_the_WRD.pdf.

Frihy, O. E. (1992). Sea-level rise and shoreline retreat of the Nile Delta promontories, Egypt. Natural Hazards, 5 (1), 65-81.

Fruchard, F. and Camenen, B. (2012). Reservoir Sedimentation: Different Type of Flushing - Friendly Flushing Example of Genissiat Dam Flushing. In: ICOLD International Symposium on Dams for a changing world. Kyoto, Japan, 5 June 2012. 6 p.
https://hal.archives-ouvertes.fr/hal-00761305/document.

Fuess, R. (2013). Meteorologische instrumente kg Evaporimeters and Evaporigraphs, Dr. Alfred müller, Königs Wusterhausen, Germany. http://www.rfuessmueller.de/231-0E.pdf

Harbeck, G. J. (1962). A Practical Field Technique for Measuring Reservoir Evaporation utilizing the Mass-Transfer Theory. U.S. Geological Survey Professional Paper 272-E, U.S. Printing Office, Washington, D.C, USA.

Halcrow (2001). Sedimentation in storage reservoirs. Final report, Halcrow Group Limited, Department of the environment transport and the regions, England.

Hargrove, W. (2008). Sedimentation in our reservoirs: Causes and solutions. Report, Department of Communications at Kansas State University, Kansas. USA.

Hassan, R. M. A., Hekal, N. T. H. and Mansor, N. M.S (2007). Evaporation reduction from Lake Naser using new environmentally safe techniques. In: Eleventh International Water Technology Conference, IWTC11, International Water Technology Association (IWTA), 15-18 March 2007, Sharm El-Sheikh, Egypt, pp. 179–194.

HDA-MWRI (2009). Lake Research Mission Reports from 1990 to 2009 in Arabic language. In: Aswan reservoir and High Dam Authority, Ministry of Water Resources and Irrigation, Aswan, Egypt.

Hegazi, A. M., Afifi, M. Y., Elwan, A. A. and El-Shorbagy, M. A. (2005). Egyptian National Action Program to Combat Desertification, Report, Desert Research Center (DRC), Ministry of Agriculture and Land Reclamation, Cairo, Egypt.
http://www.unccd.int/ActionProgrammes/egypt-eng2005.pdf.

Helfer, F., Zhang, H. and Lemckert, C. (2009). Evaporation reduction by windbreaks: overview, modelling and efficiency Urban Water Security Research Alliance. Technical Report No. 16, Urban water security research alliance, Australia's South-East Queensland (SEQ).

Ilereher, M. E. (2014). Environmental monitoring and change assessment of Toshka lakes in southern Egypt using remote sensing. Environ Earth Sci, pp. 1-10. DOI: 10.1007/s12665-014-3651-5.

Herting, A., Farmer, T. and Evans, J. (2004). Mapping of the Evaporative Loss From Elephant Butte Reservoir Using Remote Sensing and GIS Technology. Research, New Mexico State University (NMSU), Mexico. http://wrri.nmsu.edu/research/rfp/studentgrants03/reports/herting.pdf.

Hipsey, M., Sivapalan, M. and Clement, T. (2004). A numerical and field Investigation of surface heat fluxes from small wind-sheltered waterbodies in semi-arid Western Australia. Environmental Fluid Mechanics, 4 (1), 79-106.

Hurst, H. E.; Black, R. P. (1943). Monthly and annually rainfall totals and number of rainy days at stations in and near the Nile Basin. The Nile Basin Volume VI and supplements 1-11. Egyptian Government Press Cairo, Egypt.

Hurst, H.E. (1965). *The Nile*, a General account of the River and the utilization of its Waters. Constable and Company LTD, London, UK.

Hutchinson, M.F. (1988). Calculation of hydrologically sound digital elevation models. In: Third International Symposium on Spatial Data Handling, 17−19 August 1988, Sydney, Australia, pp. 117-133.

IADC (2005). Dredging: The facts. IADC publications. International Association of Dredging Companies (IADC), Netherlands.
https://www.iadc-dredging.com/ul/cms/fck-uploaded/documents/PDF%20Publications/dredging-literature-dredging-the-facts.pdf

ICID (2002). Adopted Measures to Face Major Challenges in the Egyptian Water Sector. ICID publications, The International Commission on Irrigation and Drainage (ICID), Ministry of water resources and irrigation, Egypt. World Water Council, Marseille, France, 39 p. http://www.ircwash.org/resources/adopted-measures-face-major-challenges-egyptian-water-sector

ICOLD (2009). Sedimentation and sustainable use of reservoirs and river systems. ICOLD BULLETIN, International Commission on Large Dams (ICOLD), Paris, France.
http://www.icold-cigb.org/userfiles/files/CIRCULAR/CL1793Annex.pdf

INTEA (2010). Aswan High Dam - Hydropower plant - power generation & distribution system, INTEA Automation, Aswan, Egypt.
http://www.intea.hr/Intea_pdf/05E.pdf

IPCC (2001a). Climate Change 2001: The Scientific Basis Contribution of Working Group I to the Third Assessment Report of the Intergovernmental Panel on Climate Change (IPCC), Cambridge University Press, Cambridge, United Kingdom and New York, NY, USA.
http://www.grida.no/publications/other/ipcc_tar/

IPCC (2001b). Climate Change 2001: Mitigation Contribution of Working Group III to the Third Assessment Report of the Intergovernmental Panel on Climate Change (IPCC), Cambridge University Press, Cambridge, United King-

dom and New York, NY, USA.
http://www.grida.no/publications/other/ipcc_tar/

IPCC (2007). Climate Change 2007: The Physical Science Basis. Contribution of Working Group I to the Fourth Assessment Report of the Intergovernmental Panel on Climate Change (IPCC), Cambridge University Press, Cambridge, United Kingdom and New York, NY, USA.
http://www.ipcc.ch/publications_and_data/ar4/wg1/en/contents.html

IPCC (2008). Climate change and water. Technical paper of the Intergovernmental Panel on Climate Change (IPCC): Volume 6. World Meteorological Organization (WMO), Geneva, Switzerland. Intergovernmental Panel on Climate Change. ISBN: 978-92-9169-123-4.
www.ipcc.ch/pdf/technical-papers/ccw/chapter5.pdf

IPCC (2013). Climate Change 2013: The Physical Science Basis. Contribution of Working Group I to the Fifth Assessment Report of the Intergovernmental Panel on Climate Change (IPCC). Cambridge University Press, Cambridge, United Kingdom and New York, NY, USA.
http://www.ipcc.ch/report/ar5/wg1/

IRTCES (2011). Sediment issues & sediment management in large river basins: Interim case study synthesis report International Sediment Initiative. UNESCO publications. Technical Documents in Hydrology, UNESCO, Beijing.

Ismail, A. (2001): GIS Resource Analysis for Sustainable Spatial Planning: The Case of Lake Nasser. In: CUPUM 2001, 7th International Conference on Computers in Urban Planning and Urban Management, 18–21 July 2001, Honolulu, Hawaii, USA.
https://www.academia.edu/3420036/GIS_Resource_Analysis_for_Sustainable_Spatial_Planning.

IWLP (2008). African Regional Documents. African River Basins. Nile River Basin Agreements. International Water Law Project (IWLP). Addressing the future of water law and policy in the 21st century.
http://www.internationalwaterlaw.org/documents/africa.html#Nile River Basin.

Jennison, I. (2003). Methods for reducing evaporation from storages used for urban water supplies. Final report, GHD Pty Ltd, Queensland Department Of Natural Resources and Mines, Australia.

Kaltenborn, B. P., Nellemann, C. and Vistnes, I. (2010). High mountain glaciers and climate change Challenges to human livelihoods and adaptation, UNEP publications, United Nations Environment Programme (UNEP).
http://www.unep.org/climatechange/Publications/Publication/tabid/429/language/en-US/Default.aspx?ID=4168.

Karyabwite, D.R. (2000). Water Sharing in the Nile River Valley: Using gis/remote sensing for the sustainable use of natural resources. UNEP publica-

tions, United Nations Environment Programme (UNEP).
http://www.grid.unep.ch/activities/sustainable/nile/nilereport.pdf.

Kashiwai, J. (1998). Reservoir Sedimentation and Sediment Management in Japan, Hydraulic Engineering Research Guroup,Incorporated Administrative Agency, Public Works Research Institute, Japan.

Khalil, M., Ramalho, E. and Monteiro S. F. A. (2011). Using resistivity logs to estimate hydraulic conductivity of a Nubian sandstone aquifer in southern Egypt. NSG, 9 (No 4), 349–355.

Khalil1, F., Ouda, S., Osman, N.A. and El-Hady, A. (2011). Determination of agro-climatic zones in Egypt using a robust statistical procedure. In: Fifteenth International Water Technology Conference, IWTC-15 2011, 28-30 May 2011, Alexandria, Egypt.
iwtc.info/wp-content/uploads/2011/07/G98.pdf

Klink, M. (2006). Evaporation estimation using a floating pan. Master of science civil engineering, Graduate school of clemson university.

Kim, J. and Sultan, M. (2002). Assessment of the long-term hydrologic impacts of Lake Nasser and related irrigation projects in Southwestern Egypt. Journal of Hydrology 262 (1–4), pp. 68–83. DOI: 10.1016/S0022-1694(02)00013-6.

Kohler, T. and Maselli D. (2009). Mountains and climate change from understanding to action. FAO publications. Centre for Development and Environment (CDE), Institute of Geography, University of Bern, Bern, Switzerland. ISBN: 978-3-905835-16-8.
http://www.fao.org/docrep/017/i2869e/i2869e00.pdf

Korzukhin, M. D., Kolosov, P. A. and Semenov, S. M. (2011). Applying Dalton's law of potential evaporation rate over the territory of Russia and neighboring countries using long-term observation data. Russian Meteorology and Hydrology, 36 (12), 786-793.

Li, J. and Heap, A. D. (2011). A review of comparative studies of spatial interpolation methods in environmental sciences: Performance and impact factors. Ecological Informatics. Ecological Informatics, 6 (3-4), 228–241.

LNFDC (2008). Climate change and its effects on water resources management in Egypt Lake Nassr Flood and Drought Control (LNFDC) project reports, Ministry of Water Resources and Irrigation - Planning Sector, Egypt.

Luis, L., Sidek, L. M., Desa, M. N. B. M. and Julien, P. Y. (2013). Hydropower reservoir for flood control: A case study on ringlet reservoir, Cameron Highlands, Malaysia. Journal Of Flood Engineering, 4 (0976-6219), 87–102.

Maidment, D. R. (1993). Handbook of hydrology, McGraw-Hill, New York.

Makary, A.Z. (1982), Sedimentation in the high Aswan Dam reservoir, PhD. thesis, Faculty of Engineering, Ain Shams University, Egypt.

Maliva, R. and Missimer, T. (2012). Environmental Science and Engineering, Berlin, Heidelberg, Springer Berlin Heidelberg.

MALR (2003). Deteriorated Soils in Egypt Management and Rehabilitation. Project report, Ministry of agriculture and land reclamation, Cairo, Egypt. ftp://ftp.fao.org/agl/agll/ladadocs/detsoilsegypt.doc.

MALR (2005). Egyptian National Action Program To Combat Desertification. Project report, Ministry of agriculture and land reclamation, Cairo, Egypt. http://www.unccd.int/ActionProgrammes/egypt-eng2005.pdf

Maru, M. and Teklehaimanot, G. (2012). The Nile From mistrust and sabrerattling to rapprochement. 6854 ISS Paper 238.indd, Institute for Security Studies.

Matsuura, K. D. H. (2003). Learning to combat desertification: A Teacher's Guide An activity and educational guide to understand and combat the phenomenon of desertification. UNESCO Publications, developed from the United Nations Convention to Combat Desertification. Ecological Sciences for Sustainable Development.

McCartney, M., Forkuor, G., Sood, A., Amisigo, B., Hattermann, F. (2012). The water resource implications of changing climate in the Volta River Basin. IWMI Research Report 146. International Water Management Institute (IWMI), Colombo, Sri Lanka.

Meissner, B., Wycisk, P. (1993). Geopotential and ecology: Analysis of a desert region. Cremlingen-Destedt, Germany: Catena Verlag (Catena supplement, 26).

Melesse, A. M. (2011). Nile River Basin: Hydrology, climate and water use. Dordrecht, New York. Springer.

Mcguffie, K. and Sellers, H. A. (2001). Forty years of numerical climate modelling. International Journal of Climatology, 21 (9), 1067–1109.

MDEQ (2000). Training aids for dam safety (TADS) Evaluation of seepage conditions. Report, The Mississippi Department of Environmental Quality (MDEQ), USA.

Mohamed A.F. T. and Mostafa A. K. (2009). Physicochemical Characteristics of Water Quality in Lake Nasser Water. Global Journal of Environmental Research, 3 (1990-925x), 141–148.

Mortensen, N., Hansen J. C., Badger J., Jørgensen B.H., Hasager C.B., Paulsen U.S., Hansen O.F. and Enevoldsen K. (2006). Wind atlas for egypt: measurements, micro- and mesoscale modelling. In: European Wind Energy Conference and Exhibition, 27 February to 2 March 2006, Athens, Greece. 10 pp. http://www.windatlas.dk/egypt/Download/Wind%20Atlas%20for%20Egypt%20paper%20(2006%20EWEC).pdf

Moussa, A. M. A. (2013). Predicting the deposition in the Aswan High Dam Reservoir using a 2-D model. Ain Shams Engineering Journal, 4 (2), pp. 143–153. DOI: 10.1016/j.asej.2012.08.004.

Mutua, F., Mtalo, F. and Bauwens, W. (2005). Challenges of modeling the flows of the Nile River. UNESCO. www.unesco.org/fileadmin/multimedia/field/cairo/pdf/challenges_of_modeling_ the_flows_of_the_nile_river.pdf

MWRI (2000). The Egyptian Water Strategy Till 2017. Planning Sector, Ministry of Water Resources and Irrigation, Cairo, Egypt.

MWRI (2005). National Water Policy. Report, Ministry of Water Resources and Irrigation, Egypt.

MWRI (2010). Water Resources Development and Management Strategy in Egypt – 2050. Arbaic report, Planning Sector, Ministry of Water Resources and Irrigation, Cairo, Egypt.

MWRI-NWRC (2002). Monthly average potential evapotranspiration in the main agro-climatic regions of Egypt, Arabic Report, Water Management Research Institute, National Water Research Center(NWRC), Ministry of Water Resources and Irrigation (MWRI), Cairo, Egypt.

MWRI-PS (2001). Water Resources management in Egypt. Report, Planning Sector, Ministry of Water Resources and Irrigation, Egypt.

Mynett, A. E. and Vojinovic, Z. (2009). Hydro informatics in multi colours part red: urban flood and disaster management. IWA Publishing. Journal of Hydro informatics, 11 (3–4), 166.

NAOUM, S. and TSANIS, I. (2004). Ranking spatial interpolation techniques using a GID based DSS. Global Nest the International Journal, 6, 1–20.

NBI (2012). State of the River Nile Basin 2012. Nile Basin Initiative (NBI) GIZ, Entebbe, Uganda.

Near East Foundation (2010). New Land, New Life − Project West of the High Dam Lake, Aswan, Egypt: Adaptation to Climate Change. Final Technical Report, International Development Research Center, Near East Foundation, New York, USA. http://www.neareast.org/images/uploads/NEF_Egypt_New_Lands_Final_Report_3-10.pdf.

Negm, A. M., Abdulaziz, T., Nassar, M. and Fathy, I. (2010). Predication of life time span of High Aswan Dam reservoir using CCHE2D simulation model. In: Fourteenth International Water Technology Conference, IWTC14, International Water Technology Association (IWTA), 21-23 March 2010, Cairo, Egypt. pp. 611–626.
http://www.iwtc.info/2010_pdf/10-01.pdf.

Nicol, A. and Mamdouh, S. (2003). The Nile: moving beyond cooperation. UNESCO Publications, Technical documents in hydrology, PC-CP series, Vol. 16.

Nour El-Din, M. M. (2013). Proposed Climate Change Adaptation Strategy for the Ministry of Water Resources and Irrigation in Egypt, UNESCO Publications, Cairo, Egypt.

NWRP (2005). National water resources plan facing the challenge: Integrated water resources management plan for 2017. National Water Resouces Plan (NWRP), Ministry of Water Resources and Irrigation, Cairo, Egypt.

NWS (2012). Arabic reports about Nasser Lake from 1950 till 2012. In: Nile Water Sector, Ministry of Water Resources and Irrigation, Egypt.

NYLEX (2002). AquaCap.
http://au103078256.fm.alibaba.com/product/104720126-100752418/Aqua_Cap.html.

Olea, R. A. (1999). Geostatistics for natural resources evaluation by pierre goovaerts, 31, Kluwer Academic Publishers-Plenum Publishers, Oxford University Press, Applied Geostatistics Series.

Osman, A. and Hanna, F. (1995). Agricultural land resources and the future of land reclamation and development in Egypt. Report, International Centre for Advanced Mediterranean Agronomic Studies, Paris.

Panagos, P., Jones, A., Bosco, C. and Senthil K. P.S. (2011). Map of Egypt: European digital archive on soil maps (EuDASM), Preserving important soil data for public free access. International Journal of Digital Earth, 4 (5), pp. 434-443. http://eusoils.jrc.ec.europa.eu/esdb_archive/EuDASM/africa/lists/k5_ceg.htm

Penman, H.L. (1948). Natural evaporation from open water, bare soil and grass. In: Royal Society of London, Series A, Mathematical and Physical Sciences, 22 April 1948, Vol. 193, No. 1032, pp. 120-145
http://faculty.eas.ualberta.ca/jdwilson/EAS372_13/Penman48.pdf.

Priestley, C. H. B. and Taylor R. J. (1972). On the Assessment of Surface Heat Flux and Evaporation Using Large-Scale Parameters. Monthly Weather Review, 100 (2), pp. 81–92.

OWARA (2008): Sustainable Rural Infrastructure Management in the Lake Nasser Region, Upper Egypt. Project report, Optimizing Water Allocation in Rural Areas of Lake Nasser Project (OWARA), Institute for Technologies in the Tropics, University of Applied Sciences, Cologne, Germany.
http://www.tt.fh-koeln.de/owara/Templates/10sideactivities.htm.

Ramey, V. (2004). Plant management in Florida waters. Report, Center for Aquatic and Invasive Plants, University of Florida–IFAS, Florida, USA.
http://plants.ifas.ufl.edu/guide/evaptran.html.

Rayner, N. A. (2003). Global analyses of sea surface temperature, sea ice, and night marine air temperature since the late nineteenth century. Journal of geophysical research, 108 (D14).

Rekacewicz, P., Digout, D. and UNEP/GRID-Arendal (2005). Major river basins of Africa. Map, UNEP/GRID-Arendal.
http://www.grida.no/graphicslib/detail/major-river-basins-of-africa_1ac3.

Rich, J. (2004). Integrated mass, solute, isotopic & thermal balances of a coastal wetland wetland research at perry lakes, western australia 1993-1998: Chapter 5. Water balance components. PhD thesis, Murdoch University, Perth, Western Australia.
http://perrylakes.info/wp-content/uploads/2009/11/Chapter-5.pdf.

Robert J. Z, Trabucco, A., van Straaten, O. and Bossio, D. A. (2006). Carbon, land and water A global analysis of the hydrologic dimensions of climate change mitigation through afforestation/reforestation. Research report, 101, International Water Management Institute, Colombo, Sri Lanka.

Robert, S., David, B. and Neil, C. (2007). Tree windbreaks in the wheatbelt. Report, Department of agriculture and food, State of Western Australia, Australia.

Rohwer, C. (1931). Evaporation from Free Water Surfaces. Technical Bulletin, Associate Irrigation Engineert Bureau of Agricultural Engineering, United States Department Of Agriculture, Washington, D. C, USA.

Quaider, L., Schlüter, S. and Sturm, M. (2008). Sustainable Rural Infrastructure Management in Lake Nasser Region- Upper Egypt. Report, Institute for Technologies in the Tropics, University of Applied Sciences, Cologne, Germany.
http://www.tt.fh-koeln.de/owara/Templates/10sideactivities.htm.

Said, R. (1962). The geology of Egypt. El-Sevier Publishing Company, Amesterdam, The Netherlands.

Said, R. (1993). The river Nile Geology, hydrology, and utilization, Pergamon, Oxford [England], New York.

SEKEM (2010). The SEKEM story: economics of love. the Egyptian pharmacologist and social entrepreneur Dr I.Abouleish.
http://farmhub.textileexchange.org/upload/library/Case%20studies/SEKEM%20Case%20Study.pdf.

Schleiss, A. J., Cesare, G. and Althaus, J. (2008). Reservoir sedimentation and sustainable development. In: International Workshop in Erosion, Transport and Deposition of Sediments, 28-30 April 2008, Berne, Switzerland, pp. 23–28.

Sciortino, J. A. (2010). Fishing harbour planning, construction and management. FAO fisheries and aquaculture technical paper, 539, Food and Agriculture Organization of the United Nations, Rome, Italy.

Shalash, S. (1980), Effect of sedimentation on storage capacity of high Aswan Dam Lake, Paper No. 56, El-Qanater, Egypt.

Shalash, S. (1982), Effects of sedimentation on the storage capacity of the High Aswan Dam reservoir, Hydrobiologia, 92(1), 623-639.

Shaltout, M. A. and El Housry, T. (1997). Estimating the evaporation over Nasser Lake in the upper Egypt from Meteosat observations. Elsevier, Advances in space research, 19 (3), 515–518.

Sinha, S., Kumar, L., Srivastava, R. and Thangamani, R. (2006). Evaporation Control in Reservoirs. Report, Government of India central water commission basin planning and management organisation, India, New Delhi.

Slingo, J. (2011). Climate: Observations, projections and impacts Egypt, Report, Met Office, United Kingdom.

Sokolov, A. A. and Chapman, T. G. (1974). Methods for water balance computations: An international guide for research and practice. The Unesco Press, Paris, France. A contribution to the International Hydrological Decade. ISBN 92-3-10 1227-4

Soliman, W. (2002). Environmental study on water quality assessment and prediction in lake Nasser by using monitoring networks. PhD, Cairo University, Cairo, Egypt.

Sørensen, L. S., Simonsen, S. B., Nielsen, K., Lucas-Picher, P., Spada, G., Adalgeirsdottir, G., Forsberg, R. and Hvidberg, C. S. (2011). Mass balance of the Greenland ice sheet (2003–2008) from ICES at data: the impact of interpolation, sampling and firn density. The Cryosphere, 5 (1), 173–186.

Spalding-Fecher, R., Yamba, F., Walimwipi, H., Kling, H., Tembo, B., Nyambe, I., Chapman, A. and Cuamba, B. (2014). Water Supply and Demand Scenarios for the Zambezi River Basin: Climate Change and Upstream Development Impacts on New Hydropower Projects in the Zambezi Project. Report for Climate & Development Knowledge Network. University of Cape Town, Cape Town, South Africa.
www.erc.uct.ac.za/research/hydro-zambezi/hz-water%20supply%20and%20demand%20scenarios%20report.pdf

Springuel, I. O. A. (2005). The River Nile Basin. In: The world's largest wetlands ecology and conservation, Cambridge University Press, Cambridge, New York.

Strand, R. I. and Pemberton, E. L. (1982). Reservoir sedimentation: Technical guidelines for bureau of reclamation. US Department of the Interior, Bureau of Reclamation, Sedimentation and river hydraulics section, hydrology branch, division of planning technical services, engineering and research center, Denver, Colorado.
www.uobr.gov/pmts/sediment/projects/reservoirsurveys/reports/reservoirsedimentationtechguide10_1982.pdf

Stroud, P.L. (2012). Sediment strategies: Cost analysis of sediment removal techniques from reservoir. Hydro review.
http://www.hydroworld.com/articles/hr/print/volume-31/issue-

07/articles/sediment-strategies-choosing-a-sediment-management-option-for-dam-removal.html.

Strzepek, K. M., Onyeji, C., Saleh, M. and Yates D. (1994). An assessment of integrated climate change impacts on Egypt Working Paper, International Institute for Applied Systems Analysis (IIASA), Laxenburg, Austria.

Strzepek, K. M., Yates, D. N. and El Quosy, D. E. (1996). Vulnerability assessment of water resources in Egypt to climatic change in the Nile Basin. Climate research, 6, 89–95.

Stuck, K. (2010). ECC Hollow Plastic Balls: Not just a ball but a solution. Technical documents.
http://www.eccllc.us/pdf/ECCBrochureWaterWastewater.pdf.

Sumi, T. and Hirose T. (2011). Accumulation of sediments in reservoirs: Water storage, Transport, and Distribution. UNESCO publications, Department of civil engineering, Kyoto University, Kyoto, Japan.
http://www.eolss.net/Sample-Chapters/C07/E2-12-02-05.pdf.

Sutcliffe, J.V. and Parks, Y.P. (1999). The Hydrology of the Nile. IAHS Special Publication no. 5, the International Association of Hydrological Sciences (IAHS), IAHS Press, Institute of Hydrology, Wallingford, Oxfordshire OX10 8BB, UK, ISBN 1-910502-75-9.
http://iahs.info/bluebooks/SP005/BB_005.pdf.

TechSpan (2007). Super span water conservation: Water Conservation. Online documentation.
http://www.superspan.com.au/.

Tigrek, S. and Aras, T. (2012). Reservoir sediment management, CRC Press/Balkema, London.

UN (2011). World Population Prospects the 2010 Revision: Population Division of the United Nations. Department of Economic and Social Affairs.
http://data.un.org/Data.aspx?q=egypt&d=PopDiv&f=variableID%3a12%3bcrID%3a818.

UN (2013). World Statistics Pocket book: world of information data.
http://data.un.org/Search.aspx?q=egypt.

UNESCO (1993). Biosphere Reserve Information: Egypt - WADI ALLAQUI. UNESCO publications - MAB Biosphere reserves.
http://www.unesco.org/mabdb/br/brdir/directory/biores.asp?code=EGY+02&mode=all.

UNESCO (2008). International Hydrological Programme International Sediment Initiative: Sediment in the Nile River System. UNESCO publications, UNESCO-IHP-International Sediment Initiative.
http://www.irtces.org/isi/isi_document/sediment%20in%20the%20nile%20river%20system.pdf.

UNFCCC (1999). Egyptian environmental affairs agency. The Arab Republic of Egypt. Initial national communication on climate change.

USGS (2013). Landsat Project: Online Documentation. http://landsat.usgs.gov/tools_project_documents.php.

Vente, J., Poesen, J., Verstraeten and G. (2004). Evaluation of reservoir sedimentation as a methodology for sediment yield assessment in the Mediterranean: challenges and limitations. In: Soil Conservation and Protection for Europa (SCAPE) workshop, 13-15 April 2004, Cinque Terre, Italy, pp. 139–145.

Wahby, W. S. (2004). Technologies Applied in the Toshka Project of Egypt. The Journal of Technology Studies 30 (4), pp. 86–91.
http://files.eric.ed.gov/fulltext/EJ905156.pdf.

Wang, H., Takle, E. S. and Shen, J. (2001). Shelterbelts and windbreaks: mathematical modeling and computer simulations of turbulent flows. Annual Review of Fluid Mechanics, 33 (1), 549–586.

Wang, J., Sammis, T. W. and Gutschick, V. P. (2008). A remote sensing model estimating water body evaporation. In: International Workshop on Earth Observation and Remote Sensing Applications (EORSA 2008), 30 June to 2 July 2008, Beijing, China. IEEE.

Warner, Jeroen (2011). Flood planning. The politics of water security. London: I.B. Tauris (International library of political studies, 30. ISBN 9781845118174

Wen Shen, H. (1999). Flushing sediment through reservoirs. Journal of Hydraulic Research, 37 (6), 743–757.

WMO (2010). Guide to climatological practices: Third edition, World Meteorological Organization (WMO).

Woodward, J. C., Macklin, M. G., Krom, M. D., Williams, M. A. J. (2007). Chapter 13, The Nile: Evolution, Quaternary River Environments and Material Fluxes. In: Large rivers. Geomorphology and management. Chichester, England, Hoboken, NJ. John Wiley, pp. 261–292.
http://www.researchgate.net/profile/Mark_Macklin/publication/227992119_The_Nile_Evolution_Quaternary_River_Environments_and_Material_Fluxes/links/0fcfd5069887bcc176000000.pdf.

WPDC (2010). Reducing evaporation losses. International Water Power & Dam Construction publications, UK.
http://www.waterpowermagazine.com/features/featurereducing-evaporation-losses/featurereducing-evaporation-losses-2.html

Wullschleger, S., Meinzer, F. and Vertessy, R. (1998). A review of whole-plant water use studies in trees. Tree Physiology, 18, 499-512.

Xiaoqing, Y. (2003). Manual on sediment management and measurement Operational hydrology report no. 47. WMO, 948, Secretariat of the World Meteorological Organization, Geneva, Switzerland.

Yang, C. T. (2006): Erosion and Sedimentation Manual. Sedimentation and River Hydraulics Group, Technical Service Center in Denver, USA. Hydraulic engineers of the Bureau of Reclamation, U.S. Department of the Interior.
http://www.usbr.gov/pmts/sediment/kb/ErosionAndSedimentation/Contents.pdf.

Yao. X, Zhang, H., Lemckert, C., Brook, A. and Schouten P. (2010). Evaporation reduction by suspended and floating covers: overview, modelling and efficiency Urban Water Security Research Alliance. Technical report No. 28, Description: Logan's Dam, Gatton Queensland, Australia, Urban water security research alliance, The Urban Water Security Research Alliance.

Zaghloul, S.S., Pacini N., Schwaiger, K. and Henry de Villeneuve, P. (2011). Towards a Lake Nasser management plan: results of a pilot test on integrated water resources management. International Water Technology Journal (IWTJ), 1(3), 249- 258.

Zyl, W.V. and Jager, J.D. (1987). Use of evaporimeters for estimating maximum total evaporation. Water SA, 13(4), 235-240.

Zwieten, P.A.M., Béné, C., Kolding, J., Brummett, R., Valbo-Jørgensen, J. (2011). Review of tropical reservoirs and their fisheries: The cases of Lake Nasser, Lake Volta and Indo-Gangetic Basin reservoirs. Technical Paper, No. 557, FAO Fisheries and Aquaculture Food and Agriculture Organization of the United Nations (FAO), Rome, Italy.
http://www.fao.org/docrep/015/i1969e/i1969e.pdf, checked on 11/29/2013.

Appendix A

Historical Nile River Basin Agreements (IWLP, 2008, NBI, 2012)

Protocol between Britain and Italy in 1891: The Protocol between Great Britain and Italy in 15 April 1891 (on behalf of Egypt and Ethiopia, respectively), for the demarcation of their respective spheres of influence in Eastern Africa. In its third article, the Protocol stipulates that the Italian Government engages not to construct, on the Atbara, in view of irrigation, any work which might sensibly modify its flow into the Nile.

Agreement of Brussels signed in 1894: Agreement between Great Britain and the Independent state of the Congo signed on 12 May 1894.

Treaty between United Kingdom and Ethiopia, and between United Kingdom, Italy, and Ethiopia, relative to the frontiers between the Soudan, Ethiopia, and Eretria signed at Addis Ababa on 15 May 1902: In the third article of the treaty, the Emperor of Ethiopia engages himself towards Great Britain not to construct, or allow to be constructed any work across the Blue Nile, Lake Tana, or the Sobat, which would arrest the flow of their waters into the Nile, except in agreement with the Great Britain Government and the Sudan Government.

Agreement between Britain and Congo in 1906 (Modifying the Agreement of Brussels signed in 1894): Agreement between Great Britain and the Independent state of the Congo on 9 May 1906 bringing modification to the Brussels Agreement of 12 May 1894. In its third article, the Government of the Independent state of the Congo undertakes not to construct or allow to be constructed, any work on or near the Semliki or Isango Rivers, which would diminish the volume of waters entering Lake Albert, except in agreement with the Sudanese Government.

Agreement between Britain, Italy, and Ethiopia in 1906: Concerning the territorial integrity of Ethiopia and stating that "all the previous agreements, protocols and conventions signed before will be respected." In article IV it is stated that these three countries "will unite to safeguard the interests of Egypt and Great Britain in the Nile Basin, especially in regard to the regulation of the waters of that river and its tributaries".

Exchange of Notes Between the United Kingdom and Italy Respecting Concessions for a Barrage at Lake Tana and a Railway Across Abyssinia From Eritrea To Italian Somaliland, signed in Rome on 14 and 20 December 1925: It is stated that the Italian Government recognized the prior acquired "hydraulic rights" of Egypt and the Sudan, and undertook not to construct on the head-waters of the Blue, and White Niles or their tributaries or effluents any work which might sensibly modify their flow into the Main River. it stated also that the water level of Lake Tana is not to exceed the maximum level attained during the rainy season.

Exchange of notes between His Majesty's Government in the United Kingdom (on behalf of the Sudan, Kenya, Tanganyika and Uganda) and the Egyptian Government in regard to the use of the waters of the River Nile for irrigation purposes. Signed on 7 May 1929; in force 7 May 1929: It is stated that the irrigation works on the Nile will follow those dispositions:

a) The General Inspector of the Egyptian Irrigation Service in Sudan will collaborate with the resident engineer in Sennar Dam to measure the flow and the records to assure Egyptian Government that the regulation of that dam are made in respect of the agreement.

b) No irrigation or hydro-electric projects will be built on the Nile, or its tributaries or the lakes that feed them, located in Sudan or in the countries under British administration if these works could affect the amount of waters reaching Egypt or their timing, or reducing its level in any way could have harmful effects on the Egyptian interests.

c) Egypt will enjoy all the facilities required to study and record the water of the Nile along all of its course.

d) If Egypt decides to undertake works in Sudan on the Nile or its tributaries or take measures to increase its yield, it will consult beforehand the local authorities, in view to safeguard their interests. The construction, upkeep and administration of the projects will be Egypt's responsibility.

e) The British Government will use its good offices with the governments under British influence to facilitate the execution of the works mentioned above, it is understood that if in the process of things some uncertainties might arise as to the interpretation of a principle on technical or management details, each problem will be examined in goodwill. In the case of one of the parties going against the clauses of the present agreement and if the two interested governments don't reach a solution to that problem, the case will be referred to an independent organism for arbitration.

f) The British Government recognizes Egypt's natural and historical rights over the waters of the Nile "The government of his majesty considers the respect of those rights as a fundamental principle of Great Britain's policy and affirms that this agreement will be observed at all times and under any circumstances."

Convention between Britain and Belgium in 1934. Agreement between Great Britain (on behalf of Tanganyika) and Belgium (on behalf of Rwanda and Burundi) signed in London on 22 November 1934, concerning Kagera River, one of the tributaries of Lake Victoria: The first article stipulates that the contracting parties pledge to return to the Kagera River, before it reaches the common borders of Tanganyika, Rwanda and Burundi, any quantity of water used for hydro-electric purposes. Article six states that any country planning to use the waters of the river for irrigation will notify the contracting countries six months in advance to study the projects and present any objections.

Exchange of notes constituting an agreement between the Government of the United Kingdom of Great Britain and Northern Ireland and the Government of Ethiopia amending the description of the Kenya-Ethiopia boundary, signed at Addis Ababa 29 September 1949: In force 29 September 1949.

Exchange of memoranda between Egypt and Great Britain (on behalf of Uganda) in 1949 – 1953: Exchange of memoranda between Egypt and Great Britain (on behalf of Uganda) in the period 1949 to 1953, signed on 31 May, 5 December 1949, 19 January,28 February and 20 March 1950, 16 July 1952 and 5 January 1953, regarding Egypt's participation in the construction of the Owen Falls Dam for the generation of hydro-electric power in Uganda and the control of the waters of the Nile. They included the following points:

a) Memoranda between Egypt and United Kingdom were exchanged since 1949 and it has been agreed that those memoranda between the two governments are considered to be an official agreement.

b) The erection and operation of the dam will not decrease or change the amount and timing of the amount of water reaching Egypt.

c) The government of Uganda is welcome – if the government of Egypt wishes – an Egyptian irrigation engineer to supervise the above-mentioned point.

d) Egypt and Uganda to co-operate in building the dam to generate electricity for the benefit of Uganda as well as for the sake of irrigation for Egypt.
e) The government of Egypt will contribute in the expenses of the construction of the dam.
f) The government of Uganda handles the regulation of the flow from the dam according to the instructions from the Egyptian resident Engineer.
g) The government of Egypt pays the expenses for heightening of the dam 3 meters.
h) The government of Egypt pays to Uganda the amount of L 980,000 as substitute for the electrical energy loss.

Agreement (with Annexes) between the United Arab Republic and the Republic of Sudan for the full utilization of the Nile waters. Signed at Cairo, on 8 November 1959; Came into force on 22 November 1959. Protocol to the Agreement between the United Arab Republic and the Republic of Sudan for the full utilization of the Nile waters concerning the Establishment of the Permanent Joint Technical Committee, signed at Cairo 17 January 1960; in force 17 January 1960: This agreement is considered to be supplementary for the 1929 agreement and not to cancel it. Whereas the Nile Waters Agreement concluded in 1929 has only regulated a partial use of the natural river and did not cover the future conditions of a fully controlled river supply, the two Republics have agreed to the following:

I- The present established rights

1. The quantities of water actually used by the United Arab Republic until the date of signing this Agreement constitute their established right prior to the benefit accruing them through the implementation of the control works referred to in this agreement. This established right amounts to 48 milliards of cubic meters per year measured at Aswan.

2. The quantities of water used at present by the Republic of the Sudan constitute their established right prior to the benefits accruing to them through the implementation of the afore mentioned control works. This established right amounts to 4 milliards cubic meters per year as at Aswan.

II- Nile control works and the sharing of their benefit between the Republics

1. In order to make use of the full natural river supply and stop the flow of any excess of the sea, the two Republics agree to the construction by the U.A.R. of the Sudd el Aali Reservoir at Aswan as the first of a series of over-year storage schemes on the Nile.

2. In order to enable the Republic of Sudan to exploit their share, the two Republics agree to the construction by the Sudan Republic of the Roseires Reservoir on the Blue Nile and any other works deemed necessary by the Sudan for the same purpose.

3. The net benefit from the Sudd el Aali Reservoir shall be calculated on the basis of the mean natural river supply at Aswan in the past years of this century and which amounts to 84 milliards of cubic meters per year. The established rights of the two Republics referred to in Article I, as well as the mean value of the over-years storage yearly losses in the Sudd el Aali Reservoir, shall be deducted from the above mentioned mean natural river, in order to obtain the net yearly benefit to be shared by the two Republics

4. The net benefit from the Sudd el Aali Reservoir referred to in the previous paragraph, shall be allotted between the two Republics at the ratio of 14.5 for Sudan to 7.5 for the United Arab Republic as long as the mean natural river supply remains within the limiting value mentioned in the previous paragraph. This means that as long as the computed mean natural river supply is equal to 84 milliards of cubic meters per year, and the mean value of the over year storage losses remain equal to its present estimated value of 10

milliards of cubic meters per year, then the net benefit from the Sudd el Aali reservoir is 22 milliards of cubic meters of which 14.5 milliards shall be allotted to the Republic of Sudan and 7.5 milliards to the United Arab Republic. By adding these benefits to the respective established rights, the total shares in the net mean natural supply after the working of the complete Sudd el Aali Reservoir, shall be 18.5 milliards per year for the Republic of Sudan, and 55.5 milliards per year for the United Arab Republic. If the mean natural river exceeds 84 milliards per year, then the resulting increase in the net benefit due to the increase in the mean natural river shall be equally divided between the two Republics.

5. As the net benefit from the Sudd el Aali Reservoir, referred to in paragraph (3) article II, is calculated by deducting the established rights and the mean over-year storage yearly losses, from the mean natural river supply of the past years to the present century, it is recognized that this net benefit shall be subject to revision by both parties at reasonable intervals to be agreed upon as from the date of the operation of the complete Sudd el Aali Reservoir.

6. The Government of the United Arab Republic agrees to the payment of fifteen million Egyptian pounds to the Government of the Republic of Sudan as full compensation for the damages to present Sudanese property, resulting from the storage of water in the Sudd el Aali Reservoir to a level of 182.00 meters (Survey). Such payment shall be affected as agreed upon by both parties in the Annex attached thereto.

7. The Government of the Republic of Sudan undertake to take steps to transfer the population round Halfa as well as all other Sudanese inhabitants - whose properties will be affected by the maximum storage in the Sudd el Aali Reservoir - prior to July 1963.

8. It is recognized that after the working of the complete Sudd el Aali Reservoir for over-year storage, the United Arab Republic will not require the use of Gebel Aulia Reservoir for storage. The two contracting parties shall examine all matters related to such renunciation in due time.

III - Projects for the exploitation of waters lost in the Upper Nile Basin

In view of the fact that quantities of the Nile Basin waters are wasted in the swamps of Bahr el-Zeraf, Bahr el Ghazal, River Sobat, and the conservation of these waters for increasing the present natural river supply is most vital for the future agricultural developments, the United Arab Republic and the Republic of Sudan agree to the following:

1. In agreement with the United Arab Republic, the Republic of Sudan shall carry out projects for increasing the River Nile water supply by the prevention of excess losses in the swamps of Bahr el Gebel, Bahr el Zeraf, Bahr el Gahzal and its branches, River Sobat and its branches and the White Nile. The water benefit from such projects as well as the total costs of construction shall be shared equally by the two Republics. The Republic of the Sudan shall defray the costs of the above mentioned projects and shall be reimbursed by the United Arab Republic on the basis of half the profits designated in these projects.

2. In case the United Arab republic need more water to cope with their progress in the agricultural expansion program, and therefore find it necessary to take the necessary steps to carry out one of the above mentioned schemes at a time when the need of the republic of Sudan might not have arisen, the United Arab Republic will notify the Republic of Sudan of the date on which the former intend to start the execution and in the course of two years from the date of such notification, each of the two Republics shall submit their program of expansion and the dates and quantities of their water requirements from the benefit of the scheme. Any such program shall be binding to both parties. At the expira-

tion of the two years, the United Arab Republic shall start the execution of the project at its own expense. When the Republic of the Sudan is ready to make use of its share according to the agreed program, they shall then reimburse to the United Arab Republic their share in the cost, in the same proportion to the total cost as their share in the benefit is to the total actual benefit of the scheme. The final share of either party shall not exceed 50% of the total benefit.

IV- Technical Co-operation between the two Republics

1. To insure technical co-operation between the two Republics to carry out the necessary study and research in connection with projects for the Nile Control and the increase of its supply and for the continuation of Hydrological survey work of the River in its upper reaches, the two Republics agree to constitute a Permanent Joint Technical Committee composed of an equal number of members from both Republics. This committee shall be formed after signing this agreement and shall have the following terms of reference:

    a. To draw the main lines of schemes aiming at the increase of the River supply and to supervise and direct the research work and investigations and collection of data necessary for the preparation of projects reports to be submitted to both Government for approval.

    b. To supervise the execution of the approved projects.

    c. To draw up the working arrangements for works implemented in territories outside the Sudan by agreement with their concerned authorities.

    d. To supervise the application of all aforesaid working arrangements in article (c) by means of engineers appointed for this purpose and selected from officials from the two Republics in connection with works in the Sudan and also the Sudd el Aali and Aswan Reservoir and according to agreements with other governments, in connection with works outside the Sudan.

    e. In view of the possibility of the occurrence of a series of years of low river supply causing a continuous drop in the Sudd el Aali Reservoir levels to the stage that will not enable both Republics to draw their normal quota in any year, the Committee shall put up the necessary arrangements to be followed by both parties to face the shortage of supply in such low years in a manner that will not cause any damage to either party and shall submit their proposals for approval by both Republics.

2. To enable the Committee to carry out duties referred to in paragraph 1 above, and to insure the continuation of the observation of gauges and discharges of the River in all it upper reaches, these duties shall be carried out under the supervision of the Committee within the technical field by the engineers of the Republic of Sudan and the staff of the United Arab Republic in the Sudan, and in Uganda.

3. The two Republics shall issue a joint order covering the formation of the Permanent Joint Technical Committee, the names of its members, and the necessary budget to be provided from the funds of both Republics. The Committee shall meet either in Cairo or in Khartoum according to circumstances and shall establish its own rules of procedure subject to the approval of the two Governments and which shall include the necessary regulations in connection with meetings, technical, administrative and financial activities.

V- General Provisions

1. In case any question connected with Nile water needs negotiations with the governments of any riparian territories outside the Republic of Sudan and the United Arab Republic, the two Republics shall agree beforehand on a unified view in accordance with the inves-

tigations of the problem by the Committee. This unified view shall then form the basis of instructions to be followed by the Committee in the negotiations with the governments concerned. Should such negotiations result in an agreement to construct works on the Nile in territories outside the two Republics, the Permanent Joint Committee shall then assume the responsibility to contact the concerned authorities in those territories, in order to lay down all the technical details in connection with the execution as well as the Working Arrangements and maintenance of the works in question. After agreement on these points with the governments concerned, the Committee shall supervise the execution of the technical provisions of such agreements.

2. Since other riparian countries on the Nile besides the Republic of Sudan and the United Arab Republic claim a share in the Nile waters, both Republics agree to study together these claims and adopt a unified view thereon. If such studies result in the possibility of allotting an amount of the Nile water to one or the other of these territories, then the value of this amount as at Aswan shall be deducted in equal shares from the share of each of the two Republics. The Permanent Joint Technical Committee shall make arrangements with the concerned authorities in other territories in connection with the control and checking of the agreed amounts of Nile water consumption.

VI- Transition period before the working of the complete Sudd el Aali

Whereas both Republics shall benefit from their respective shares in the net benefit of the Sudd el Aali Reservoir only when the latter shall be complete and shall yield its benefit, both parties shall agree on their interim program of expansion in the transition period - from now until the working of the complete Sudd el Aali - in a manner that shall not affect their present water requirements.

a) At the time of the agreement, 48 and 4 milliards of cubic meters per year estimated at Aswan were considered as acquire rights for Egypt and Sudan, respectively.

b) The approval for the High Aswan Dam erection in Egypt.

c) The approval for the Roseires dam erection in the Sudan on Blue Nile and any other works to exploit her share.

d) On the basis of the average annual natural flow at Aswan being estimated at 84 milliards; the benefit from the High Aswan Dam amounted to 22 km^3 annually will be distributed as 14.5 and 7.5 for the Sudan and Egypt, respectively.

e) The execution of the Upper Nile projects, namely Bahr El Jebel, Bahr El Zeraf, Bahr El Gazal, River Sobat are ready to start. The net benefit estimated at Aswan as well as the costs will be shared equally between the two countries.

f) As the riparian states, other than the two republics, claim a share in the Nile waters, the two republics have agreed that they will jointly consider and reach one unified view regarding the said claim. And if the said consideration results in the acceptance allotting an amount of the Nile waters to one or the other of the said states, the accepted amount shall be deducted from the shares of the two republics in equal parts, as calculated at Aswan.

g) The Technical Commission mentioned in this agreement shall make the necessary arrangements with the states concerned, in order to ensure that their water consumption shall not exceed the amounts agreed upon.

h) To establish the Permanent Joint Technical Commission for Nile Waters (PJTC) to control the river and conduct all hydrological studies necessary for the increase of the Nile Yield. The operation of the gauging station covering the Main Nile and its tributaries is conducted by the Egyptian Irrigation Department in the Sudan and the Hydrological Division of the Sudan Ministry of Irrigation. All data collected pours into the secretariat office of the Commission in Khartoum for analyses and studies.

i) The 1959 Nile Water Agreement stipulated the necessity for realization of the technical co-operation in the problems of the Nile waters. To assure this spirit of co-operation, at all times, a joint technical commission had to be established. Each country is represented on this commission by an equal number of their experts in the Nile matters. It is very satisfying that now the activities of the Commission extended to participate in the Hydromet Survey of the catchments of Lakes Victoria, Kioga and Albert with the assistance of the United Nation Special Fund. The participation of the concerned governments in this important project is a manifestation of international co-operation. It is realized that technical co-operation and co-ordination of studies and collection of basic data in this great river is vital to all the Nile countries.

Exchange of memoranda between Egypt and Uganda in 1991. Exchange of memoranda between the Egyptian Government and the Government of Uganda signed in Cairo on 12th May, 1991: Regarding the construction of the extension of Hydro-Electric Station on Owen Falls Dam in Uganda. It is agreed that the Owen Falls extension project can go forward, subject to and taking into consideration the following:

a) The pattern of the out-flow from lake Victoria shall be in accordance with the present operating policy of the agreed curve. This will take into account the full 90 years-hydrological regime.

b) With the extension of Owen Falls Power project, Uganda will respect the hitherto operational storage reserve policy "as much as full 3 meters range" which was agreed upon at the time of the construction of the Owen Falls Dam (1953).

c) In order to enable Uganda to benefit from the available water storage and increased flow in the Nile, Lake Victoria could be regulated whenever needed by Uganda, on the understanding that such regulation policies should be discussed and reviewed by the two parties, within safe limits not adversely affecting the water needs of the downstream countries.

d) The average discharge in the period 1900-1990 as determined by Uganda's consultants was about 800 m3/sec. Uganda's decision to design the Owen Falls extension with a discharge of 1200 m3/sec is based on expected sustainable high natural discharge values experienced since 1960.

e) However, the discharge utilized for power should be consistent with the natural flow mentioned in Para (1) and (3).

Agreement between Egypt and Uganda for controlling water hyacinth in 1998: In the framework of the interest of both the two countries to reinforce and enhance the ties of friendship and fraternity of the Nile Riparian and properly manage the Upper Nile water catchments, the two countries, Egypt and Uganda agreed in Cairo on 22 March, 1998 that Egypt makes available an amount of 13.9 million US $, to support, Uganda on grant basis. The grant is allocated to assist the Government of Uganda, in combating and controlling the aquatic weeds, specially the water hyacinth in the outlets and inlets of lakes Victoria, Kioga, Albert and in the Nile, through purchasing and delivering suitable equipment and machinery to the sites of concern.

Agreement on the Nile River Basin Cooperative Framework, opened for signature on 14 May 2010: in Entebbe, Uganda (not in force), as Egypt and Sudan did not agree to sign it. Article 3 stated the general principles. The Nile River System and its waters shall be protected, used, conserved and developed in accordance with the following general principles.

a) Cooperation: The principle of cooperation between States of the Nile River Basin on the basis of sovereign equality, territorial integrity, mutual benefit and good faith in order to attain optimal utilization and adequate protection and conservation of the Nile River Basin and to promote joint efforts to achieve social and economic development.

b) Sustainable development: The principle of sustainable development of the Nile River Basin.

c) Subsidiarity: The principle of subsidiarity, whereby development and protection of the Nile River Basin water resources is planned and implemented at the lowest appropriate level.

d) Equitable and reasonable utilization: The principle of equitable and reasonable utilization of the waters of the Nile River System.

e) Prevention of the causing of significant harm: The principle of preventing the causing of significant harm to other States of the Nile River Basin.

f) The right of Nile Basin States to use water within their territories: The principle that each Nile Basin State has the right to use, within its territory, the waters of the Nile River System in a manner that is consistent with the other basic principles referred to herein.

g) Protection and conservation: The principle that Nile Basin States take all appropriate measures, individually and, where appropriate, jointly, for the protection and conservation of the Nile River Basin and its ecosystems.

h) Information concerning planned measures: The principle that the Nile Basin States exchange information on planned measures through the Nile River Basin Commission.

i) Community of interest: The principle of the community of interest of the Nile Basin

j) States in the Nile River System.

k) Exchange of data and information: The principle of the regular and reciprocal exchange among States of the Nile River Basin of readily available and relevant data and information on existing measures and on the condition of water resources of the Basin, where possible in a form that facilitates its utilization by the States to which it is communicated.

l) Environmental impact assessment and audits: The principle of environmental impact assessment and audits.

m) Peaceful resolution of disputes: The principle of the peaceful resolution of disputes.

n) Water as a finite and vulnerable resource: The principle that fresh water is a finite and vulnerable resource, essential to sustain life, development and the environment, and must be managed in an integrated and holistic manner, linking social and economic development with protection and conservation of natural ecosystems.

o) Water has social and economic value: The principle that water is a natural resource having social and economic value, whose utilization should give priority to its most economic use, taking into account the satisfaction of basic human needs and the safeguarding of ecosystems.

p) Water security: The principle of water security for all Nile Basin States

Appendix B

Meteorological Stations along the High Aswan Dam Reservoir (NWS, 2012)

Aswan Shore station.

The Aswan Shore station was constructed at the end of 1986 on the North-West bank of the lake, one kilometer from the HAD and about 50 m from the water bed when the lake level is at 182 m AMSL. It exists at latitude 23° 59`N and longitude 32° 48`E with an elevation of 192 m AMSL. The meteorological enclosure contains the instruments of aspirated dry and wet-bulb temperatures, maximum and minimum temperatures, a thermo-hygrograph, an electric anemograph at 10 m height, two cup anemometers at two and four meters height, and a Class A pan evaporimeter. Hourly observations during the period from 0600 to 1800 UT are recorded for dry and wet-bulb temperatures, relative humidity, wind speed and direction. Daily observations for the extreme temperatures are recorded at 0600 and 1800 UT. Daily wind run at the two heights and pan evaporation are measured at 0600 UT.

The HADR Shore Station

Aswan raft station.

This station was in operation for 14 years, from 1977 to 1990. It was located at one kilometer upstream the HAD at latitude 23°58`N and longitude 31°51`E. The average cross section of the lake at the raft site was about six km, and the raft was fixed approximately in the center. Water depth at this fairly open site was about 24 m. The raft is a boat-anchorage with a platform of 6.5 m by 14.5 m, rising above the water surface by 0.6 m. Mechanical recordings and visual instruments were fixed above the platform to be at two and four meters heights above the water surface. The measured parameters were air temperature, relative humidity, water surface temperature and wind speed. Observations of air temperature and relative humidity were taken daily every three hours during the period between 0900 and 1800 UT in order to correct the daily records of air temperature and relative humidity. Daily wind run at two and four meters heights and daily maximum and minimum air temperatures above the water surface were recorded at 0600 UT. Also, water temperatures at 0.20 m and 1.0 m depths were recorded.

Old Aswan Raft Station (1977-1990)

The conventional observing system stopped in 1990 and resumed with a new automatic system in 1995 at a new location, about two km south of the HAD. The automatic system records hourly data of maximum, minimum and mean air temperatures, relative humidity, water temperature at the surface level as well as at different depths, wind speed and direction, wind run and peak gust, and atmospheric air pressure. The sensors were fixed on a mast at two and four meters heights above the water surface. The system is operated by rechargeable batteries through solar panels.

Aswan airport station

The Aswan airport station is situated about four kilometers to the west of the HAD at an elevation of about 193 m AMSL, latitude 23°8`N and longitude 32° 47`E. The meteorological enclosure contains the instruments of dry and wet-bulb temperatures, maximum and minimum temperatures, mechanical recording thermo-graph, a Dines anemograph, a mercury barometer, mechanical barograph, and a piche evaporimeter. Hourly observations during 24 hours are taken for dry and wet-bulb temperatures, relative humidity, wind speed and direction and atmospheric air pressure. The extreme temperatures are measured daily at 0600 and 1800 Universal Time (UT) and piche evaporation at 0600 UT.

New Aswan Raft Station (1995- now)

New Aswan Raft Station automatic system (1995- now)

North El-Alaky automatic recording buoy station

This station was established in 1995. It started to operate in 1998 at 75 km upstream the HAD. It is exposed north of khor El-Alaky at latitude 23°19`N and longitude 32°52`E. The average cross section of the HADR at the buoy site is about 10 km, and the buoy lies at about three km from the eastern shore and seven km from the western shore. Average water depth at this open site is about 70 m. Measuring sensors for air temperature, relative humidity, wind speed and direction are fixed on a mast above the buoy at two and four meters above the water surface. The pressure sensor is fixed inside the buoy. Water temperatures at the surface level and at different depths are also measured. The system is operated by rechargeable batteries with solar panels. It records hourly data of maximum, minimum and mean air temperatures, relative humidity, water temperatures, wind speed and direction, wind run and peak gust, and atmospheric air pressure.

North El-Alaky Automatic Station

Toshka station

Toshka station is a floating telemetric meteorological station close to the shore. It is located at 280 km upstream the HAD. The available raw data of the station are hourly wind speed, wind direction, average air temperature, minimum air temperature, maximum air temperature, average relative humidity, maximum relative humidity, minimum relative humidity, and average net radiation.

Toshka Meteorological Station

Khor-Kalabsha station

Khor-Kalabsha Station is a floating meteorological station close to the shore. It is located at 40 km upstream the HAD. The available raw data of the station are daily wind speed, wind direction, average air temperature, minimum air temperature, maximum air temperature, average relative humidity, maximum relative humidity, minimum relative humidity, and evaporation.

Kalabsha Meteorological Station

Wadi El-Arab (Amada) automatic recording buoy station.

This station is a new hydro-meteorological automatic station installed in January 2000 south of the Wadi El-Arab region at latitude 22°41`N and longitude 32°25`E and 185 km from the HAD. The average cross section of the HADR at the buoy site is about three km, and the buoy is located about one km from the eastern shore and two km from the western shore. In this station, hourly values of air temperature, relative humidity, wind speed and direction, atmospheric air pressure, and water surface temperature are measured. The measuring sensors for air temperature, relative humidity, wind speed and direction, are fixed on a tower above the buoy at two and four meters above the water surface. The system operates by rechargeable batteries with solar panels and records the hourly data by a data logger.

Wadi El-Arab (Amada) automatic recording buoy station

Abu Simbel automatic recording buoy station.

This station was installed in October 1986 and is located 280 km south of the HAD. It was installed in the middle of the HADR, 1500 m from both eastern and western shores, adjacent to the Abu Simbel Temples. In December 1999, the buoy site was shifted south by about 200 m. The average of the lake's cross section at the new site is about 2.4 km, and the buoy is about 700 m away from the eastern shore and 1700 m from the western shore. The average water depth at this open site is about 55 m. The measuring sensors for air temperature, relative humidity, wind speed and direction are installed on a tower on the buoy at four meters height above the water surface, and the pressure sensor is installed inside the buoy.

Water temperatures at the surface and at different depths are also measured. The instruments are operated by rechargeable batteries through solar panels. Hourly data are recorded and stored on a magnetic cassette. The parameters obtained from these recorded data are maximum, minimum and mean air temperatures, relative humidity, water surface temperature, wind speed and direction, wind run, peak gust, air pressure, and water temperature at 10 different depths. Daily mean values are obtained from the average of 24-hour values for each parameter. At the end of 1998, the system was upgraded to include meteorological parameters at two and four meters heights above the water surface, and the recording system was started to be operated through a data logger instead of a magnetic cassette.

Abu Simbel Automatic Recording Buoy Station

Appendix C

Samples of reports of measured data at Aswan Raft stations

Samples of monthly data measured at Aswan raft station in May 2005

HIGH ASWAN DAM RESERVOIR – HYDROMETIOROLOGICAL DATA AND EVAPORATION FROM RAFT STATION 2 KM U.S. HIGH DAM

مايو ٢٠٠٥

DAY	BP (mbar)	AT2 (Deg C)	AT4 (Deg C)	RH2 (%)	RH4 (%)	WST (Deg C)	NR max (W/m2)	NR min (W/m2)	WS2 (m/s)	WS4 (m/s)	WD4 (Deg)	evap. mm / day
1	987.8	29.0	29.9	40	33	23.1	821.0	-73.7	3.9	4.5	109.2	7.87
2	991.7	28.3	28.8	25	22	22.6	819.0	-70.3	4.8	5.4	236.6	10.79
3	993.4	23.4	23.6	28	25	20.2	826.0	-91.9	5.5	6.1	216.8	10.89
4	992.7	22.9	23.2	28	25	19.1	839.0	-83.0	4.6	5.0	209.4	8.38
5	994.0	24.6	24.9	29	26	20.0	860.0	-74.8	2.5	2.9	204.6	4.78
6	993.6	26.2	26.8	28	22	20.5	852.0	-75.3	2.3	2.7	230.6	4.47
7	991.7	26.4	27.1	39	30	22.3	884.0	-79.9	1.7	2.1	144.0	2.82
8	991.5	28.6	29.3	38	30	24.2	874.0	-87.5	1.9	2.2	142.2	4.54
9	993.0	29.2	29.6	24	21	23.4	834.0	-75.9	4.3	4.8	247.4	10.45
10	993.3	26.8	27.3	20	16	21.0	880.0	-77.4	4.6	5.1	229.5	10.48
11	993.0	25.9	26.4	24	21	20.4	855.0	-83.6	4.5	5.0	220.8	9.27
12	992.3	25.6	26.1	26	22	20.4	876.0	-94.5	3.4	3.8	219.2	6.92
13	993.1	27.7	28.4	25	21	20.6	785.0	-68.5	3.8	4.3	229.0	7.32
14	992.3	30.0	31.2	23	17	21.0	784.0	-47.0	3.7	4.3	230.4	7.50
15	991.0	29.4	30.5	25	18	21.4	860.0	-66.3	2.9	3.5	230.8	6.37
16	991.3	28.9	29.7	24	19	21.7	866.0	-74.0	3.0	3.4	238.7	6.35
17	991.3	29.3	30.1	24	20	22.0	839.0	-70.8	3.7	4.1	239.8	7.78
18	989.6	29.4	30.1	20	16	22.0	862.0	-79.8	3.4	3.8	238.6	7.74
19	988.4	30.7	31.7	22	16	22.6	882.0	-68.3	3.0	3.5	255.8	6.95
20	988.6	32.0	33.4	25	19	23.2	861.0	-69.8	2.4	2.9	262.9	5.50
21	988.2	32.1	33.3	33	27	23.7	882.0	-57.8	2.2	2.6	233.8	4.36
22	989.6	32.1	33.0	27	24	23.5	854.0	-54.8	4.7	5.3	247.6	9.50
23	992.0	29.9	30.5	21	18	21.8	850.0	-70.1	4.9	5.5	243.5	10.81
24	991.4	28.1	28.8	25	22	21.3	864.0	-72.0	3.9	4.5	233.8	8.09
25	991.7	28.4	29.1	25	21	21.8	864.0	-75.7	2.9	3.3	244.8	6.20
26	991.9	28.9	29.8	25	18	22.7	793.0	-79.2	2.3	2.7	235.5	5.43
27	993.0	30.6	31.5	24	18	23.8	892.0	-87.3	2.1	2.5	254.4	5.66
28	993.5	32.1	33.2	21	15	23.9	849.0	-76.2	2.9	3.5	252.5	7.69
29	992.8	32.6	33.8	26	17	24.9	872.0	-71.8	1.7	2.1	223.6	4.66
30	994.4	32.7	33.5	22	19	24.0	832.0	-60.8	4.2	4.7	253.2	10.29
31	994.0	32.9	33.9	20	16	24.0	867.0	-58.4	3.3	3.8	258.5	8.42
Mean	991.8	28.9	29.6	26	21	22.2	850.9	-73.4	3.4	3.9		7.36
Min	987.8	22.9	23.2	20	15	19.1	784.0	-94.5	1.7	2.1		2.82
Max	994.4	32.9	33.9	40	33	24.9	892.0	-47.0	5.5	6.1		10.89

Samples of hourly data measured at Aswan stations

	WS2 (m/s)	WD2 (Deg)	SG2 (Deg)	WS4 (m/s)	WD4 (Deg)	SG4 (Deg)	AT2 (Deg C)	AT2mx (Deg C)	AT2mn (Deg C)	AT4 (Deg C)	AT4mx (Deg C)	AT4mn (Deg C)	RH2 (%)	RH2mx (%)	RH2mn (%)	RH4 (%)	RH4mx (%)	RH4mn (%)
01:00	2.7	299.1	4.9	0.4	272.9	4.8	23.3	23.7	23	23.7	24.2	23.4	41	46.5	32.4	35.1	40.7	28.1
02:00	2.3	309.5	9.5	0.1	289.4	2.8	22.9	23.3	22.6	23.4	23.8	23	43.2	52.4	34.7	37	45.7	31
03:00	3.4	301	9.6	1.9	260.2	20.6	22.4	22.8	21.8	22.7	23.6	22	42.7	50.6	35.4	38.3	46.4	31.6
04:00	3.9	295.4	5.7	2.8	241.9	9.9	21.3	21.9	20.7	21.5	22.1	20.8	46.3	52.4	41.9	43.5	48.9	39.6
05:00	4.4	293	4.1	3.5	232.9	8.1	20.4	20.7	20.1	20.4	20.8	20.2	47.2	52.8	41.3	44.5	50.4	39.7
06:00	4.2	293.7	6.1	3.2	237.8	12.4	19.9	20.4	19.5	20	20.3	19.6	46.1	52.8	40.3	43.8	49	39.7
07:00	2.6	300.1	14.6	1.2	253.8	8.9	19.4	19.6	19	19.4	19.7	19.1	46.3	52.9	42	44.6	51.7	41
08:00	1.9	296.1	31.3	0.3	245.6	13.9	18.8	19.1	18.6	18.9	19.1	18.7	44	51.9	38	42.9	50.3	38.9
09:00	3.3	299.6	10.2	3.1	226.4	14.2	18.9	19.3	18.6	18.9	19.3	18.6	44.6	53.8	39.5	43.2	49.9	38.9
10:00	3.4	308.9	5.6	3.2	233.4	20.2	20	20.8	19.2	20	20.8	19.2	44.3	53.3	39	42.5	49.2	38.2
11:00	3.8	313.3	6	3.5	236.8	12.7	21.6	22.6	20.7	21.7	22.7	20.8	39.5	49.1	31.5	36.8	44.8	31.2
12:00	3.3	314.8	5.5	3.2	223.9	17.6	23.6	24.6	22.5	23.8	24.9	22.7	35.3	42.2	27.4	31.4	37.8	26.6
13:00	3	328.9	8	1.8	280.1	16.6	25.9	27.8	24.5	26.5	28.5	24.9	32.7	41.6	20.4	26.7	35.1	19.1
14:00	2.3	352.7	17.1	1.1	238.4	21.9	28.3	29.8	27.2	29	30.1	28	25.2	39.2	16.1	21	34.2	13.8
15:00	2.5	57.8	21.8	1.2	303.8	23.2	25.5	30.5	28.6	30.5	31.3	29.8	24.1	35	13.3	18.1	28.4	11.2
16:00	3.2	58.5	15.6	2.4	326	19.8	25.6	30.4	28.6	31.2	32.2	30.1	25	34.3	15	17.1	25.7	11.2
17:00	1.7	47.3	31.8	1.9	328.2	26.8	29.4	30.6	28	31.1	31.9	29.4	27.8	39.9	13.2	19.6	33.2	12.2
18:00	1.9	343.5	11.3	0.9	274.6	15.7	30.2	31.3	29.1	31.5	32	30.7	22.1	34.3	14.7	17.1	27.8	13.9
19:00	3.2	314.8	9.2	1.7	274.2	12.8	29.1	30.1	28.4	30.4	31	29.8	24.1	32.1	16.5	18.4	23.5	15.8
20:00	2.8	313.2	6.8	0.9	276.8	9.6	27.9	28.7	27.4	29.3	29.9	28.7	26.9	35.7	20.6	20.2	25.6	18.1
21:00	2.8	302.8	5.7	1.1	269	9.7	26.9	27.7	26.3	28.1	28.9	27.5	30.2	36.6	21.2	22.2	27.7	19.1
22:00	3.3	293.7	4.5	1.9	252.5	11.3	25	26.8	25.3	27.1	27.7	26.4	32.4	40	23.5	24.5	30.6	20.9
23:00	2.5	298	7.4	0.6	261.9	8.4	24.6	25.4	24.1	25.6	26.5	24.8	39.6	46.3	32.7	30.9	39.1	23.2
24:00:00	2.6	303.7	9.3	1.1	255.8	11	24.2	24.5	24	25	25.3	24.7	39.2	48.5	29.4	31	43.3	25.8

	DP2 (Deg C)	NR (W/m2)	BP (mbar)	WTs (Deg C)	WT2 (Deg C)	WT5 (Deg C)	WT10 (Deg C)	WT15 (Deg C)	WT20 (Deg C)	WTsad (Deg C)	Ept (mm/hr)	Eh (mm/hr)	Batt (VDC)	EVAP. mm/hr
01:00	9.2	-79.3	993	21.10	20.3	21	20.7	21	20.6	0.44	-0.143	-0.1	11.96	0.2
02:00	9.8	-81.2	992	21.10	20.3	21	20.7	21	20.6	0.43	-0.148	-0.035	11.96	0.2
03:00	9.1	-81.5	992	21.00	20.2	21	20.7	21	20.6	0.39	-0.151	-0.036	11.95	0.2
04:00	9.3	-83.5	992	21.00	20.2	21	20.8	21	20.6	0.37	-0.149	-0.009	11.94	0.3
05:00	8.7	-83.9	992	21.00	20.2	21	20.8	21	20.6	0.33	-0.15	0.021	11.93	0.3
06:00	8	-86.1	992	20.90	20.1	20.9	20.8	21	20.6	0.31	-0.15	0.033	11.92	0.3
07:00	7.6	-89.4	992	20.90	20.1	20.9	20.7	21	20.6	0.27	-0.152	0.03	11.92	0.3
08:00	6.3	-70.9	992	20.80	20	20.8	20.7	21	20.6	0.2	-0.157	0.028	11.92	0.1
09:00	6.6	49.6	993	20.80	20	20.8	20.8	21	20.6	0.17	-0.124	0.047	11.96	0.3
10:00	7.5	237.1	995	20.80	20	20.8	20.8	21	20.6	0.19	0.089	0.021	12	0.3
11:00	7.2	399.8	995	20.90	20	20.8	20.8	21	20.6	0.2	0.435	-0.024	12.04	0.3
12:00	7.4	513.3	995	21.00	20.1	20.8	20.8	21	20.6	0.33	0.754	-0.076	12.06	0.3
13:00	8.1	572.6	994	21.10	20.1	20.8	20.8	21	20.6	0.48	0.997	-0.13	12.08	0.2
14:00	6.3	562.4	994	21.40	20.1	20.8	20.8	21	20.6	0.78	1.146	-0.157	12.09	0.2
15:00	6.6	466	993	21.60	20.1	20.8	20.8	21	20.6	0.94	1.139	-0.2	12.09	0.2
16:00	7.3	422.5	993	21.60	20.1	20.8	20.7	21	20.6	0.99	0.945	-0.267	12.09	0.3
17:00	8.6	133.5	993	21.50	20.1	20.8	20.8	21	20.6	0.85	0.855	-0.134	12.06	0.1
18:00	6	-19.9	993	21.30	20.1	20.8	20.8	21	20.6	0.72	0.272	-0.176	12.02	0.2
19:00	6.4	-64.6	992	21.20	20.2	20.8	20.8	21.1	20.7	0.56	-0.04	-0.255	11.98	0.3
20:00	7.1	-67.3	992	21.10	20.2	20.8	20.8	21.1	20.7	0.44	-0.129	-0.183	11.96	0.2
21:00	7.5	-67.5	993	21.00	20.2	20.8	20.8	21.1	20.6	0.39	-0.132	-0.153	11.95	0.2
22:00	8.2	-68.5	993	21.00	20.2	20.9	20.8	21.1	20.7	0.31	-0.131	-0.153	11.94	0.2
23:00	9.9	-72.8	993	20.90	20.1	20.9	20.8	21.1	20.7	0.26	-0.131	-0.083	11.93	0.2
24:00:00	9.4	-72.6	993	20.90	20.1	20.8	20.8	21.1	20.6	0.25	-0.138	-0.074	11.92	0.2

Appendix D

Samples of Hydrological data

Samples of daily reports at HAD gauge station from 1964 to 2010
Region: Main Nile in Egypt
River: Main Nile in Egypt

Year	Month	day		High Dam U.S. Level M.t.	Diff. In Level cm	High Dam Contents M.m3	D.S. Aswan Release M.m3	spilled in Toshka M.m3	Arriving Aswan M.m3
1964	AUG.	1	1-Aug-1964	111,89	128	0	505		505
		2	2-Aug-1964	112,93	104	0	565		565
		3	3-Aug-1964	113,52	59	0	578		578
		4	4-Aug-1964	113,77	25	0	598		598
		5	5-Aug-1964	113,89	12	0	619		619
		6	6-Aug-1964	114,07	18	0	605		605
		7	7-Aug-1964	114,37	30	0	618		618
		8	8-Aug-1964	114,71	34	0	631		631
		9	9-Aug-1964	115,13	42	0	648		648
		10	10-Aug-1964	115,64	51	0	673		673
		11	11-Aug-1964	115,73	9	0	683		683
		12	12-Aug-1964	116,17	44	0	705		705
		13	13-Aug-1964	116,27	10	0	732		732
		14	14-Aug-1964	116,2	-7	0	726		726
	AUG.	15	15-Aug-1964	116,23	3	0	737		737
		16	16-Aug-1964	116,35	12	0	750		750
		17	17-Aug-1964	116,43	8	0	762		762
		18	18-Aug-1964	116,48	5	0	747		747
		19	19-Aug-1964	116,51	3	0	761		761
		20	20-Aug-1964	116,62	11	0	760		760
		21	21-Aug-1964	116,8	18	0	782		782
		22	22-Aug-1964	117	20	0	803		803
		23	23-Aug-1964	117,13	13	0	821		821
		24	24-Aug-1964	117,32	19	0	829		829
		25	25-Aug-1964	117,5	18	0	846		846
		26	26-Aug-1964	117,72	22	0	848		848
		27	27-Aug-1964	118	28	0	859		859
		28	28-Aug-1964	118,24	24	0	862		862
		29	29-Aug-1964	118,38	14	0	896		896
		30	30-Aug-1964	118,49	11	0	907		907
		31	31-Aug-1964	118,54	5	0	911		911
	SEP.	1	1-Sep-1964	118,72	18	0	887		887
		2	2-Sep-1964	118,97	25	0	896		896
		5	5-Jul-2010	170,23	-0,01	98589	250		207
		6	6-Jul-2010	170,21	-0,02	98503	250		164
		7	7-Jul-2010	170,19	-0,02	98417	250		164
		8	8-Jul-2010	170,16	-0,03	98288	250		121
		9	9-Jul-2010	170,14	-0,02	98202	250		164
		10	10-Jul-2010	170,11	-0,03	98073	250		121
		11	11-Jul-2010	170,08	-0,03	97944	245		116
		12	12-Jul-2010	170,06	-0,02	97858	245		159
		13	13-Jul-2010	170,03	-0,03	97729	245		116
		14	14-Jul-2010	170	-0,03	97600	245		116
	JUL	15	15-Jul-2010	169,98	-0,02	97514	245		159
		16	16-Jul-2010	169,97	-0,01	97471	245		202
		17	17-Jul-2010	169,97	0	97471	245		245
		18	18-Jul-2010	169,97	0	97471	240		240
		19	19-Jul-2010	169,97	0	97471	240		240
		20	20-Jul-2010	169,97	0	97471	240		240
		21	21-Jul-2010	169,97	0	97471	240		240
		22	22-Jul-2010	169,96	-0,01	97428	240		197
		23	23-Jul-2010	169,94	-0,02	97342	235		149
		24	24-Jul-2010	169,92	-0,02	97256	235		149
		25	25-Jul-2010	169,9	-0,02	97170	235		149
		26	26-Jul-2010	169,88	-0,02	97084	235		149
		27	27-Jul-2010	169,87	-0,01	97041	230		187
		28	28-Jul-2010	169,86	-0,01	96998	225		182
		29	29-Jul-2010	169,85	-0,01	96955	225		182
		30	30-Jul-2010	169,82	-0,03	96826	235		106
		31	31-Jul-2010	169,79	-0,03	96697	230		101

Monthly mean gauge readings in meters at Wadi Halfa about 345 Kilometers upstream the HAD from 1968 to 2007
Region: Main Nile in Sudan
River: Main Nile in Sudan

Year	Jan	Feb	Mar	Apr	May	Jun	Jul	Aug	Sep	Oct	Nov	Dec	Max	Min
1968								151.2	153.3	155.4	156.5	156.4	156.50	151.17
1969	156.3	155.7	154.9	154.4	154.2	152.8	151.1	152.9	158.9	161.2	161.1	160.8	161.17	151.12
1970	160.6	160.2	159.3	158.6	158.0	156.5	154.6	155.4	161.3	164.1	164.7	164.8	164.78	154.57
1971	164.6	164.3	163.8	163.2	162.7	161.5	160.0	160.9	165.1	167.2	167.6	167.6	167.61	160.01
1972	167.5	167.2	166.6	166.0	165.5	164.3	162.9	162.8	164.3	165.0	165.2	164.8	167.48	162.82
1973	164.4	163.8	163.0	162.2	161.5	160.3	158.8	159.3	163.2	165.5	166.3	166.2	166.29	158.76
1974	165.8	165.6	164.8	164.0	163.3	162.3	161.2	163.4	167.6	170.1	170.6	170.5	170.60	161.24
1975	170.3	169.9	169.3	168.6	168.2	167.2	165.9	166.9	171.6	175.0	175.7	175.7	175.69	165.93
1976	175.6	175.4	175.0	174.5	174.0	173.4	172.6	173.1	175.5	176.4	176.4	176.2	176.44	172.58
1977	175.8	175.4	174.8	174.2	173.7	172.9	171.9	172.9	175.7	176.8	177.1	177.0	177.11	171.93
1978	176.6	176.3	175.7	174.8	174.2	173.4	172.6	173.5	175.8	177.0	177.4	177.1	177.40	172.57
1979	176.8	176.5	175.9	175.5	175.1	174.3	173.3	173.5	175.1	175.8	175.9	175.6	176.78	173.34
1980	175.3	174.8	174.2	173.6	173.1	172.4	171.4	172.4	175.2	176.1	176.2	175.9	176.17	171.42
1981	175.5	175.1	174.5	173.9	173.5	172.6	171.5	171.9	174.2	175.6	175.9	175.5	175.88	171.49
1982	175.1	174.6	174.1	173.4	173.0	172.0	170.8	170.4	171.8	172.3	172.5	172.0	175.09	170.41
1983	171.5	170.8	170.1	169.5	168.9	167.9	166.5	165.9	168.3	169.4	169.8	169.6	171.50	165.94
1984	169.1	168.7	168.1	167.3	166.7	165.5	164.0	163.9	164.6	164.5	164.0	163.3	169.11	163.33
1985	162.7	161.9	161.0	160.3	159.5	158.0	156.4	157.3	161.7	164.2	164.2	163.9	164.22	156.39
1986	163.5	162.9	162.1	161.2	160.5	159.1	157.4	158.2	161.1	162.5	162.6	162.0	163.50	157.36
1987	161.5	160.8	159.9	158.9	158.0	156.5	155.1	154.8	157.6	158.1	158.4	158.2	161.46	154.82
1988	157.8	157.3	156.4	155.6	154.8	152.9	151.3	155.2	162.6	166.7	168.4	168.7	168.75	151.34
1989	168.6	168.5	168.0	167.5	167.0	165.9	164.6	165.1	167.6	169.3	169.8	169.7	169.77	164.61
1990	169.5	169.1	168.5	167.8	167.3	166.1	164.4	164.2	166.4	168.0	168.4	168.1	169.45	164.18
1991	167.8	167.3	166.6	165.9	165.2	163.8	162.5	163.6	167.3	169.2	169.3	169.1	169.25	162.50
1992	169.0	168.7	168.0	167.4	166.9	165.7	164.3	164.4	167.7	169.5	170.5	170.7	170.69	164.28
1993	170.6	170.3	169.8	169.2	168.8	167.9	167.4	168.5	171.6	173.6	174.2	174.3	174.32	167.38
1994	174.3	173.9	173.2	172.5	172.0	171.0	169.8	171.1	174.9	177.1	177.1	177.0	177.13	169.77
1995	176.9	176.5	176.0	175.4	174.8	173.8	172.7	173.0	175.4	176.2	176.1	175.9	176.86	172.66
1996	175.8	175.5	174.9	174.3	173.7	172.8	172.4	173.9	176.8	178.3	178.5	178.5	178.52	172.40
1997	178.4	178.2	177.7	177.3	176.9	176.1	175.5	176.1	177.6	177.9	178.2	178.5	178.48	175.49
1998	178.5	178.2	177.7	177.2	176.8	175.8	174.9	176.1	179.4	181.0	181.2	180.8	181.21	174.87
1999	180.2	179.7	179.0	178.2	177.5	176.6	175.8	176.8	179.6	180.8	181.5	181.2	181.53	175.81
2000	180.6	180.0	179.3	178.7	178.0	177.0	176.1	176.6	179.1	180.3	180.6	180.4	180.60	176.06
2001	179.9	179.5	179.0	178.4	177.8	176.8	175.9	177.2	180.0	180.5	180.1	180.0	180.47	175.91
2002	179.8	179.6	179.2	178.5	177.8	176.7	175.6	175.6	177.0	177.6	177.5	177.2	179.84	175.56
2003	177.0	176.5	175.9	175.3	174.7	173.5	172.3	173.2	176.2	177.7	177.8	177.5	177.75	172.27
2004	177.3	176.9	176.3	175.6	174.6	173.4	172.1	172.4	174.3	175.1	175.4	175.1	177.30	172.07
2005	174.9	174.5	173.8	172.9	172.0	170.8	169.7	170.5	173.0	174.6	174.6	174.5	174.94	169.70
2006	174.3	173.9	173.3	172.3	171.3	170.1	168.9	170.0	173.9	176.3	176.4	176.4	176.44	168.88
2007	176.3	176.0	175.5	174.8	174.0	173.0	172.6	174.8	178.0	179.9	179.8	179.6	179.92	172.60

Monthly average discharge in hm³ at Dongola about 776 Kilometers from the HAD from 1962 to 2007
Region: Main Nile in Sudan
River: Main Nile in Sudan

Year	Jan	Feb	Mar	Apr	May	Jun	Jul	Aug	Sep	Oct	Nov	Dec	Max	Min
1962					9.74	9.93	10.62	13.15	14.00	12.81	10.84	10.29	14.00	9.74
1963	9.99	9.90	9.63	9.75	9.88	10.04	11.01	13.80	13.78	12.15	10.48	10.30	13.80	9.63
1964	10.19	9.87	9.62	9.57	9.91	10.13	11.09	14.32	14.58	13.20	11.95	10.94	14.58	9.57
1965	10.80	10.76	10.38	10.12	10.26	10.45	10.84	12.82	13.21	12.25	11.12	10.45	13.21	10.12
1966	10.04	9.80	9.42	9.79	9.90	9.74	10.72	12.59	13.52	11.31	10.52	10.49	13.52	9.42
1967	10.40	9.96	9.34	9.70	9.80	9.72	10.62	13.61	14.11	12.71	10.99	10.63	14.11	9.34
1968	10.18	10.00	9.63	9.80	9.80	9.72	11.03	13.34	12.31	11.98	10.45	10.05	13.34	9.63
1969	9.62	9.36	9.46	10.09	10.01	9.46	10.76	13.49	13.50	10.98	10.21	9.90	13.50	9.36
1970	9.76	9.53	9.25	9.88	9.72	9.33	9.79	13.46	13.94	11.80	10.90	10.12	13.94	9.25
1971	9.96	10.04	9.42	10.02	9.90	9.42	10.51	13.28	13.62	11.59	10.65	10.17	13.62	9.42
1972	10.00	9.83	9.36	9.89	9.73	9.41	10.42	11.88	11.47	11.11	9.94	9.45	11.88	9.36
1973	9.30	9.02	8.95	9.47	9.57	9.57	10.03	13.22	13.32	11.93	10.45	9.74	13.32	8.95
1974	9.66	9.41	9.26	9.64	9.66	9.58	11.42	13.88	13.79	12.24	10.71	10.02	13.88	9.26
1975	9.90	9.55	9.24	9.83	9.85	9.33	10.38	13.82	14.85	12.90	11.09	10.50	14.85	9.24
1976	10.13	9.88	9.49	9.70	10.11	9.98	10.69	13.19	12.92	11.32	10.44	10.06	13.19	9.49
1977	9.73	9.41	9.28	9.66	9.75	9.61	10.56	13.85	13.13	11.68	11.13	10.08	13.85	9.28
1978	9.94	9.60	9.34	9.58	9.79	9.78	10.86	13.51	12.92	12.35	10.62	10.03	13.51	9.34
1979	10.01	9.67	9.61	9.91	9.99	9.72	10.52	12.77	12.31	11.33	10.33	9.91	12.77	9.61
1980	9.53	9.19	9.16	9.62	9.87	9.60	10.63	13.82	13.26	10.94	10.39	9.75	13.82	9.16
1981	9.56	9.26	9.11	9.64	9.68	9.32	10.18	13.12	13.29	11.96	10.29	9.86	13.29	9.11
1982	9.68	9.56	9.18	9.72	9.55	9.39	9.79	12.00	12.11	11.38	10.13	9.68	12.11	9.18
1983	9.26	9.08	9.08	9.66	9.39	9.34	9.50	12.20	12.99	11.21	10.46	9.63	12.99	9.08
1984	9.60	9.47	9.12	9.62	9.53	9.17	10.25	11.80	10.83	10.24	9.43	9.09	11.80	9.09
1985	9.07	8.94	8.94	9.36	9.43	9.34	10.75	12.98	13.95	11.15	10.07	9.73	13.95	8.94
1986	9.40	9.22	9.09	9.49	9.59	9.05	10.67	12.84	12.89	10.99	9.83	9.25	12.89	9.05
1987	9.10	8.94	8.85	9.21	9.42	9.76	10.41	12.15	11.99	10.65	10.03	9.45	12.15	8.85
1988	9.30	9.05	9.03	9.53	9.16	9.01	11.31	14.96	15.02	13.14	11.54	10.24	15.02	9.01
1989	10.17	9.71	9.46	9.86	10.03	9.64	10.76	12.81	13.32	11.78	10.40	9.69	13.32	9.46
1990	9.55	9.26	9.05	9.78	9.80	9.12	9.58	12.62	13.02	11.30	9.99	9.50	13.02	9.05
1991	9.25	8.90	8.95	9.50	9.49	9.31	10.82	13.62	13.79	10.88	10.28	9.87	13.79	8.90
1992	9.63	9.28	9.17	9.77	9.61	9.24	10.26	12.88	13.68	11.97	11.14	10.05	13.68	9.17
1993	9.64	9.51	9.35	9.74	9.87	10.27	11.38	13.84	14.26	12.18	10.79	10.11	14.26	9.35
1994	9.49	9.21	9.09	9.52	9.71	9.41	10.67	14.65	15.07	11.88	10.28	9.89	15.07	9.09
1995	9.64	9.31	9.27	9.24	9.78	9.31	10.31	13.74	13.02	11.01	10.03	9.53	13.74	9.24
1996	9.48	9.22	9.20	9.45	9.80	10.05	12.31	14.29	14.18	12.36	10.58	9.97	14.29	9.20
1997	9.81	9.70	9.44	9.98	9.96	9.93	11.56	13.63	12.04	10.97	11.36	10.36	13.63	9.44
1998	9.69	9.48	9.35	9.68	9.90	9.47	10.70	14.41	15.59	13.48	12.14	10.70	15.59	9.35
1999	9.98	9.88	9.54	9.78	9.95	10.00	11.39	14.60	14.77	13.28	11.93	10.69	14.77	9.54
2000	10.04	9.88	9.68	10.13	9.88	9.81	10.91	14.24	13.87	12.30	11.42	10.34	14.24	9.68
2001	9.95	9.64	9.43	9.69	9.92	9.73	11.31	15.32	14.38	11.55	11.08	10.26	15.32	9.43
2002	9.91	9.79	9.76	9.93	10.02	9.62	10.27	13.41	12.51	11.01	9.92	9.64	13.41	9.62
2003	9.49	9.27	9.20	9.77	9.59	9.23	10.95	14.30	13.94	12.01	10.22	9.90	14.30	9.20
2004	9.46	9.24	9.31	9.62	9.57	9.46	10.68	13.50	12.31	11.95	10.11	9.70	13.50	9.24
2005	9.45	9.15	8.97	9.23	9.60	9.73	11.55	13.50	13.79	11.64	10.17	9.76	13.79	8.97
2006	9.47	9.23	9.03	9.07	9.80	9.75	11.01	14.50	15.07	12.20	10.93	10.23	15.07	9.03
2007	9.75	9.74	9.51	9.53	10.05	10.16	13.17	15.07	15.38	12.71	10.69	10.13	15.38	9.51

Appendix E

Samples of Topographic data

An example of raw data measured at Adindan in November 2000

#	Time	Alongline1 (805)	Off-line1 (8	X (original)	Y (original)	Z (interp)
1	Nov 21, 2000 11:36:48.00	331.92	2082.73	348909.10	2454924.88	8.13
2	Nov 21, 2000 11:36:49.00	331.72	2073.85	348902.84	2454918.58	8.14
3	Nov 21, 2000 11:36:51.00	331.65	2068.81	348899.27	2454915.03	8.16
4	Nov 21, 2000 11:36:53.00	331.58	2057.00	348890.82	2454906.78	8.16
5	Nov 21, 2000 11:36:55.00	331.68	2049.50	348885.35	2454901.64	8.14
6	Nov 21, 2000 11:36:57.00	331.90	2041.80	348879.65	2454896.46	8.13
7	Nov 21, 2000 11:37:00.00	332.20	2033.85	348873.72	2454891.15	8.11
8	Nov 21, 2000 11:37:01.00	332.51	2026.47	348868.19	2454886.25	8.11
9	Nov 21, 2000 11:37:03.00	332.86	2018.90	348862.50	2454881.25	8.05
10	Nov 21, 2000 11:37:05.00	333.20	2011.66	348857.06	2454876.46	8.00
11	Nov 21, 2000 11:37:07.00	333.27	2004.42	348851.79	2454871.49	7.88
12	Nov 21, 2000 11:37:09.00	333.18	1997.22	348846.67	2454866.43	7.82
13	Nov 21, 2000 11:37:11.00	333.05	1990.04	348841.60	2454861.35	7.73
14	Nov 21, 2000 11:37:13.00	333.02	1982.87	348836.46	2454856.35	7.65
15	Nov 21, 2000 11:37:14.00	333.12	1979.30	348833.81	2454853.95	7.58
16	Nov 21, 2000 11:37:17.00	333.56	1968.65	348825.85	2454846.86	7.52
17	Nov 21, 2000 11:37:19.00	333.85	1961.53	348820.52	2454842.13	7.19
18	Nov 21, 2000 11:37:20.00	334.04	1957.90	348817.77	2454839.75	6.94
19	Nov 21, 2000 11:37:23.00	334.71	1947.32	348809.69	2454832.88	7.08
20	Nov 21, 2000 11:37:24.00	335.02	1943.72	348806.89	2454830.61	7.06
21	Nov 21, 2000 11:37:26.00	335.79	1936.59	348801.22	2454826.22	6.91
22	Nov 21, 2000 11:37:28.00	336.69	1929.53	348795.52	2454821.96	6.75
23	Nov 21, 2000 11:37:30.00	337.60	1922.45	348789.79	2454817.70	6.60
24	Nov 21, 2000 11:37:32.00	338.49	1915.30	348784.02	2454813.38	6.35
25	Nov 21, 2000 11:37:34.00	339.43	1908.28	348778.32	2454809.18	6.20
26	Nov 21, 2000 11:37:36.00	340.40	1901.14	348772.51	2454804.92	5.92
27	Nov 21, 2000 11:37:38.00	341.47	1893.98	348766.61	2454800.73	5.76
28	Nov 21, 2000 11:37:40.00	342.66	1886.94	348760.72	2454796.69	5.70
29	Nov 21, 2000 11:37:42.00	344.00	1879.91	348754.73	2454792.78	5.53
30	Nov 21, 2000 11:37:44.00	345.28	1872.95	348748.83	2454788.87	5.28
31	Nov 21, 2000 11:37:47.00	346.74	1862.45	348740.26	2454782.63	5.02
32	Nov 21, 2000 11:37:50.00	347.55	1855.38	348734.61	2454778.31	4.53
33	Nov 21, 2000 11:37:51.00	348.19	1848.48	348729.20	2454773.97	4.05
34	Nov 21, 2000 11:37:53.00	348.63	1841.38	348723.78	2454769.37	3.42
35	Nov 21, 2000 11:37:55.00	348.91	1834.29	348718.48	2454764.64	2.70
36	Nov 21, 2000 11:37:57.00	348.90	1827.25	348713.42	2454759.75	1.70

#	Time	Alongline1	Offline1	X	Y	Z
13197	Nov 20, 2000 12:56:11.00	4585.45	19697.15	347363.41	2453343.15	6.40
13198	Nov 20, 2000 12:56:13.00	4591.63	19695.07	347369.22	2453340.19	5.28
13199	Nov 20, 2000 12:56:15.00	4597.76	19692.88	347375.07	2453337.35	5.22
13200	Nov 20, 2000 12:56:17.00	4603.88	19690.53	347381.04	2453334.62	5.14
13201	Nov 20, 2000 12:56:19.00	4610.06	19688.22	347387.01	2453331.83	5.03
13202	Nov 20, 2000 12:56:21.00	4616.23	19686.07	347392.87	2453328.93	4.85
13203	Nov 20, 2000 12:56:23.00	4622.32	19683.59	347398.90	2453326.32	4.71
13204	Nov 20, 2000 12:56:25.00	4628.29	19680.83	347405.05	2453323.99	4.71
13205	Nov 20, 2000 12:56:27.00	4634.32	19678.32	347411.07	2453321.44	4.59
13206	Nov 20, 2000 12:56:29.00	4640.39	19675.85	347417.08	2453318.83	4.42
13207	Nov 20, 2000 12:56:30.00	4643.53	19674.54	347420.22	2453317.51	4.34
13208	Nov 20, 2000 12:56:33.00	4652.40	19670.49	347429.31	2453314.01	4.25
13209	Nov 20, 2000 12:56:34.00	4655.41	19669.02	347432.47	2453312.88	4.10
13210	Nov 20, 2000 12:56:37.00	4663.99	19664.79	347441.50	2453309.72	4.01
13211	Nov 20, 2000 12:56:39.00	4669.29	19662.01	347447.20	2453307.88	3.85
13212	Nov 20, 2000 12:56:41.00	4674.40	19659.61	347452.49	2453305.92	3.67
13213	Nov 20, 2000 12:56:43.00	4678.93	19657.61	347457.08	2453304.09	3.45
13214	Nov 20, 2000 12:56:45.00	4682.53	19655.92	347460.81	2453302.70	3.39
13215	Nov 20, 2000 12:56:47.00	4686.44	19654.20	347464.78	2453301.11	3.31
13216	Nov 20, 2000 12:56:49.00	4689.88	19652.64	347468.30	2453299.74	3.23
13217	Nov 20, 2000 12:56:50.00	4691.36	19651.98	347469.81	2453299.15	3.20
13218	Nov 20, 2000 12:56:52.00	4694.61	19650.33	347473.26	2453297.98	3.15
13219	Nov 20, 2000 12:56:55.00	4699.22	19648.21	347478.00	2453296.18	3.08
13220	Nov 20, 2000 12:56:57.00	4702.22	19646.60	347481.25	2453295.16	3.00
13221	Nov 20, 2000 12:56:58.00	4703.78	19645.84	347482.88	2453294.58	2.96
13222	Nov 20, 2000 12:57:01.00	4708.39	19643.62	347487.70	2453292.84	2.90
13223	Nov 20, 2000 12:57:02.00	4709.89	19642.77	347489.35	2453292.37	2.83
13224	Nov 20, 2000 12:57:05.00	4714.50	19640.50	347494.21	2453290.66	2.74
13225	Nov 20, 2000 12:57:07.00	4717.54	19639.01	347497.40	2453289.53	2.71
13226	Nov 20, 2000 12:57:09.00	4720.64	19637.64	347500.55	2453288.28	2.57

List of the downloaded images from Landsat 123 MMS.

#	File name	#	File name
1	LM1_187_044_19722 59AAA04.	30	LM2_187_043_1976013AAA01.
2	LM1_187_044_1972313AAA04.	31	LM2_187_044_1975180AAA02.
3	LM1_187_044_1973181AAA03.	32	LM2_187_044_1977259AAA02.
4	LM1_188_045_1972278AAA01.	33	LM2_188_044_1977224AAA01.
5	LM1_188_045_1973128AAA01.	34	LM2_188_044_1977242AAA01.
6	LM1_188_044_1972314AAA03.	35	LM2_188_044_1977260AAA05.
7	LM1_188_044_1973164AAA03.	36	LM2_188_044_1977278AAA01.
8	LM2_188_045_1975181AAA04.	37	LM2_188_045_1977260AAA03.
9	LM2_188_045_1977224AAA01.	38	LM2_188_045_1977278AAA01.
10	LM2_188_045_1977242AAA01.	39	LM2_188_045_1979016AAA03.
11	LM2_188_044_197 5181AAA04.	40	LM2_188_045_1979052AAA03.
12	LM2_188_044_1976014AAA01.	41	LM2_188_045_1979070AAA04.
13	LM1_187_043_1972259AAA04.	42	LM2_188_045_1979088AAA04.
14	LM1_187_043_1972313AAA04.	43	LM2_188_045_1979106AAA05.
15	LM1_187_043_1972349AAA01.	44	LM2_188_045_1979214AAA03.
16	LM1_187_043_1973019AAA01.	45	LM2_188_045_1979268AAA04.
17	LM1_187_043_1973055AAA04.	46	LM2_188_045_1979304AAA03.
18	LM1_187_043_1973181AAA03.	47	LM3_188_045_1979025AAA03.
19	LM1_187_044_1972277AAA02.	48	LM3_188_045_1979043AAA07.
20	LM1_187_044_1972349AAA02.	49	LM3_188_045_1979133AAA04.
21	LM1_187_044_1973037AAA01.	50	LM3_188_045_1979151AAA03.
22	LM1_187_044_1973271AAA03.	51	LM3_188_045_1979223AAA03.
23	LM1_188_045_1973164AAA03.	52	LM3_188_045_1979241AAA03.
24	LM1_188_044_1972278AAA01.	53	LM3_188_045_1979259AAA03.
25	LM1_188_044_1972296AAA01.	54	LM3_188_045_1979277AAA03.
26	LM1_188_044_1973290AAA03.	55	LM3_188_045_1979295AAA03.
27	LM1_188_045_1972296AAA04.	56	LM3_188_045_1979313AAA03.
28	LM1_188_045_1972314AAA02.	57	LM3_188_045_1979349AAA05.
29	LM2_187_043_1975180AAA03.		

List of the downloaded images from Landsat 45 MMS.

#	File name	#	File name
1	LM5_174_044_1984314AAA03.	39	LM5_174_043_1985092AAA03.
2	LM5_174_044_1984154AAA03.	40	LM5_174_043_1985108AAA03.
3	LM5_174_043_1986063AAA03.	41	LM5_174_043_1986015AAA05.
4	LM5_174_043_1985140AAA03.	42	LM5_174_044_1985028AAA03.
5	LM5_174_044_1984218AAA03.	43	LM5_174_044_1985060AAA03.
6	LM5_174_044_1984346AAA03.	44	LM5_174_044_1985076AAA03.
7	LM5_174_044_1984138AAA03.	45	LM5_174_044_1985092AAA03.
8	LM5_174_043_1984346AAA03.	46	LM5_174_044_1985108AAA03.
9	LM5_174_043_1984138AAA03.	47	LM5_174_044_1985140AAA03.
10	LM5_174_043_1984154AAA03.	48	LM5_175_044_1984273AAA03.
11	LM5_174_043_1984218AAA03.	49	LM5_175_044_1984289AAA03.
12	LM5_174_043_1984282AAA03.	50	LM5_175_044_1984321AAA03.
13	LM5_174_043_1984298AAA03.	51	LM5_175_044_1985035AAA03.
14	LM5_174_043_1984330AAA03.	52	LM5_175_044_1985083AAA03.
15	LM5_174_043_1986031AAA03.	53	LM5_175_044_1985147AAA03.
16	LM5_174_043_1986047AAA03.	54	LM5_175_044_1985307AAA03.
17	LM5_174_043_1987274AAA04.	55	LM5_175_044_1986006AAA03.
18	LM5_174_043_1987290AAA04.	56	LM5_175_044_1986022AAA03.
19	LM5_174_044_1984282AAA03.	57	LM5_175_044_1986038AAA03.
20	LM5_174_044_1984298AAA03.	58	LM5_175_044_1986054AAA03.
21	LM5_174_044_1984330AAA03.	59	LM5_175_044_1987281AAA03.
22	LM5_174_044_1986015AAA03.	60	LM5_175_044_1990241AAA03.
23	LM5_174_044_1986031AAA03.	61	LM5_175_045_1984273AAA03.
24	LM5_174_044_1986047AAA03.	62	LM5_175_045_1984289AAA03.
25	LM5_174_044_1987274AAA04.	63	LM5_175_045_1984305AAA03.
26	LM5_174_044_1987290AAA04.	64	LM5_175_045_1984321AAA03.
27	LM5_175_044_1984145AAA03.	65	LM5_175_045_1985035AAA03.
28	LM5_175_045_1984145AAA03.	66	LM5_175_045_1985051AAA03.
29	LM4_174_043_1983287AAA03.	67	LM5_175_045_1985083AAA03.
30	LM4_174_043_1990242AAA09.	68	LM5_175_045_1985147AAA03.
31	LM4_174_043_1990274AAA03.	69	LM5_175_045_1985307AAA03.
32	LM4_174_044_1983287AAA03.	70	LM5_175_045_1986006AAA03.
33	LM4_174_044_1990242AAA08.	71	LM5_175_045_1986022AAA03.
34	LM4_174_044_1990274AAA03.	72	LM5_175_045_1986038AAA03.
35	LM4_175_045_1990281AAA08.	73	LM5_175_045_1986054AAA03.
36	LM5_174_043_1985012AAA03.	74	LM5_175_045_1987281AAA03.
37	LM5_174_043_1985028AAA03.	75	LM5_175_045_1990241AAA03.
38	LM5_174_043_1985060AAA03.		

List of the downloaded images from Landsat 45 TM.

#	File name	#	File name	#	File name
1	LT4_174_043_1988029_XXX03.	83	LT5_174_044_1984218_AAA0 7.	165	LT5_175_044_200012 5_XXX01.
2	LT4_174_043_1988061_AAA05.	84	LT5_174_044_1984234_XXX08.	166	LT5_175_044_2000253_XXX03.
3	LT4_174_043_1988301_AAA03.	85	LT5_174_04419 84 2 5 0_XXX01.	167	LT5_175_044_2000269_AAA02.
4	LT4_174_043_1988333_XXX06.	86	LT5_174_044_1984298_XXX02.	168	LT5_175_044_2002002_MTI00.
5	LT4_174_043_1988836 5_XXX04.	87	LI 5_174_04419 84 314_XXX0 3.	169	LT5_175_044_2002146_MTI00.
6	LT4_174_043_1989319_AAA02.	88	LT5_174_044_1984330_XXX02.	170	LT5_175_044_2002162_MTI00.
7	LT4_174_043_1990114_XXX01.	89	LT5_174_044_1984346_XXX02.	171	LT5_175_044_2002178_MTI00.
8	LT4_174_043_1990210_XXX03.	90	LT5_174_044_198 5028_XXX03.	172	LT5_175_044_2003181_MTI01.
9	LT4_174_043_1990242_XXX04.	91	LT5_174_044_198 5060_AAA03.	173	LT5_175_044_2003197_MTI01.
10	LT4_174_043_1990274_XXX03.	92	LT5_174_044_198 5076_AAA03.	174	LT5_175_044_2003213_MTI01.
11	LT4_174_044_1988029_XXX03.	93	LT5_174_044_198 5092_AAA03.	175	LT5_175_044_2003229_MTI01.
12	LT4_174_044_1988061_AAA05.	94	LT5_174_044_198 5140_AAA03.	176	LT5_175_044_2003245_MTI01.
13	LT4_174_044_1988301_AAA03.	95	LT5_174_044_1986015_XXX04.	177	LT5_175_044_2003261_MTI01.
14	LT4_174_044_1988333_XXX07.	96	LT5_174_044_1986031_XXX04.	178	LT5_175_044_2003277_MTI01.
15	LT4_174_044_1988365_XXX04.	97	LT5_174_044_198 6604 7_XXX03.	179	LT5_175_044_2003293_MTI01.
16	LT4_174_044_1989319_AAA0 2.	98	LT5_174_044_198627_XXX01.	180	LT5_175_045_1984161_XXX03.
17	LT4_174_044_1990114_XXX01.	99	LT5_174_044_1986319_AAA03.	181	LT5_175_04 5_1984257_XXX02.
18	LT4_174_044_1990242_XXX05.	100	LT5_174_044_198633 5_XXX05.	182	LT5_175_045_1984273_XXX02.
19	LT4_174_044_1990274_XXX01.	101	LT5_174_044_1987002_AAA03.	183	LT5_175_04 51984430 5_XXX06.
20	LT4_175_044_1987321_XXX08.	102	LT5_174_044_1987018_AAA03.	184	LT5_175_04 5_1985035_XXX04.
21	LT4_175_044_1988020_XXX03.	103	LT5_174_044_1987034_AAA02.	185	LT5_175_045_1985051_AAA03.
22	LT4_175_044_1988084_XXX03.	104	LT5_174_044_1987226_XXX01.	186	LT5_175_045_198 5083_XXX04.
23	LT4_175_044_1989006_XXX02.	105	LT5_174_044_1987274_XXX03.	187	LT5_175_045_198 5307_AAA04.
24	LT4_175_044_198934 2_XXX02.	106	LT5_174_044_1998080_XXX05.	188	LT5_175_045_1986006_XXX03.
25	LT4_175_045_1987321_XXX08.	107	LT5_174_044_1998096_AAA03.	189	LT5_175_045_1986022_XXX03.
26	LT4_175_045_1988020_XXX03.	108	LT5_174_044_1998112_XXX01.	190	LT5_175_045_1986038_AAA03.
27	LT4_175_045_1988084_XXX03.	109	LT5_174_04419 9 8 2 24_XXX01.	191	LI 5_175_045_1986054_XXX03.
28	LT4_175_045_1989006_XXX02.	110	LT5_174_044_1998288_AAA03.	192	LT5_175_045_1986310_XXX03.
29	LT4_175_045_198934 2_XXX02.	111	LT5_174_044_1998320_XXX01.	193	LT5_175_045_1986326_AAA04.
30	LT5_174_043_1984154_XXX03.	112	LT5_174_044_1999019_AAA02.	194	LT5_175_04 5_1986342_XXX03.
31	LT5_174_043_1984218_AAA03.	113	LT5_174_044_1999067_XXX02.	195	LT5_175_045_1986358_AAA03.
32	LT5_174_043_1984234_XXX03.	114	LT5_174_044_1999099_XXX03.	196	LT5_175_045_1987009_XXX04.
33	LT5_174_043_19842 50_XXX01.	115	LT5_174_044_1999115_AAA03.	197	LT5_175_04 5_198704 I_XXX02.
34	LT5_174_043_1984298_XXX02.	116	LT5_174_044_1999307_XXX02.	198	LT5_175_045_1987281_XXX03.
35	LT5_174_043_1984330_XXX02.	117	LT5_174_044_2000022_XXX02.	199	LT5_175_045_199024 I_AAA04.
36	LT5_174_043_1984 346_XXX02.	118	LT5_174_044_2000038_AAA03.	200	LT5_175_045_1995031_XXX00.
37	LT5_174_043_198 5012_XXX02.	119	LT5_174_044_2000070_XXX03.	201	LT5_175_045_1995047_XXX01.
38	LT5_174_043_1985028_XXX03.	120	LT5_174_044_2000278_XXX02.	202	LT5_175_045_1998119_AAA01.
39	LT5_174_043_198 5060_AAA04.	121	LT5_174_044_2002171_MTI00.	203	LT5_175_045_1998135_AAA01.
40	LT5_174_043_1985092_AAA03.	122	LT5_174_044_2003174_MTI02.	204	LT5_175_045_1998183_AAA01.
41	LT5_174_043_198 5140_AAA03.	123	LT5_174_044_2003190_MTI01.	205	LT5_175_045_1998199_AAA01.
42	LT5_174_043_1986015_XXX04.	124	LT5_174_044_2003238_MTI01.	206	LT5_175_045_1998263_XXX01.
43	LT5_174_043_1986031_XXX04.	125	LT5_174_044_2009030_MTI00.	207	LT5_175_045_1998279_AAA01.
44	LT5_174_043_1986047_XXX03.	126	LT5_174_044_2010337_MTI00.	208	LT5_175_045_1998295_AAA02.
45	LT5_174_043_1986063_XXX03.	127	LT5_175_044_1984161_XXX02.	209	LT5_175_045_1998311_AAA01.
46	LT5_174_043_1986271_XXX01.	128	LT5_175_044_19842 57_XXX02.	210	LT5_175_04 51998343_XXX01.
47	LT5_174_043_1986319_AAA09.	129	LT5_175_044_1984273_XXX02.	211	LT5_175_045_1999026_AAA02.
48	LT5_174_043_1986335_XXX08.	130	LT5_175_044_1984305_XXX01.	212	LT5_175_045_199904 2_XXX01.
49	LT5_174_043_1987002_AAA03.	131	LT5_175_044_1985035_XXX04.	213	LT5_175_045_19990 58_XXX01.
50	LT5_174_043_1987018_AAA03.	132	LT5_175_044_1985083_XXX04.	214	LT5_175_045_1999090_XXX01.
51	LT5_174_043_1987034_AAA02.	133	LT5_175_044_1985307_AAA04.	215	LT5_175_045_1999106_AAA01.
52	LT5_174_043_1987226_XXX01.	134	LT5_175_044_1986006_XXX03.	216	LT5_175_045_1999218_XXX02.
53	LT5_174_043_1987274_XXX03.	135	LT5_175_044_1986038_AAA03.	217	LT5_175_045_1999234_AAA02.
54	LT5_174_043_1987290_AAA04.	136	LT5_175_044_19860 54_XXX03.	218	LT5_175_045_1999314_XXX02.
55	LT5_174_043_1998048_AAA01.	137	LT5_175_044_1986310_XXX03.	219	LT5_175_045_2000029_XXX02.
56	LT5_174_043_1998096_AAA03.	138	LT5_175_044_1986326_AAA04.	220	LT5_175_045_2000077_AAA04.
57	I_T 5_174_04 319 9 8112_XXX01.	139	LT5_175_044_198634 2_XXX03.	221	LT5_175_045_2000125_XXX01.
58	LT5_174_043_1998224_XXX01.	140	LT5_175_044_1986358_AAA03.	222	LT5_175_045_2000269_AAA02.
59	LT5_174_043_1998240_XXX01.	141	LT5_175_044_1987009_XXX10.	223	LT5_175_045_2002002_MTI00.
60	LT5_174_043_1998288_AAA03.	142	LT5_175_044_198704_raX02.	224	LT5_175_045_2002146_MTI00.
61	LT5_174_043_1998320_XXX01.	143	LT5_175_044_1987281_XXX03.	225	LT5_175_045_2002162_MTI00.
62	LT5_174_043_1999003_XXX03.	144	LT5_175_044_199024 I_AAA04.	226	LT5_175_045_2002178_MTI00.
63	LT5_174_043_1999019_AAA02.	145	LT5_175_044_1998039_XXX01.	227	LT5_175_045_2003197_MTI01.
64	LT5_174_043_1999067_XXX02.	146	LT5_175_044_1998119_AAA01.	228	LT5_175_045_2003213_MTI01.
65	LT5_174_043_1999081_XXX03.	147	LT5_175_044_19981 5_AAA01.	229	LT5_175_044_1999081_XXX01.
66	LT5_174_043_1999115_AAA03.	148	LT5_175_044_1998183_AAA0 1.	230	LT5_176_044_1999225_AAA02.
67	LT5_174_043_1999307_XXX02.	149	LT5_175_0441998199_AAA0 1.	231	LT5_176_044_2000004_AAA03.
68	LT5_174_043_2000038_AAA03.	150	LT5_175_044_1998263_XXX01.	232	LT5_176_044_2000036_AAA02.
69	LT5_174_043_2000054_AAA02.	151	LT5_175_044_1998279_AAA01.	233	LT5_176_044_2000052_XXX03.
70	LT5_174_043_2000070_XXX03.	152	LT5_175_044_1998295_AAA02.	234	LT5_176_044_2000084_XXX02.
71	LT5_174_043_2000166_XXX01.	153	LT5_175_044_1998311_AAA01.	235	LT5_176_044_2000100_XXX02.
72	LT5_174_043_2000278_XXX02.	154	LT5_175_044_199834 3_XXX01.	236	LT5_176_044_2000212_XXX02.
73	LT5_174_043_2iD02171_MTI00.	155	LT5_175_044_1999026_AAA02.	237	LT5_176_044_2000228_XXX04.
74	LT5_174_043_2003174_MTI02.	156	LT5_175_044_199904 2_XXX01.	238	LT5_176_044_2000260_AAA02.
75	LT5_174_043_2003190_MTI01.	157	LT5_175_044_19990 58_XXX00.	239	LT5_176_044_2002137_MTI00.
76	LT5_174_043_2003222_MTI01.	158	LT5_175_044_1999090_XXX01.	240	LT5_176_044_2002169_MTI00.
77	LT5_174_043_2003238_MTI01.	159	LT5_175_044_1999106_AAA0 1.	241	LT5_176_044_2003188_MTI01.
78	LT5_174_043_2003254_MTI01.	160	LT5_175_044_1999218_XXX02.	242	LT5_176_044_2003204_MTI01.
79	LT5_174_043_2003270_MTI01.	161	LT5_175_044_1999234_AAA04.	243	LT5_176_044_2003220_MTI01.
80	LT5_174_043_2003286_MTI01.	162	LT5_175_044_1999314_XXX02.	244	LT5_176_044_2003252_MTI01.
81	LT5_174_043_2009030_MTI00.	163	LT5_175_044_2000029_XXX02.	245	LT5_176_044_2003268_MTI01.
82	LT5_174_044_1984154_XXX03.	164	LT5_175_044_2000077_AAA04.	246	LT5_176_044_2003284_MTI01.

List of the downloaded images from Landsat 4,5 TM.

#	File name	#	File name	#	File name
1	LE7_174_043_1999219EDCOO.	46	LE7_174_044_2002275SGS01.	91	LE7_175_045_2000245SGSOO.
2	LE7_174_043_1999283EDCOO.	47	LE7_174_044_2002291SGS01.	92	LE7_175_045_2000293SGSOO.
3	LE7_174_043_1999331SGSOO.	48	LE7_174_044_2002355SGSOO.	93	LE7_175_045_2000309EDCOO.
4	LE7_174_043_2000014SGSOO.	49	LE7_174_044_2003022SGSOO.	94	LE7_175_045_2000341SGSOO.
5	LE7_174_043_2000062SGSOO.	50	LE7_174_044_2003038SGSOO.	95	LE7_175_045_2001039SGSOO.
6	LE7_174_043_2000142SGSOO.	51	LE7_174_044_2003070SGSOO.	96	LE7_175_045_2001087SGSOO.
7	LE7_174_043_2000174SGS01.	52	LE7_174_044_2003118ASNOO.	98	LE7_175_045_2001231SGSOO.
8	LE7_174_043_2000254SGSOO.	53	LE7_174_044_2005171ASNOO.	99	LE7_175_045_2001327SGSOO.
9	LE7_174_043_2000334EDCOO.	54	LE7_174_044_2005187ASN01.	100	LE7_175_045_2002010SGSOO.
10	LE7_174_043_2001064SGSOO.	55	LE7_174_044_2005219ASNOO.	101	LE7_175_045_2002074SGSOO.
11	LE7_174_043_2001128SGSOO.	56	LE7_174_044_2005235ASNOO.	102	LE7_175_045_2002138AGS01.
12	LE7_174_043_2001272EDCOO.	57	LE7_174_044_2006126ASNOO.	103	LE7_175_045_2002202SGSOO.
13	LE7_174_043_2001320SGSOO.	58	LE7_174_044_2006158ASNOO.	104	LE7_175_045_2002314SGSOO.
14	LE7_174_043_2002019SGSOO.	59	LE7_174_044_2006174ASNOO.	105	LE7_175_045_2002362SGSOO.
15	LE7_174_043_2002051SGSOO.	60	LE7_174_044_2006190ASNOO.	106	LE7_175_045_2003045SGSOO.
16	LE7_174_043_2002067SGSOO.	61	LE7_174_044_2006222ASNOO.	107	LE7_175_045_2003077SGSOO.
17	LE7_174_043_2002083SGSOO.	62	LE7_174_044_2006238ASNOO.	108	LE7_175_045_2003125ASNOO.
18	LE7_174_043_2002099SGSOO.	63	LE7_175_044_1999242EDCOO.	109	LE7_175_045_2005210ASNOO.
19	LE7_174_043_2002163SGSOO.	64	LE7_175_044_2000085EDCOO.	110	LE7_175_045_2005210ASNOO.
20	LE7_174_043_2002259SGSOO.	65	LE7_175_044_2000149SGSOO.	111	LE7_175_045_2005226ASNOO.
21	LE7_174_043_2002275SGSOO.	66	LE7_175_044_2000245SGSOO.	112	LE7_175_045_2006165ASNOO.
22	LE7_174_043_2002291SGSOO.	67	LE7_175_044_2000293SGSOO.	113	LE7_175_045_2006197ASNOO.
23	LE7_174_043_2002307SGSOO.	68	LE7_175_044_2000309EDCOO.	114	LE7_176_044_1999313EDCOO.
24	LE7_174_043_2003006EDC03.	69	LE7_175_044_2000341SGSOO.	115	LE7_176_044_2000012SGSOO.
25	LE7_174_043_2003054SGSOO.	70	LE7_175_044_2001039SGSOO.	116	LE7_176_044_2000156EDCOO.
26	LE7_174_043_2003118ASNOO.	71	LE7_175_044_2001087SGSOO.	117	LE7_176_044_2000236EDCOO.
27	LE7_174_044_1999219EDCOO.	72	LE7_175_044_2001103SGSOO.	118	LE7_176_044_2000364SGSOO.
28	LE7_174_044_1999331SGSOO.	73	LE7_175_044_2001231SGSOO.	119	LE7_176_044_2001046SGSOO.
29	LE7_174_044_2000014SGSOO.	74	LE7_175_044_2001359SGSOO.	120	LE7_176_044_2001078EDCOO.
30	LE7_174_044_2000142SGSOO.	75	LE7_175_044_2002074SGSOO.	121	LE7_176_044_2001110EDCOO.
31	LE7_174_044_2000254SGSOO.	76	LE7_175_044_2002138AGSOO.	122	LE7_176_044_2001174EDCOO.
32	LE7_174_044_2000302SGSOO.	77	LE7_175_044_2002202SGSOO.	123	LE7_176_044_2001238SGS01.
33	LE7_174_044_2000334EDCOO.	78	LE7_175_044_2002314SGSOO.	124	LE7_176_044_2001350SGSOO.
34	LE7_174_044_2001064SGSOO.	79	LE7_175_044_2002362SGSOO.	125	LE7_176_044_2002033SGSOO.
35	LE7_174_044_2001128SGSOO.	80	LE7_175_044_2003029SGSOO.	126	LE7_176_044_2002081EDCOO.
36	LE7_174_044_2001224SGSOO.	81	LE7_175_044_2003045SGSOO.	127	LE7_176_044_2002097SGSOO.
37	LE7_174_044_2001304SGSOO.	82	LE7_175_044_2003077SGSOO.	128	LE7_176_044_2002161EDCOO.
38	LE7_174_044_2001320SGSOO.	83	LE7_175_044_2003093ASNOO.	129	LE7_176_044_2002241SGSOO.
39	LE7_174_044_2002019SGSOO.	84	LE7_175_044_2005210ASNOO.	130	LE7_176_044_2002337SGS01.
40	LE7_174_044_2002051SGSOO.	85	LE7_175_044_2005226ASNOO.	131	LE7_176_044_2003036SGSOO.
41	LE7_174_044_2002067SGSOO.	86	LE7_175_044_2006165ASNOO.	132	LE7_176_044_2003068SGSOO.
42	LE7_174_044_2002083SGSOO.	87	LE7_175_044_2010336ASNOO.	133	LE7_176_044_2003116ASNOO.
43	LE7_174_044_2002099SGS01.	88	LE7_175_045_1999242EDCOO.	134	LE7_174_044_1988061AAA05.
44	LE7_174_044_2002163SGSOO.	89	LE7_175_045_2000085EDCOO.	135	LE7_174_044_2009030MTIOO.
45	LE7_174_044_2002259SGSOO.	90	LE7_175_045_2000149SGSOO.		

Appendix F

HADR Bed Soil Sedimentary Categories Database (HADRBSDB)

Sample of reports for collecting bed soil data by the survey mission in 2007
Primary report for mechanical analysis of bed material sediments along inside the Sudanese limits (January- February 2007).

Med.= Medium, Mid.= Middle, Avg.= Average, Loc.= Location

No.	Name of Cross Section	Distance from High Dam (Km.)	Loc.	Gravel				Sand				Silt and Clay
				Coarse	Med.	Fine	Total	Coarse	Med.	Fine	Total	
1	Dobrosa	337.5	East	0.00	0.00	0.00	0.00	0.00	0.00	0.07	0.07	99.93
			Mid.	0.00	0.00	0.00	0.00	0.00	0.00	0.08	0.08	99.92
			West	0.00	0.00	0.00	0.00	0.00	6.50	10.97	17.47	82.53
			Avg.	0.00	0.00	0.00	0.00	0.00	2.17	3.71	5.87	94.13
2	Halfa Doghaim	347	East	0.00	0.00	0.00	0.00	0.00	0.00	0.09	0.09	99.91
			Mid.	0.00	0.00	0.00	0.00	0.00	0.00	0.10	0.10	99.90
			West	0.00	0.00	0.00	0.00	0.00	0.00	0.09	0.09	99.91
			Avg.	0.00	0.00	0.00	0.00	0.00	0.00	0.09	0.09	99.91
3	Abd El Kader	352	East	0.00	0.00	0.00	0.00	0.00	0.00	0.13	0.13	99.87
			Mid.	0.00	0.00	0.00	0.00	0.00	0.00	0.12	0.12	99.88
			West	0.00	0.00	0.00	0.00	0.00	0.00	0.11	0.11	99.89
			Avg.	0.00	0.00	0.00	0.00	0.00	0.00	0.12	0.12	99.88
4	Second Cataract	357	East	0.00	0.00	0.00	0.00	0.00	0.00	0.10	0.10	99.90
			Mid.	0.00	0.00	0.00	0.00	0.00	0.00	0.04	0.04	99.96
			West	0.00	0.00	0.00	0.00	0.00	0.00	0.14	0.14	99.86
			Avg.	0.00	0.00	0.00	0.00	0.00	0.00	0.09	0.09	99.91
5	Amaka	364	East	0.00	0.00	0.00	0.00	0.00	0.00	0.06	0.06	99.94
			Mid.	0.00	0.00	0.00	0.00	0.00	0.00	0.07	0.07	99.93
			West	0.00	0.00	0.00	0.00	0.00	0.00	0.13	0.13	99.87
			Avg.	0.00	0.00	0.00	0.00	0.00	0.00	0.09	0.09	99.91
6	Amaka Strait	368	East				No Sample					
			Mid.	0.00	0.00	0.00	0.00	0.00	0.00	0.60	0.60	99.40
			West	0.00	0.00	0.00	0.00	0.00	0.00	1.53	1.53	98.47
			Avg.	0.00	0.00	0.00	0.00	0.00	0.00	1.07	1.07	98.94
7	Gomay	372	East	0.00	0.00	0.00	0.00	0.00	0.00	0.20	0.20	99.80
			Mid.	0.00	0.00	0.00	0.00	0.00	0.00	0.26	0.26	99.74
			West	0.00	0.00	0.00	0.00	0.00	0.00	1.70	1.70	98.30
			Avg.	0.00	0.00	0.00	0.00	0.00	0.00	0.72	0.72	99.28
8	Morshed	378.5	East	0.00	0.00	0.00	0.00	0.00	0.00	0.14	0.14	99.86
			Mid.	0.00	0.00	0.00	0.00	0.00	0.00	0.26	0.26	99.74
			West	0.00	0.00	0.00	0.00	0.00	0.00	3.69	3.69	96.31
			Avg.	0.00	0.00	0.00	0.00	0.00	0.00	1.36	1.36	98.64
9	North of Kajnarity (W-W)	384	East	0.00	0.00	0.00	0.00	0.00	0.00	1.94	1.94	98.06
			Mid.	0.00	0.00	0.00	0.00	0.00	0.00	0.37	0.37	99.63
			West	0.00	0.00	0.00	0.00	0.00	0.00	0.53	0.53	99.47
			Avg.	0.00	0.00	0.00	0.00	0.00	0.00	0.95	0.95	99.05
10	Kajnarity	394	East	0.00	0.00	0.00	0.00	0.00	0.00	0.18	0.18	99.82
			Mid.	0.00	0.00	0.00	0.00	0.00	0.00	0.64	0.64	99.36
			West	0.00	0.00	0.00	0.00	0.00	0.00	0.09	0.09	99.91
			Avg.	0.00	0.00	0.00	0.00	0.00	0.00	0.30	0.30	99.70

#	Name	Dist.	Dir.									
11	Semna	403.5	East	0.00	0.00	0.00	0.00	0.00	0.00	0.59	0.59	99.41
			Mid.	0.00	0.00	0.00	0.00	0.00	0.00	1.47	1.47	98.53
			West	0.00	0.00	0.00	0.00	0.00	0.00	0.19	0.19	99.81
			Avg.	0.00	0.00	0.00	0.00	0.00	0.00	0.75	0.75	99.25
12	Ateery	415.5	East	0.00	0.00	0.00	0.00	0.02	0.28	19.69	19.99	80.01
			Mid.	0.00	0.00	0.00	0.00	0.00	0.00	0.34	0.34	99.66
			West	0.00	0.00	0.00	0.00	0.00	0.20	79.50	79.70	20.30
			Avg.	0.00	0.00	0.00	0.00	0.01	0.16	33.18	33.34	66.66
13	North of El Dowaishat (H-H)	422	East	0.00	0.00	0.00	0.00	0.00	0.00	0.76	0.76	99.24
			Mid.	0.00	0.00	0.00	0.00	0.00	0.80	58.11	58.91	41.09
			West	0.00	0.00	0.00	0.00	0.00	0.00	4.22	4.22	95.78
			Avg.	0.00	0.00	0.00	0.00	0.00	0.27	21.03	21.30	78.70
14	El Dowaishat	431	East	0.00	0.00	0.00	0.00	0.00	0.00	2.78	2.78	97.22
			Mid.	0.00	0.00	0.00	0.00	0.00	19.00	57.59	76.59	23.41
			West	0.00	0.00	0.00	0.00	0.00	0.00	1.84	1.84	98.16
			Avg.	0.00	0.00	0.00	0.00	0.00	6.33	20.74	27.07	72.93
15	North of El Malek El Nasser (C-C)	438	East	0.00	0.00	0.00	0.00	0.00	25.00	71.74	96.74	3.26
			Mid.	0.00	0.00	0.00	0.00	0.00	14.50	83.48	97.98	2.02
			West	0.00	0.00	0.00	0.00	0.00	0.00	0.00	0.00	100.00
			Avg.	0.00	0.00	0.00	0.00	0.00	13.17	51.74	64.91	35.09
16	El Malek El Nasser	448	East	0.00	0.00	0.00	0.00	0.00	0.00	4.30	4.30	95.70
			Mid.	0.00	0.00	0.00	0.00	0.00	22.00	56.11	78.11	21.89
			West	0.00	0.00	0.00	0.00	0.00	16.00	80.38	96.38	3.62
			Avg.	0.00	0.00	0.00	0.00	0.00	12.67	46.93	59.60	40.40
17	North of Okama (B-B)	456	East	0.00	0.00	0.00	0.00	0.00	39.00	42.83	81.83	18.17
			Mid.	0.00	0.00	0.00	0.00	0.20	34.80	59.78	94.78	5.22
			West	0.00	0.00	0.00	0.00	0.00	19.00	58.95	77.95	22.05
			Avg.	0.00	0.00	0.00	0.00	0.07	30.93	53.85	84.85	15.15
18	Okama	466	East	0.00	0.00	0.00	0.00	0.00	3.70	52.20	55.90	44.10
			Mid.	0.00	0.00	0.00	0.00	0.00	46.00	46.52	92.52	7.48
			West				No Sample					
			Avg.	0.00	0.00	0.00	0.00	0.00	24.85	49.36	74.21	25.79
19	Akasha (A-A)	477	East	0.00	0.00	0.00	0.00	0.00	76.00	23.73	99.73	0.27
			Mid.	0.00	0.00	0.00	0.00	0.02	71.98	27.92	99.92	0.08
			West	0.00	0.00	0.00	0.00	0.09	58.91	39.50	98.50	1.50
			Avg.	0.00	0.00	0.00	0.00	0.04	68.96	30.38	99.38	0.62
20	Daka	487.5	East	0.00	0.00	0.00	0.00	0.30	90.70	7.77	98.77	1.23
			Mid.	0.00	0.00	0.48	0.48	1.73	90.29	6.24	98.26	1.26
			West	0.00	0.00	0.00	0.00	0.09	91.71	7.47	99.27	0.73
			Avg.	0.00	0.00	0.16	0.16	0.71	90.90	7.16	98.77	1.07

Appendix F

Sediment accumulation and erosion volumes at the different sections along Lake Nubia from 1964 till 2010

1978 - 1979		1977 - 1978		1973 - 1977		1964 - 1973		Distance around section (km)	Distance from the HAD (km)	Section
Change in Section Volume (hm³)	Change in Section area (m²)	Change in Section Volume (hm³)	Change in Section area (m²)	Change in Section Volume (hm³)	Change in Section area (m²)	Change in Section Volume (hm³)	Change in Section area (m²)			
-0.65	-38.24	-0.65	-25.29	-	-	-	-	17.0	487.0	Daka
1.26	64.62	1.26	56.92	-	-	-	-	19.5	466.0	Okma
12.14	693.71	12.14	679.43	44.05	2517.14	6.41	366.29	17.5	448.0	Malk Nasser
8.89	547.08	8.89	614.15	110.42	6795.08	60.65	3732.31	16.3	431.0	Diwaishat
23.27	1692.36	23.27	1765.09	124.94	9086.55	48.03	3493.09	13.8	415.5	Ateery
10.74	681.90	10.74	504.13	52.83	3354.29	30.12	1912.38	15.8	403.5	Samanh
21.86	1748.80	21.86	1955.20	-	-	58.42	4673.60	12.5	394.0	Kaganarty
16.19	1471.82	16.19	1409.09	101.39	9217.27	80.87	7351.82	11.0	378.5	Morshid
9.18	1748.57	9.18	1771.43	47.58	9062.86	43.31	8249.52	5.3	372.0	Gomy
4.55	1137.50	4.55	1130.00	-	-	-	-	4.0	368.0	Madik Amka
5.00	909.09	5.00	898.18	-	-	-	-	5.5	364.0	Amka
23.22	3870.00	23.22	3841.67	-	-	-	-	6.0	357.0	Second Cataract
-	-	-	-	-	-	-	-	5.0	352.0	Abd El kadir
22.87	3154.48	22.87	3111.72	-	-	-	-	7.3	347.0	Halfa Dighaim
0.24	50.53	0.24	44.21	-	-	-	-	4.8	338.0	Dabrosa
0.77	128.33	0.77	123.33	-	-	-	-	6.0	331.1	Arkeen
9.72	304.50	9.72	296.12	-	-	-	-	32.3	325.0	Sara
169.25		169.25		481.21		327.81			Total Change in Volume hm³	

1984 - 1985		1983 - 1984			1982 - 1983		1981 - 1982		1980 - 1981		1979 - 1980	
Change in Section Volume (hm³)	Change in Section area (m²)	Change in Section Volume (hm³)	Change in Section area (m²)	Change in Section Volume (hm³)	Change in Section area (m²)	Change in Section Volume (hm³)	Change in Section area (m²)	Change in Section Volume (hm³)	Change in Section area (m²)	Change in Section Volume (hm³)	Change in Section area (m²)
-1.79	-93.53	-1.26	-90.59	-1.26	-83.53	-1.26	-74.12	-0.96	-56.47	-0.78	-45.88
1.30	121.03	2.94	124.62	2.94	138.97	2.94	150.77	1.78	91.28	1.42	72.82
5.59	369.71	7.96	407.43	7.96	429.14	7.96	454.86	13.07	746.86	12.21	697.71
2.12	192.00	4.22	240.62	4.22	256.00	4.22	259.69	6.79	417.85	7.20	443.08
5.68	560.73	11.15	719.27	11.15	752.00	11.15	810.91	16.03	1165.82	20.77	1510.55
9.46	690.16	13.84	756.83	13.84	781.59	13.84	878.73	15.06	956.19	12.95	822.22
8.68	829.60	14.99	884.80	14.99	1040.00	14.99	1199.20	17.56	1404.80	19.89	1591.20
7.91	999.09	14.04	1101.82	14.04	1230.00	14.04	1276.36	15.38	1398.18	17.07	1551.82
2.36	689.52	5.1	920.00	5.1	946.67	5.1	971.43	7.16	1363.81	7.45	1419.05
5.43	1292.50	4.7	1250.00	4.7	1217.50	4.7	1175.00	4.68	1170.00	4.59	1147.50
21.13	2987.27	6.1	2350.91	6.1	1621.82	6.1	1109.09	3.75	681.82	4.47	812.73
23.09	3968.33	24.22	4061.67	24.22	4085.00	24.22	4036.67	23.87	3978.33	23.34	3890.00
-	-	-	-	-	-	-	-	-	-	-	-
20.08	3089.66	23.92	3259.31	23.92	3358.62	23.92	3299.31	23.63	3259.31	23.27	3209.66
0.29	75.79	0.48	82.11	0.48	92.63	0.48	101.05	0.35	73.68	0.31	65.26
0.77	133.33	0.98	136.67	0.98	143.33	0.98	163.33	0.92	153.33	0.81	135.00
18.17	496.74	13.20	452.40	13.20	422.95	13.20	410.23	11.00	341.09	10.26	318.14
130.27		146.58		146.58		146.58		160.07		165.23	

Appendix F

1990 - 1991		1989 - 1990		1988 - 1989		1987 - 1988		1986 - 1987		1985 - 1986	
Change in Section Volume (hm³)	Change in Section area (m²)	Change in Section Volume (hm³)	Change in Section area (m²)	Change in Section Volume (hm³)	Change in Section area (m²)	Change in Section Volume (hm³)	Change in Section area (m²)	Change in Section Volume (hm³)	Change in Section area (m²)	Change in Section Volume (hm³)	Change in Section area (m²)
-0.85	-59.41	-2.08	-110.00	-2.08	-122.35	-1.91	-112.35	-1.79	-105.29	-1.79	-94.71
-0.94	-43.08	1.06	49.23	1.06	54.36	0.93	47.69	1.30	66.67	1.30	101.03
3.05	152.29	2.78	146.29	2.78	158.86	2.82	161.14	5.59	319.43	5.59	338.29
2.16	134.00	1.84	51.69	1.84	113.23	1.13	69.54	2.12	130.46	2.12	150.15
5.29	362.55	5.20	379.64	5.20	378.18	5.48	398.55	5.68	413.09	5.68	461.09
5.83	358.25	5.59	366.98	5.59	354.92	5.49	348.57	9.46	600.63	9.46	638.73
5.21	449.20	6.51	480.00	6.51	520.80	5.59	447.20	8.68	694.40	8.68	726.40
6.91	540.91	5.02	512.73	5.02	456.36	6.97	633.64	7.91	719.09	7.91	829.09
3.86	643.33	2.78	485.71	2.78	529.52	2.07	394.29	2.36	449.52	2.36	531.43
5.26	1302.50	5.16	1292.50	5.16	1290.00	4.24	1060.00	5.43	1357.50	5.43	1302.50
17.19	3505.45	21.58	3923.64	21.58	3923.64	20.01	3638.18	21.13	3841.82	21.13	3465.45
22.60	3803.33	24.22	3993.33	24.22	4036.67	26.21	4368.33	23.09	3848.33	23.09	3910.00
23.65	5250.00										
15.61	2626.21	26.73	2995.86	26.73	3686.90	26.23	3617.93	20.08	2769.66	20.08	3028.97
		0.22		0.22	46.32	0.25	52.63	0.29	61.05	0.29	71.58
		0.75		0.75	125.00	0.76	126.67	0.77	128.33	0.77	130.00
		13.23		13.23	410.23	15.02	465.74	18.17	563.72	18.17	531.47
114.83		120.59		120.59		121.29		130.27		130.27	

1996- 1997		1995 - 1996		1994 - 1995		1993 - 1994		1992 - 1993		1991 - 1992	
Change in Section Volume (hm³)	Change in Section area (m²)	Change in Section Volume (hm³)	Change in Section area (m²)	Change in Section Volume (hm³)	Change in Section area (m²)	Change in Section Volume (hm³)	Change in Section area (m²)	Change in Section Volume (hm³)	Change in Section area (m²)	Change in Section Volume (hm³)	Change in Section area (m²)
3.97	233.53	-0.65	114.12	-0.65	-38.24	-0.62	-36.47	-0.70	-41.18	-0.85	-50.00
3.68	188.72	-1.78	147.69	-1.78	-91.28	-2.61	-133.85	-1.15	-58.97	-0.94	-48.21
3.51	200.57	-0.74	67.43	-0.74	-42.29	1.07	61.14	3.39	193.71	3.05	174.29
8.95	550.77	1.29	260.92	1.29	79.38	2.04	125.54	2.49	153.23	2.16	132.92
5.09	370.18	3.97	234.91	3.97	288.73	4.69	341.09	5.32	386.91	5.29	384.73
6.69	424.76	4.50	249.52	4.50	285.71	5.42	344.13	5.86	372.06	5.83	370.16
10.83	866.40	6.82	756.80	6.82	545.60	5.79	463.20	4.49	359.20	5.21	416.80
7.55	686.36	6.37	672.73	6.37	579.09	6.30	572.73	7.39	671.82	6.91	628.18
2.72	518.10	2.43	508.57	2.43	462.86	1.94	369.52	4.02	765.71	3.86	735.24
7.12	1780.00	5.67	1520.00	5.67	1417.50	5.41	1352.50	5.31	1327.50	5.26	1315.00
12.86	2338.18	8.12	1123.64	8.12	1476.36	10.06	1829.09	16.42	2985.45	17.19	3125.45
9.45	1575.00	15.93	1925.00	15.93	2655.00	20.42	3403.33	22.52	3753.33	22.60	3766.67
14.13	2826.00	18.89	3464.00	18.89	3778.00	20.97	4194.00	27.86	5572.00	23.65	4730.00
13.06	1801.38	12.27	1558.62	12.27	1692.41	13.21	1822.07	13.84	1908.97	15.61	2153.10
-	-	-	-	-	-	-	-	-	-	-	-
-	-	-	-	-	-	-	-	-	-	-	-
109.61		83.09		83.09		94.09		117.06		114.83	

Appendix F

2002-2003		2001 - 2002		2000- 2001		1999- 2000		1998 - 1999		1997- 1998	
Change in Section Volume (hm³)	Change in Section area (m²)	Change in Section Volume (hm³)	Change in Section area (m²)	Change in Section Volume (hm³)	Change in Section area (m²)	Change in Section Volume (hm³)	Change in Section area (m²)	Change in Section Volume (hm³)	Change in Section area (m²)	Change in Section Volume (hm³)	Change in Section area (m²)
7.66	450.59	7.66	450.59	-3.39	-199.41	9.89	581.76	7.95	467.65	6.98	410.59
16.07	823.85	16.07	823.85	10.26	526.15	9.91	508.21	7.49	384.10	6.16	315.90
4.88	278.86	4.88	278.86	8.73	498.86	7.39	422.29	6.93	396.00	4.36	249.14
-0.87	-53.23	-0.87	-53.23	11.93	734.15	12.19	750.15	13.97	859.69	11.71	720.62
0.71	51.27	0.71	51.27	6.34	461.09	7.81	568.00	8.99	653.82	6.87	499.64
2.29	45.40	2.29	145.40	5.61	356.19	8.44	535.87	9.17	582.22	8.64	548.57
4.22	-37.20	4.22	337.20	10.33	826.40	11.13	890.40	19.98	1598.40	13.87	1109.60
10.78	579.55	10.78	979.55	2.52	229.09	7.40	672.73	9.74	885.45	8.06	732.73
0.97	183.81	0.97	183.81	5.54	1055.24	5.38	1024.76	5.60	1066.67	2.96	563.81
-0.95	-236.25	-0.95	-236.25	4.98	1245.00	8.18	2045.00	10.10	2525.00	7.28	1820.00
9.03	1640.91	9.03	1640.91	18.21	3310.91	18.41	3347.27	19.25	3500.00	15.20	2763.64
25.05	4174.17	25.05	4174.17	4.97	828.33	8.34	1390.00	13.94	2323.33	10.98	1830.00
23.90	4780.00	23.90	4780.00	14.28	2856.00	15.13	3026.00	16.96	3392.00	15.32	3064.00
8.02	1105.52	8.02	1105.52	5.16	711.72	12.32	1699.31	15.09	2081.38	14.17	1954.48
-14.13	-2974.74	-14.13	-2974.74	72.45	15252.63	-	-	-	-	-	-
-	-	-	-	-	-	-	-	-	-	-	-
-	-	-	-	-	-	-	-	-	-	-	-
97.60		97.60		177.92		141.92		165.16		132.56	

Appendix F

2008- 2009		2007- 2008		2006- 2007		2005- 2006		2004 - 2005		2003- 2004	
Change in Section Volume (hm³)	Change in Section area (m²)	Change in Section Volume (hm³)	Change in Section area (m²)	Change in Section Volume (hm³)	Change in Section area (m²)	Change in Section Volume (hm³)	Change in Section area (m²)	Change in Section Volume (hm³)	Change in Section area (m²)	Change in Section Volume (hm³)	Change in Section area (m²)
-1.84	-349.97	0.48	91.00	-0.69	-131.00	0.29	55.50	0.29	55.50	-1.67	-317.16
9.34	889.28	0.47	45.00	-6.95	-662.00	-1.24	-118.00	-1.24	-118.00	2.10	200.40
0.30	32.80	2.02	224.00	-3.81	-423.40	1.40	155.50	1.40	155.50	-3.65	-405.89
-1.64	-204.80	4.31	539.00	0.22	28.00	-0.76	-94.50	-0.76	-94.50	-1.56	-194.50
-0.01	-0.90	0.44	48.00	0.74	80.50	-0.77	-82.50	-0.77	-82.50	2.29	247.46
-1.35	-125.36	-1.66	-154.00	-1.69	-157.00	1.02	95.00	1.02	95.00	6.56	610.00
-7.43	761.60	7.96	816.00	-0.48	-49.00	-0.87	-89.00	-0.87	-89.00	-9.98	-1023.68
4.21	702.20	-0.39	-65.00	-4.63	-771.00	-0.06	-9.00	-0.06	-9.00	5.98	997.27
-0.95	181.50	2.79	532.00	-0.49	-93.00	-0.79	-149.50	-0.79	-149.50	-2.68	-510.00
-0.98	-245.30	4.68	1169.00	-1.08	-270.00	1.05	261.00	1.05	261.00	-4.34	-1085.80
0.90	163.90	24.83	4515.00	-3.12	-568.00	3.03	551.00	3.03	551.00	6.37	1158.50
21.74	3622.80	80.04	13340.00	30.10	5016.00	2.00	333.00	2.00	333.00	124.04	20672.80
20.70	4139.10	41.77	8354.00	83.06	16612.00	48.75	9749.50	48.75	9749.50	105.74	21148.00
9.02	1244.70	14.84	2047.00	54.79	7557.00	9.27	1278.50	9.27	1278.50	86.70	11958.50
23.82	2996.60	9.91	1246.00	69.81	8781.00	11.32	1424.00	11.32	1424.00	63.00	7924.50
-	-	-	-	-	-	-	-	-	-	-	-
-	-	-	-	-	-	-	-	-	-	-	-
56.83		192.49		215.78		73.66		73.66		378.90	

Appendix F

Average *Change in Section Volume (hm³)	Maximum *Erosion Volume (hm³)	Maximum *Sediment Volume (hm³)	Change in Section Volume (hm³)	Total Sediment Volume (hm³)	Erosion Volume (hm³)	2009-2010 Change in Section Volume (hm³)	2009-2010 Change in Section area (m²)
0.40	-3.39	9.89	13.16	43.38	-30.22	-1.84	-349.97
2.84	-6.95	16.07	93.74	112.37	-18.63	9.34	889.28
4.43	-3.81	13.07	196.77	205.71	-8.94	0.30	32.80
3.70	-1.64	13.97	293.3	301.38	-8.08	-1.64	-204.80
6.57	-0.77	23.27	389.64	391.19	-1.55	-0.01	-0.90
6.48	-1.69	15.06	296.63	302.68	-6.05	-1.35	-125.36
7.90	-9.98	21.86	319.02	345.21	-26.19	-7.43	761.60
7.70	-4.63	17.07	436.29	436.79	-0.5	4.21	702.20
3.08	-2.68	9.18	192.69	198.55	-5.855	-0.95	181.50
3.97	-4.34	10.10	131.06	140.33	-9.27	-0.98	-245.30
11.46	-3.12	24.83	378.1	381.22	-3.12	0.90	163.90
24.54		124.04	809.66	809.66		21.74	3622.80
31.35		105.74	627	627.00		20.70	4139.10
19.88		86.70	655.88	655.88		9.02	1244.70
13.82		72.45	317.85	317.85		23.82	2996.60
0.83		0.98	10.78	10.78		-	-
13.56		18.17	176.29	176.29		-	-
137.24	-23.88	402.78	5337.86	5456.27	-126.48	75.83	

* The maximum sediment, the maximum erosion and the average were estimated from 1978 to 2010.

Appendix G

Results of the Analysis of HADR Meteorological Database (HADRMTDB)

Air Temperature

Kalabsha Station:

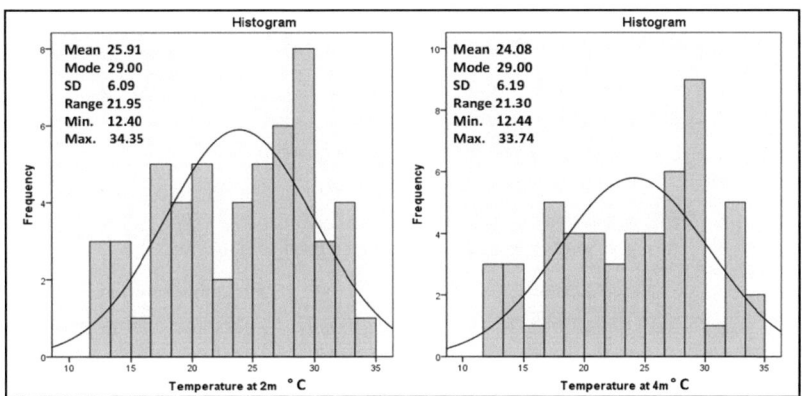

Histogram of the monthly mean air temperatures measured at two and four meters above surface water level at Kalabsha station °C

The average, median, maximum and minimum monthly air temperatures measured at two and four meters above surface water level at Kalabsha station

Month	Air temperature at 2 m (°C)				Air temperature at 4 m (°C)			
	Mean	Median	Maximum	Minimum	Mean	Median	Maximum	Minimum
January	14.16	13.68	16.82	12.47	14.11	13.52	16.93	12.44
February	15.39	14.82	19.50	12.40	15.59	15.02	19.83	12.51
March	18.96	18.40	21.87	17.17	19.22	18.57	22.34	17.42
April	22.45	21.46	26.31	20.58	22.85	21.75	27.00	20.91
May	26.18	25.46	29.13	24.66	26.45	25.60	29.75	24.83
June	29.28	27.85	31.89	27.53	29.62	28.09	32.36	27.80
July	30.38	29.24	32.90	28.22	30.66	29.37	33.50	28.32
August	30.63	29.33	34.35	28.91	30.83	29.39	33.74	29.13
September	28.95	28.45	31.48	26.38	29.12	28.33	31.86	26.28
October	26.30	25.07	29.06	24.43	26.36	25.11	29.38	24.41
November	20.89	21.56	23.85	17.75	20.85	21.47	23.67	17.52
December	17.60	17.70	21.20	13.77	17.59	17.73	21.35	13.52
Annual	25.91	24.84	34.35	12.40	24.08	24.95	33.74	12.44

Aswan station:

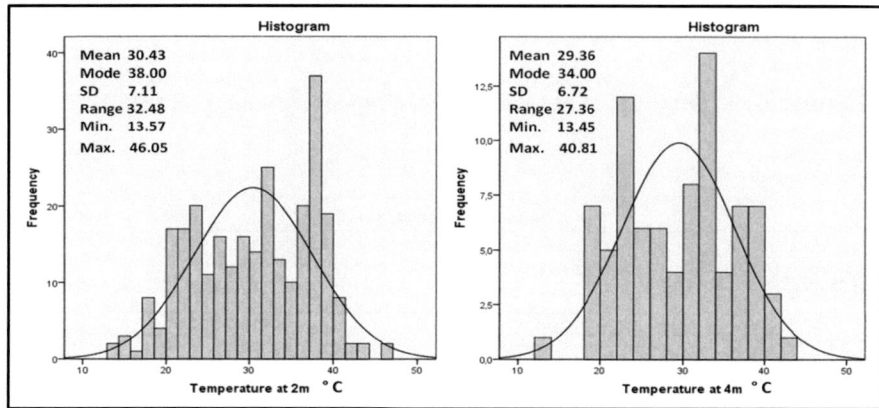

Histogram of the monthly mean air temperature measured at two and four meters above surface water level at Aswan station in °C

The average, median, maximum and minimum monthly air temperatures measured at two and four meters above surface water level at Aswan station

Month	Air temperature at 2 m (°C)				Air temperature at 4 m (°C)			
	Mean	Median	Maximum	Minimum	Mean	Median	Maximum	Minimum
January	20.81	21.25	26.23	14.86	21.06	21.14	23.73	18.21
February	22.77	22.46	40.72	16.89	21.65	22.05	25.76	18.07
March	26.57	26.35	43.04	21.19	24.94	23.99	30.16	21.63
April	31.64	32.40	39.50	25.22	29.84	30.29	33.56	25.84
May	35.24	36.10	41.29	28.62	34.24	35.82	39.81	29.63
June	37.16	38.00	46.05	30.62	35.14	35.29	37.68	32.49
July	37.51	38.05	45.72	31.24	36.80	36.49	40.45	33.48
August	36.90	38.00	42.12	29.32	37.35	38.48	40.81	33.77
September	35.19	36.10	40.58	24.51	35.05	35.30	38.35	31.22
October	31.48	32.27	36.24	22.48	30.56	30.98	34.36	24.11
November	26.02	26.20	32.37	17.81	25.01	27.00	29.12	18.72
December	21.72	22.80	28.41	13.57	20.90	22.72	24.01	13.45
Annual	30.43	31.24	46.05	13.57	29.36	30.20	40.81	13.45

The average, median, maximum and minimum monthly air temperatures measured at two and four meters above surface water level at Aswan station for different decades

Decade	Temperature at 2 m (°C)				Temperature at 4 m (°C)			
	Mean	Median	Maximum	Minimum	Mean	Median	Maximum	Minimum
1970s	28.75	29.32	38.60	13.98	no data	no data	no data	no data
1980s	31.36	32.60	40.24	17.60	no data	no data	no data	no data
1990s	32.10	34.59	40.90	15.94	31.30	32.42	40.45	14.14
2000s	33.49	35.27	48.89	13.57	32.04	33.16	40.81	13.45
Average	30.48	31.24	48.89	13.57	29.36	30.2	40.81	13.45

El-Alaky Station:

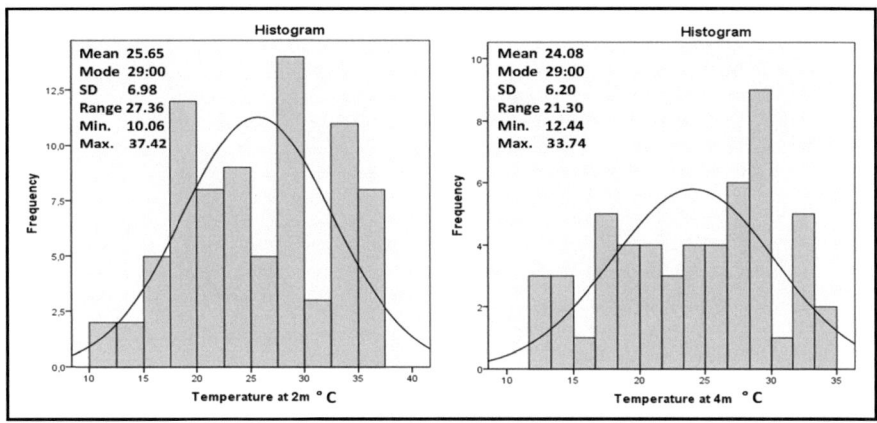

Histogram of the monthly mean air temperatures measured at two and four meters above surface water level at El-Alaky station °C

The average, median, maximum and minimum monthly air temperatures measured at two and four meters above surface water level at El-Alaky station

Month	Air temperature at 2 m (°C)				Air temperature at 4 m (°C)			
	Mean	Median	Maximum	Minimum	Mean	Median	Maximum	Minimum
January	17.01	17.97	20.59	10.98	14.11	13.52	16.93	12.44
February	17.10	17.80	21.75	10.06	15.59	15.02	19.83	12.51
March	21.44	22.40	31.43	13.94	19.22	18.57	22.34	17.42
April	24.62	28.22	29.49	17.81	22.85	21.75	27.00	20.91
May	28.97	31.72	34.74	19.68	26.45	25.60	29.75	24.83
June	31.45	33.70	37.42	24.98	29.62	28.09	32.36	27.80
July	32.65	35.57	36.52	27.65	30.66	29.37	33.50	28.32
August	32.73	34.68	35.80	28.83	30.83	29.39	33.74	29.13
September	29.86	30.51	33.56	23.66	29.12	28.53	31.86	26.28
October	25.47	25.22	29.12	20.70	26.36	25.11	29.38	24.41
November	21.92	21.84	26.57	17.57	20.85	21.47	23.67	17.52
December	19.45	21.10	21.52	14.10	17.59	17.73	21.35	13.52
Annual	25.65	25.31	37.42	10.06	24.08	24.95	33.74	12.44

The average, median, maximum and minimum monthly air Temperature measured at two and four meters above surface water level by El-Alaky station for different decades

Decade	Temperature at 2 m (°C)				Temperature at 4 m (°C)			
	Mean	Median	Maximum	Minimum	Mean	Median	Maximum	Minimum
1990s	28.25	29.16	36.52	19.51	no data	no data	no data	no data
2000s	29.19	29.98	37.42	10.06	28.08	24.95	33.74	12.44
Average	25.87	25.31	37.42	10.06	28.08	24.95	33.74	12.44

Abu -Simbel Station:

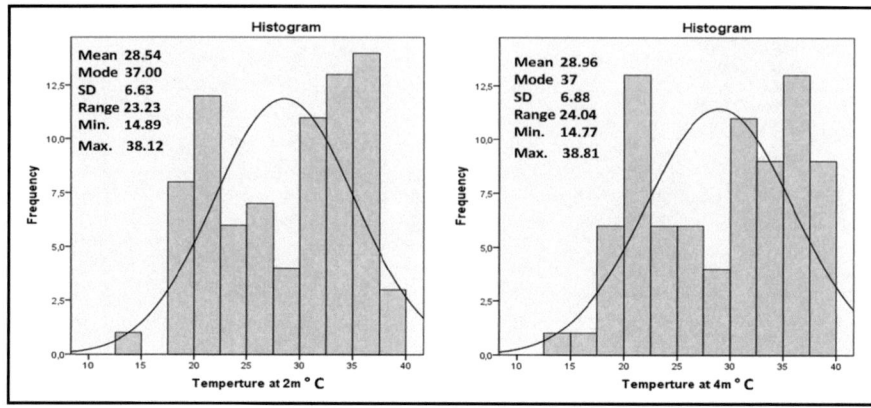

Histogram of the monthly mean air temperatures measured at two and four meters above surface water level at Abu -Simbel station °C

The average, median, maximum and minimum monthly air temperatures measured at two and four meters above surface water level at Abu -Simbel station

Month	Air temperature at 2 m (°C)				Air temperature at 4 m (°C)			
	Mean	Median	Maximum	Minimum	Mean	Median	Maximum	Minimum
January	18.59	18.95	20.73	14.89	18.71	19.14	20.92	14.77
February	20.16	20.39	22.79	17.79	20.40	20.68	23.17	17.78
March	24.46	23.30	27.76	21.05	24.94	23.65	29.05	21.11
April	28.72	30.09	31.11	23.56	29.37	31.02	31.83	23.54
May	33.08	33.33	34.50	30.35	33.97	34.14	35.92	30.88
June	35.60	35.20	36.46	35.00	36.45	35.95	37.60	35.64
July	36.06	37.20	37.56	32.78	36.50	37.66	38.18	32.60
August	36.17	37.04	38.12	32.97	36.64	37.58	38.81	32.71
September	33.91	34.75	35.77	30.77	34.34	35.30	36.31	30.92
October	30.36	30.25	35.10	27.35	30.79	30.74	36.55	27.39
November	24.12	25.07	26.90	20.73	24.21	25.11	27.09	20.92
December	20.41	21.15	21.63	17.54	20.43	21.32	21.69	17.34
Annual	28.54	30.09	38.12	14.89	28.96	30.92	38.81	14.77

The average, median, maximum and minimum monthly air temperatures measured at two and four meters above surface water level at Abu -Simbel station for different decades

Decade	Temperature at 2 m (°C)				Temperature at 4 m (°C)			
	Mean	Median	Maximum	Minimum	Mean	Median	Maximum	Minimum
1990s	28.77	29.90	37.37	19.65	28.31	30.58	38.14	19.89
2000s	29.49	30.35	38.12	14.89	28.90	30.92	38.81	14.77
Average	28.54	30.09	38.12	14.89	28.96	30.92	38.81	14.77

Toshka Station:

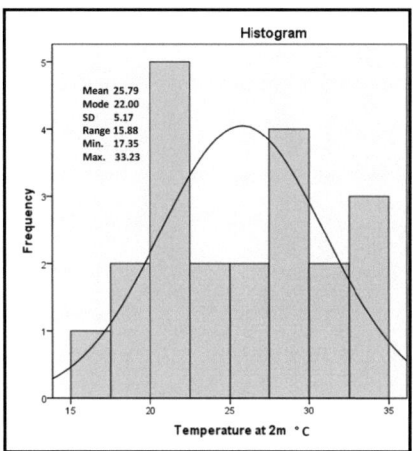

Histogram of the monthly mean air temperatures measured at two meters above surface water level at Toshka station °C

The average, median, maximum and minimum monthly air temperatures measured at two meters above surface water level at Toshka station

Month	Air temperature at 2 m (°C)			
	Mean	Median	Maximum	Minimum
January	18.24	18.24	19.13	17.35
February	20.33	20.33	20.52	20.14
March	22.80	22.80	23.25	22.36
April	26.39	26.39	27.04	25.74
May	29.77	29.77	29.94	29.60
June	32.61	32.61	33.13	32.10
July	32.76	32.76	32.76	32.76
August	33.23	33.23	33.23	33.23
September	31.18	31.18	31.18	31.18
October	28.25	28.25	28.40	28.09
November	23.14	23.14	24.14	22.14
December	20.72	20.72	21.83	19.62
Annual	25.79	25.74	33.23	17.35

Amada Station:

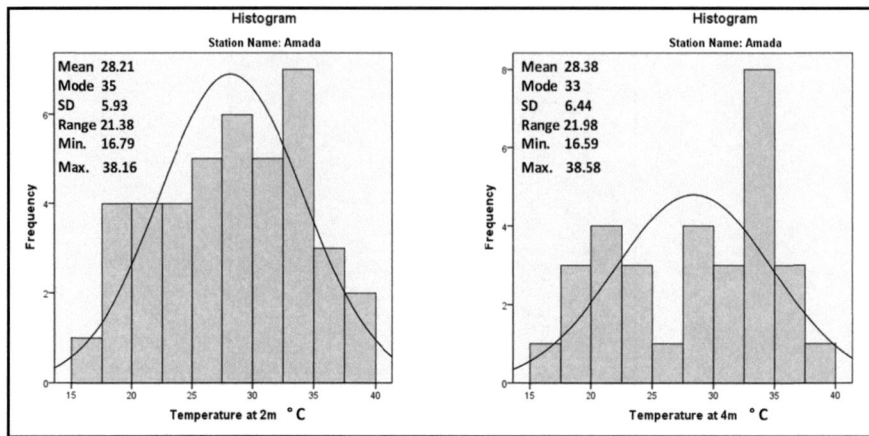

Histogram of the monthly mean air temperatures measured at two and four meters above surface water level at Amada station °C

The average, median, maximum and minimum monthly air temperatures measured at two and four meters above surface water level at Amada station

Month	Air temperature at 2 m (°C)				Air temperature at 4 m (°C)			
	Mean	Median	Maximum	Minimum	Mean	Median	Maximum	Minimum
January	17.88	17.88	18.97	16.79	17.86	17.86	19.12	16.59
February	19.42	19.42	19.87	18.96	19.60	19.60	20.16	19.04
March	21.61	21.61	22.18	21.04	21.98	21.98	22.53	21.43
April	27.29	27.29	27.66	26.92	27.95	27.95	28.36	27.54
May	32.30	30.87	36.44	29.60	32.82	31.46	36.82	30.17
June	33.31	32.47	36.60	31.69	34.36	33.58	36.93	32.56
July	32.08	32.79	38.16	26.47	35.24	33.94	38.58	33.21
August	33.78	33.83	37.60	26.47	35.25	34.21	37.49	34.04
September	31.34	32.39	34.06	26.24	32.50	32.85	34.02	30.64
October	26.96	27.92	29.00	22.56	27.81	28.22	29.23	25.96
November	23.27	23.38	24.38	21.94	23.54	24.38	24.54	21.70
December	20.11	20.11	20.68	19.55	20.19	20.19	20.78	19.60
Annual	28.21	28.92	38.16	16.79	28.38	29.23	38.58	16.59

Relative Humidity

Kalabsha Station:

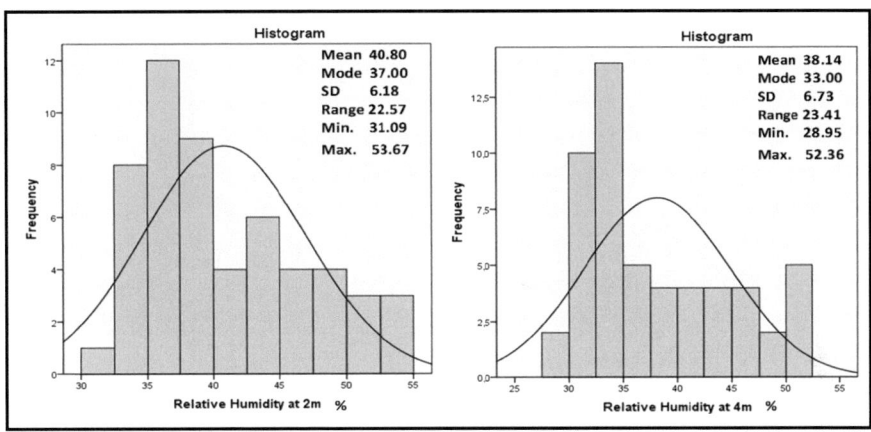

Histogram of the monthly mean relative humidity measured at two and four meters above surface water level at Kalabsha station in percentage

The average, median, maximum and minimum monthly relative humidity measured at two and four meters above surface water level at Kalabsha station

Month	Relative humidity at 2 m (%)				Relative humidity at 4 m (%)			
	Mean	Median	Maximum	Minimum	Mean	Median	Maximum	Minimum
January	52.72	53.02	53.67	51.15	51.29	51.88	52.36	49.03
February	47.45	47.12	49.96	45.61	44.79	44.44	47.96	42.30
March	40.94	40.72	43.97	38.33	37.23	37.10	39.75	34.98
April	38.18	38.17	38.62	37.74	34.11	34.01	35.15	33.25
May	35.41	35.44	36.49	34.29	31.52	31.70	32.35	30.34
June	34.79	35.06	37.28	32.59	31.27	31.78	33.50	28.95
July	36.04	36.35	37.27	34.60	33.30	33.20	34.86	32.18
August	36.08	36.05	38.76	32.63	33.54	33.47	35.93	31.45
September	36.68	37.49	39.97	31.09	34.18	34.64	37.80	30.38
October	40.56	40.58	44.03	33.98	38.05	37.87	40.51	33.70
November	45.53	46.07	47.77	42.98	44.04	44.07	45.65	42.23
December	49.11	50.34	52.28	43.47	48.02	48.61	51.09	43.76
Annual	40.81	38.48	53.67	31.09	38.14	35.08	52.36	28.95

Aswan Station:

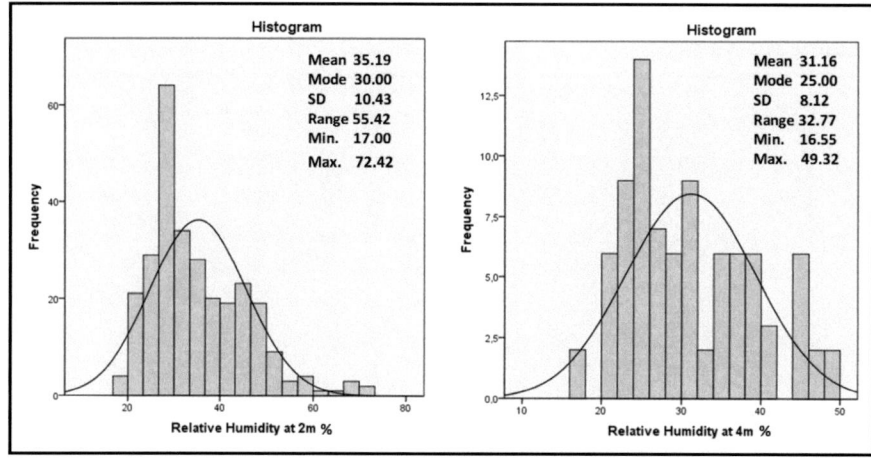

Histogram of the monthly mean relative humidity measured at two and four meters above surface water level at Aswan station

The average, median, maximum and minimum monthly relative humidity means measured at two and four meters above surface water level at Aswan station

Month	Relative humidity at 2 m (%)				Relative humidity at 4 m (%)			
	Mean	Median	Maximum	Minimum	Mean	Median	Maximum	Minimum
January	47.24	45.81	69.00	39.00	44.67	45.74	47.97	38.32
February	40.66	38.38	69.34	33.00	36.90	36.41	40.86	32.69
March	34.92	33.05	61.84	23.00	30.22	30.77	32.74	26.69
April	29.83	28.96	49.53	22.00	24.32	24.48	27.67	20.49
May	26.69	25.51	49.26	17.00	22.80	22.52	25.71	20.46
June	27.71	25.85	50.23	19.00	23.49	23.23	28.70	17.40
July	30.24	27.47	51.39	21.00	24.54	24.19	28.00	21.65
August	30.52	28.00	57.29	21.74	25.93	25.27	37.42	16.55
September	30.52	28.95	50.70	25.00	27.74	26.02	36.30	24.86
October	35.33	33.90	59.48	28.82	32.28	31.59	35.48	28.46
November	42.63	42.42	69.30	23.28	39.37	39.71	41.40	36.57
December	47.61	46.55	72.42	19.38	44.90	45.48	49.32	36.12
Annual	35.20	32.55	72.42	17.00	31.16	29.12	49.32	16.55

The average, median, maximum and minimum monthly relative humidity means measured at two and four meters above surface water level at Aswan station for different decades

Decade	Relative humidity at 2 m (%)				Relative humidity at 4 m (%)			
	Mean	Median	Maximum	Minimum	Mean	Median	Maximum	Minimum
1970s	40.48	38.00	72.42	19.38	no data	no data	no data	no data
1980s	32.20	30.00	51.00	17.00	no data	no data	no data	no data
1990s	35.21	32.13	50.23	27.42	32.57	28.85	49.06	23.65
2000s	34.32	32.55	51.50	21.45	30.94	29.12	49.32	16.55
Average	35.20	32.55	72.42	17.00	31.16	29.12	49.32	16.55

El-Alaky Station:

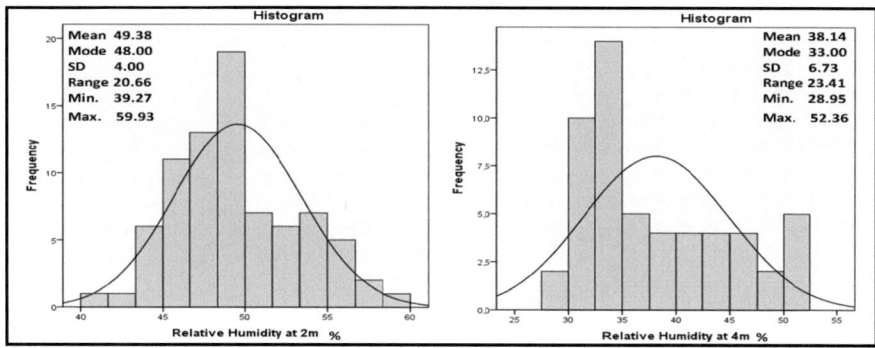

Histogram of the monthly mean relative humidity measured at two and four meters above surface water level at El-Alaky station in percentage

The average, median, maximum and minimum monthly mean relative humidity measured at two and four meters above surface water level at El-Alaky station

Month	Relative humidity at 2 m (%)				Relative humidity at 4 m (%)			
	Mean	Median	Maximum	Minimum	Mean	Median	Maximum	Minimum
January	55.79	55.55	57.87	54.55	51.29	51.88	52.36	49.03
February	54.49	53.84	59.93	51.12	44.79	44.44	47.96	42.30
March	50.59	51.02	54.39	48.04	37.23	37.10	39.75	34.98
April	49.88	49.90	52.93	47.84	34.11	34.01	35.15	33.25
May	47.68	48.40	49.58	43.85	31.52	31.70	32.35	30.34
June	45.34	45.93	48.47	39.27	31.27	31.78	33.50	28.95
July	45.41	45.68	47.52	42.92	33.30	33.20	34.86	32.18
August	46.59	47.10	48.03	44.16	33.54	33.47	35.93	31.45
September	47.55	47.95	49.13	45.31	34.18	34.64	37.80	30.38
October	48.98	48.74	53.37	44.95	38.05	37.87	40.51	33.70
November	49.76	50.74	53.70	41.23	44.04	44.07	45.65	42.23
December	53.60	53.59	56.55	50.65	48.02	48.61	51.09	43.76
Annual	49.38	48.68	59.93	39.27	38.14	35.08	52.36	28.95

The average, median, maximum and minimum monthly relative humidity measured at two and four meters above surface water level at El-Alaky station for the 1990s and 2000s

Decade	Relative humidity at 2 m (%)				Relative humidity at 4 m (%)			
	Mean	Median	Maximum	Minimum	Mean	Median	Maximum	Minimum
1990s	50.33	49.52	55.65	46.60	no data	no data	no data	no data
2000s	49.31	48.57	59.93	39.27	38.14	35.08	52.36	28.95
Average	49.38	48.68	59.93	39.27	38.14	35.08	52.36	28.95

Abu -Simbel Station:

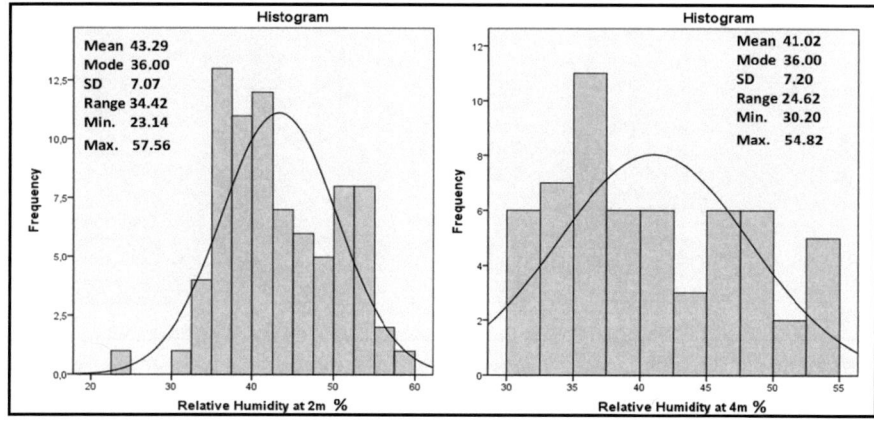

Histogram of the monthly mean relative humidity measured at two and four meters above surface water level at Abu -Simbel station in percentage

The average, median, maximum and minimum monthly relative humidity measured at two and four meters above surface water level at Abu-Simbel station

Month	Relative humidity at 2 m (%)				Relative humidity at 4 m (%)			
	Mean	Median	Maximum	Minimum	Mean	Median	Maximum	Minimum
January	54.58	54.98	57.11	52.25	53.09	52.65	54.82	51.74
February	49.05	48.45	53.25	44.52	47.05	47.32	49.69	44.47
March	44.50	44.95	49.45	42.05	41.31	40.43	45.29	38.46
April	41.30	40.67	44.80	39.66	36.96	37.05	40.03	34.45
May	39.69	39.83	42.32	36.08	34.46	35.23	36.50	30.87
June	35.63	35.44	36.77	34.96	31.64	31.35	33.37	30.20
July	35.37	36.19	37.74	31.98	33.18	32.71	36.92	30.61
August	34.71	35.44	38.90	23.14	34.48	35.45	36.61	31.19
September	38.57	38.88	42.28	35.25	36.70	37.32	38.46	34.62
October	44.94	44.13	53.04	39.57	40.45	40.63	43.47	36.39
November	49.06	49.37	50.72	46.89	46.75	47.10	48.91	45.14
December	53.00	52.30	57.56	50.17	51.18	49.55	54.63	47.96
Annual	43.29	42.28	57.56	23.14	41.02	39.56	54.82	30.20

The average, median, maximum and minimum monthly relative humidity measured at two and four meters above surface water level at Abu-Simbel station for the 1990s and 2000s

Decade	Relative humidity at 2 m (%)				Relative humidity at 4 m (%)			
	Mean	Median	Maximum	Minimum	Mean	Median	Maximum	Minimum
2000s	45.88	44.51	57.56	36.77	42.26	40.30	54.63	33.37
2010s	42.83	42.22	57.11	23.14	40.69	38.83	54.82	30.20
Average	43.29	42.28	57.56	23.14	41.02	39.56	54.82	30.20

Toshka Station:

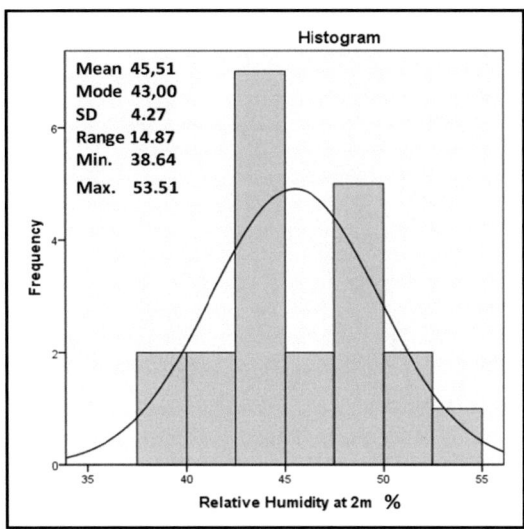

Histogram of the monthly mean relative humidity measured at two above surface water level at Toshka station

The average, median, maximum and minimum monthly relative humidity measured at two and four meters above surface water level at Toshka station

Month	Relative humidity at 2 m (%)			
	Mean	Median	Maximum	Minimum
January	52.39	52.39	53.51	51.27
February	48.71	48.71	48.89	48.53
March	44.44	44.44	45.27	43.61
April	42.91	42.91	43.13	42.69
May	40.17	40.17	40.85	39.50
June	39.56	39.56	40.49	38.64
July	42.83	42.83	42.83	42.83
August	42.99	42.99	42.99	42.99
September	43.24	43.24	43.24	43.24
October	46.84	46.84	48.89	44.80
November	47.85	47.85	49.85	45.85
December	50.44	50.44	51.08	49.79
Annual	45.51	44.80	53.51	38.64

Amada Station:

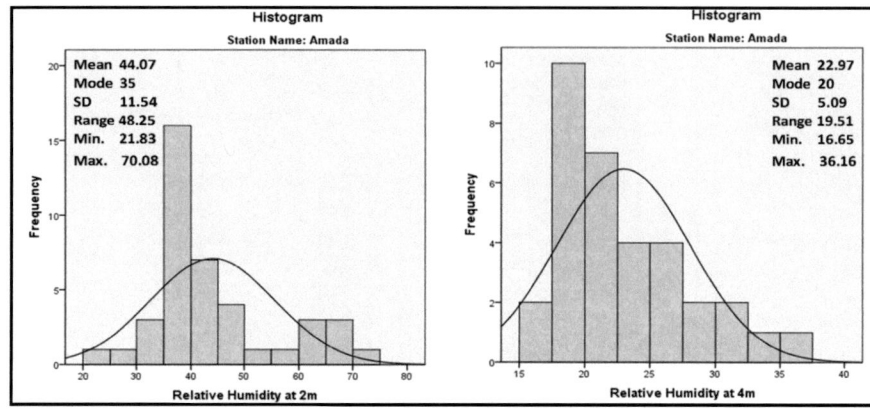

Histogram of the monthly mean relative humidity measured at two and four meters above surface water level at Amada station

The average, median, maximum and minimum monthly relative humidity measured at two and four meters above surface water level at Amada station

Month	Relative humidity at 2 m (%)				Relative humidity at 4 m (%)			
	Mean	Median	Maximum	Minimum	Mean	Median	Maximum	Minimum
January	47.73	47.73	51.16	44.30	30.24	30.24	33.54	26.94
February	43.80	43.80	46.23	41.38	26.10	26.10	28.54	23.67
March	38.86	38.86	39.45	38.26	21.52	21.52	21.82	21.22
April	36.88	36.88	37.34	36.43	19.61	19.61	20.04	19.19
May	36.60	36.45	38.32	35.02	19.22	18.86	20.62	18.18
June	42.96	35.85	66.21	33.91	18.18	18.04	19.69	16.81
July	38.84	36.93	63.96	21.83	18.56	18.88	19.83	16.65
August	41.80	36.26	69.96	29.51	23.57	19.90	36.16	18.34
September	48.01	39.55	65.50	37.82	21.57	21.20	22.51	21.00
October	51.36	43.49	70.08	40.03	24.78	25.39	26.17	22.78
November	49.07	46.68	60.82	42.12	27.89	26.84	32.26	24.57
December	47.10	47.10	48.32	45.89	29.83	29.83	30.87	28.80
Annual	44.07	39.55	70.08	21.83	22.97	21.20	36.16	16.65

Wind speed

Kalabsha Station:

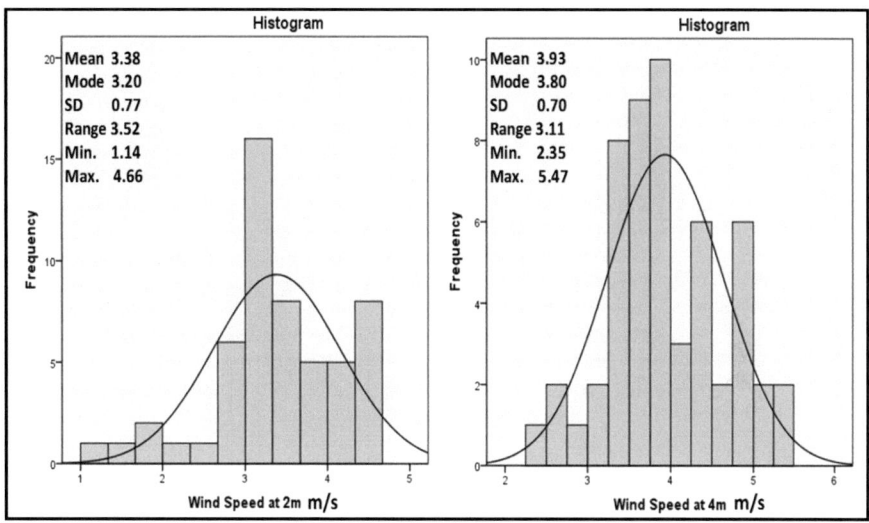

Histogram of the monthly mean wind speeds measured at two and four meters above surface water level at Kalabsha station

The average, median, maximum and minimum monthly wind speeds measured at two and four meters above surface water level at Kalabsha station

Month	Wind speed at 2 m (m/s)				Wind speed at 4 m (m/s)			
	Mean	Median	Maximum	Minimum	Mean	Median	Maximum	Minimum
January	4.14	4.14	4.43	3.85	4.71	4.81	5.16	4.07
February	3.67	3.68	4.15	3.18	4.23	4.01	5.37	3.52
March	3.43	3.42	3.81	3.09	3.99	3.98	4.20	3.81
April	3.38	3.24	3.94	3.10	3.99	3.98	4.49	3.52
May	3.19	3.22	3.33	2.99	3.76	3.77	3.94	3.57
June	3.07	3.03	3.49	2.82	3.58	3.66	3.93	3.19
July	2.88	3.00	3.04	2.64	3.30	3.33	3.44	3.02
August	2.90	3.00	3.20	2.27	3.30	3.34	3.53	2.85
September	3.15	3.46	3.56	1.83	3.71	3.93	4.28	2.64
October	3.64	3.75	4.47	1.99	4.31	4.34	4.93	3.41
November	3.73	4.22	4.59	1.40	4.37	4.51	5.47	2.71
December	3.61	4.33	4.66	1.14	4.19	4.69	5.04	2.35
Annual	3.38	3.31	4.66	1.14	3.93	3.83	5.47	2.35

Aswan Station:

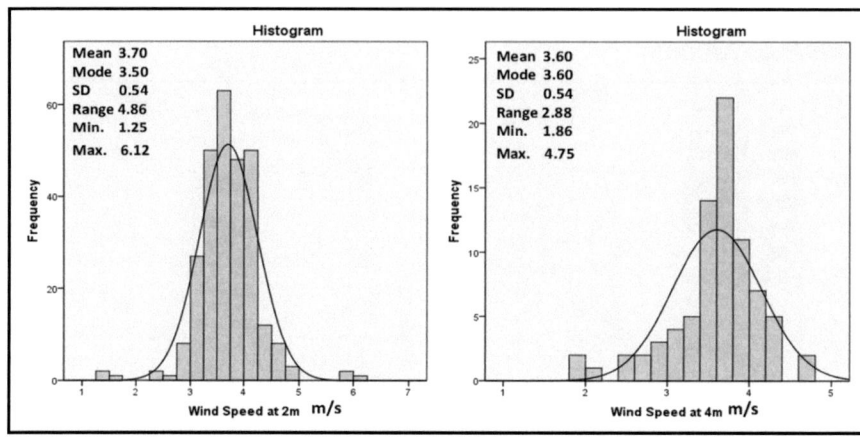

Histogram of the monthly mean wind speeds measured at two and four meters above surface water level at Aswan station

The average, median, maximum and minimum monthly wind speeds measured at two and four meters above surface water level at Aswan station

Month	Wind speed at 2 m (m/s)				Wind speed at 4 m (m/s)			
	Mean	Median	Maximum	Minimum	Mean	Median	Maximum	Minimum
January	3.89	3.93	4.68	2.91	3.73	3.80	4.34	2.79
February	3.56	3.70	4.43	1.52	3.15	2.98	4.37	2.05
March	3.66	3.75	4.97	1.25	3.55	3.65	4.38	1.94
April	3.45	3.54	4.36	1.29	3.31	3.56	3.80	1.86
May	3.52	3.60	4.17	3.07	3.70	3.74	3.90	3.49
June	3.63	3.67	4.21	3.14	3.87	3.75	4.36	3.56
July	3.39	3.40	4.05	2.76	3.38	3.38	3.79	3.00
August	3.55	3.50	4.35	2.96	3.70	3.67	4.06	3.36
September	3.95	3.89	4.53	3.36	3.85	3.86	4.11	3.54
October	4.07	4.03	6.12	3.20	4.10	3.85	4.75	3.70
November	3.92	3.89	5.89	3.17	3.69	3.73	4.11	3.17
December	3.93	3.95	5.80	2.96	3.36	3.51	3.80	2.44
Annual	3.70	3.70	6.12	1.25	3.60	3.70	4.75	1.86

The average, median, maximum and minimum monthly wind speed measured at two and four meters above surface water level at Aswan station for different decades

Decade	Wind speed at 2 m (m/s)				Wind speed at 4 m (m/s)			
	Mean	Median	Maximum	Minimum	Mean	Median	Maximum	Minimum
1970s	3.98	4.00	4.97	3.07	no data	no data	no data	no data
1980s	3.75	3.70	4.90	2.89	no data	no data	no data	no data
1990s	3.61	3.53	4.16	3.16	3.85	3.77	4.38	3.46
2000s	3.38	3.36	6.12	1.25	3.56	3.67	4.75	1.86
Average	3.70	3.70	6.12	1.25	3.60	3.70	4.75	1.86

El-Alaky Station:

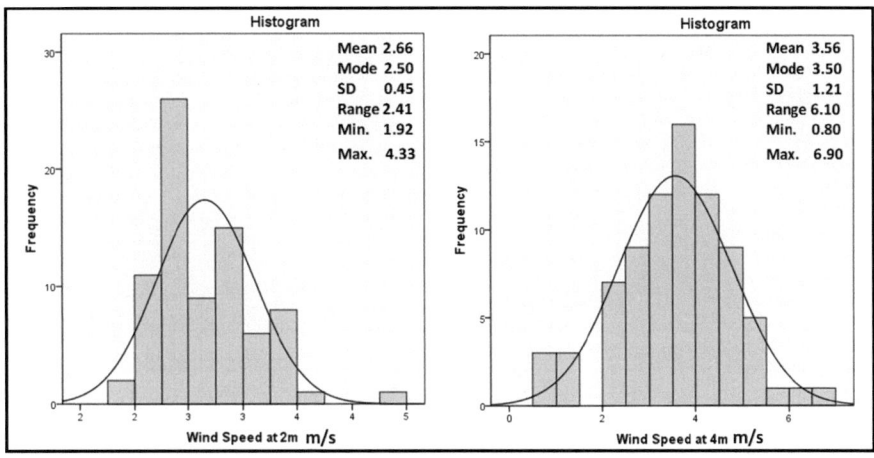

Histogram of the monthly mean wind speeds measured at two and four meters above surface water level at El-Alaky station in m/s

The average, median, maximum and minimum monthly wind speeds measured at two and four meters above surface water level at El-Alaky station

Month	Wind speed at 2 m (m/s)				Wind speed at 4 m (m/s)			
	Mean	Median	Maximum	Minimum	Mean	Median	Maximum	Minimum
January	2.94	2.96	3.34	2.25	3.43	3.51	5.09	.99
February	2.78	2.62	3.49	2.35	3.43	3.65	5.35	.86
March	2.57	2.56	2.86	2.20	3.42	4.13	4.53	1.11
April	2.41	2.37	2.86	2.06	3.23	3.74	4.70	1.07
May	2.37	2.35	2.68	2.11	3.03	3.18	4.39	.80
June	2.58	2.38	4.33	2.03	3.37	3.19	6.90	1.22
July	2.15	2.18	2.35	1.92	3.09	3.41	3.65	2.11
August	2.37	2.43	2.51	2.11	3.38	3.51	3.93	2.65
September	2.76	2.76	3.05	2.41	3.92	4.17	4.72	2.64
October	3.05	2.99	3.43	2.70	4.30	4.43	5.35	3.13
November	3.14	3.20	3.56	2.53	4.58	4.96	6.20	2.79
December	3.14	3.10	3.47	2.91	4.24	4.11	5.53	3.20
Annual	2.66	2.51	4.33	1.92	3.56	3.67	6.90	.80

The average, median, maximum and minimum monthly wind speeds measured at two and four meters above surface water level at El-Alaky station for the 1990s and 2000s

Decade	Wind speed at 2 m (m/s)				Wind speed at 4 m (m/s)			
	Mean	Median	Maximum	Minimum	Mean	Median	Maximum	Minimum
1990s	2.55	2.47	3.22	2.06	2.81	2.72	3.54	2.28
2000s	2.67	2.51	4.33	1.92	3.70	3.77	6.90	.80
Average	2.66	2.51	4.33	1.92	3.56	3.67	6.90	.80

Abu -Simbel Station:

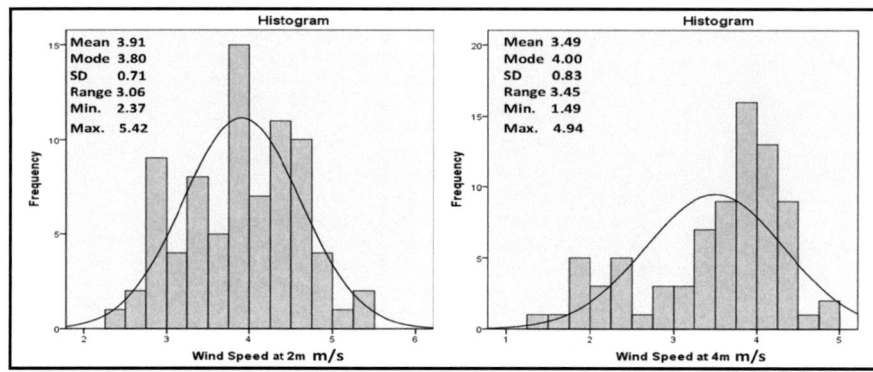

Histogram of the monthly mean wind speeds measured at two and four meters above surface water level at Abu -Simbel station in m/s

The average, median, maximum and minimum monthly wind speeds measured at two and four meters above surface water level at Abu -Simbel station

Month	Wind speed at 2 m (m/s)				Wind speed at 4 m (m/s)			
	Mean	Median	Maximum	Minimum	Mean	Median	Maximum	Minimum
January	4.34	4.35	5.40	3.42	3.83	3.85	4.93	2.44
February	4.05	4.46	5.08	2.65	3.72	4.21	4.35	1.85
March	3.74	3.46	4.66	2.81	3.36	3.24	4.29	2.16
April	3.67	3.95	4.35	2.88	3.40	3.78	4.08	1.77
May	3.91	4.03	4.46	2.77	3.41	3.68	4.14	1.49
June	3.80	3.99	4.31	2.72	3.45	3.73	4.00	2.38
July	3.29	3.33	3.90	2.37	3.02	3.27	3.56	1.52
August	3.77	3.99	4.51	2.77	3.40	3.57	4.12	1.85
September	4.09	4.47	4.88	2.82	3.72	4.04	4.46	1.78
October	4.08	4.06	5.42	2.98	3.60	4.08	4.94	1.92
November	4.12	3.88	4.85	3.69	3.53	3.91	4.39	2.37
December	4.09	4.00	4.99	3.45	3.51	3.97	4.51	2.16
Annual	3.91	3.95	5.42	2.37	3.49	3.76	4.94	1.49

The average, median, maximum and minimum monthly wind speed measured at two and four meters above surface water level at Abu -Simbel station for the last the 1990s and 2000s

Decade	Wind speed at 2 m (m/s)				Wind speed at 4 m (m/s)			
	Mean	Median	Maximum	Minimum	Mean	Median	Maximum	Minimum
1990s	4.15	4.07	5.08	3.08	3.68	3.69	4.32	2.93
2000s	3.87	3.92	5.42	2.37	3.46	3.77	4.94	1.49
Average	3.91	3.95	5.42	2.37	3.49	3.76	4.94	1.49

Toshka Station:

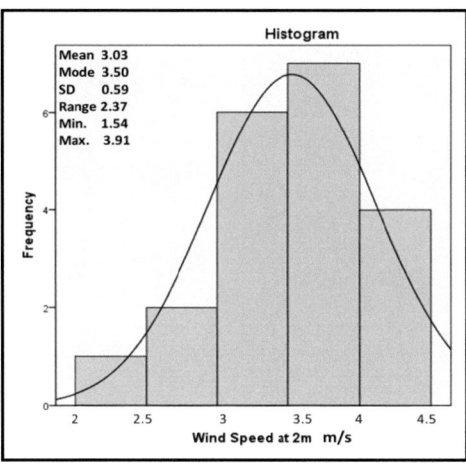

Histogram of the monthly mean wind speed measured at two meters above surface water level at Toshka station in m/s

The average, median, maximum and minimum monthly wind speed measured at two meters above surface water level at Toshka station

Month	Wind speed at 2 m (m/s)			
	Mean	Median	Maximum	Minimum
January	3.37	3.37	3.76	2.99
February	2.84	2.84	3.02	2.66
March	2.91	2.91	3.18	2.65
April	2.90	2.90	2.90	2.90
May	2.77	2.77	3.14	2.39
June	2.90	2.90	3.01	2.78
July	1.54	1.54	1.54	1.54
August	2.19	2.19	2.19	2.19
September	3.44	3.44	3.44	3.44
October	3.33	3.33	3.80	2.87
November	3.58	3.58	3.91	3.25
December	3.52	3.52	3.84	3.20
Annual	3.03	3.02	3.91	1.54

Amada Station:

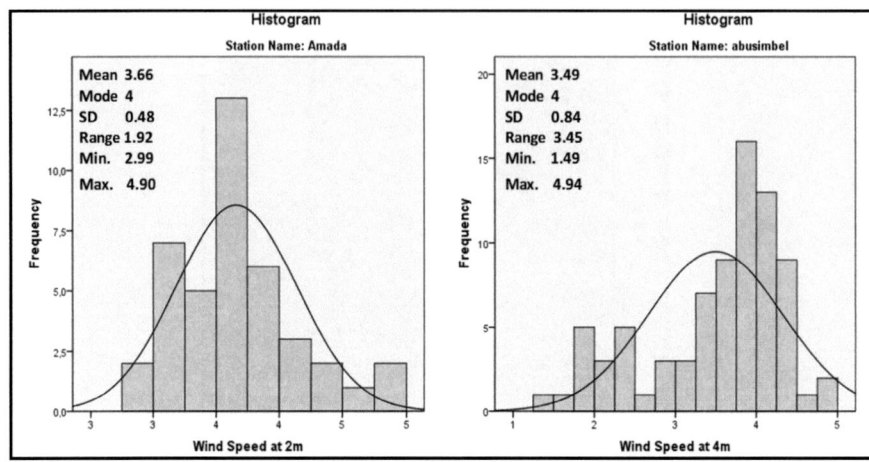

Histogram of the monthly mean wind speeds measured at two and four meters above surface water level at Abu -Simbel station in m/s

The average, median, maximum and minimum monthly wind speeds measured at two and four meters above surface water level at Abu -Simbel station

Month	Wind speed at 2 m (m/s)				Wind speed at 4 m (m/s)			
	Mean	Median	Maximum	Minimum	Mean	Median	Maximum	Minimum
January	4.10	4.10	4.50	3.69	3.83	3.85	4.93	2.44
February	3.61	3.61	3.78	3.45	3.72	4.21	4.35	1.85
March	3.83	3.83	3.96	3.70	3.36	3.24	4.29	2.16
April	3.47	3.47	3.55	3.38	3.40	3.78	4.08	1.77
May	3.55	3.57	3.58	3.51	3.41	3.68	4.14	1.49
June	3.50	3.48	4.05	3.01	3.45	3.73	4.00	2.38
July	3.26	3.22	3.72	2.99	3.02	3.27	3.56	1.52
August	3.35	3.25	3.74	3.04	3.40	3.57	4.12	1.85
September	3.65	3.77	4.48	2.99	3.72	4.04	4.46	1.78
October	3.93	3.82	4.81	3.29	3.60	4.08	4.94	1.92
November	3.99	3.76	4.90	3.55	3.53	3.91	4.39	2.37
December	4.21	4.21	4.37	4.05	3.51	3.97	4.51	2.16
Annual	3.91	3.95	5.42	2.37	3.49	3.76	4.94	1.49

Water Temperature

Kalabsha Station:

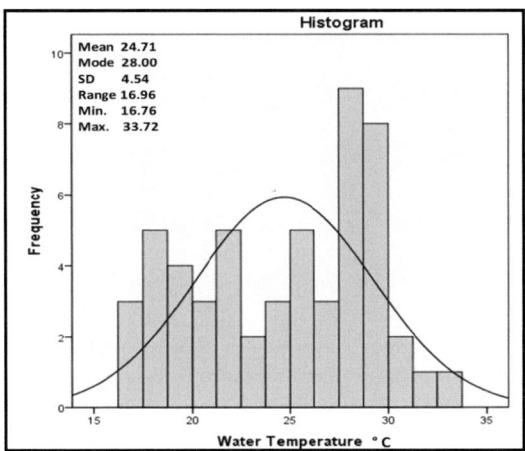

Histogram of the monthly means water temperature measured at Kalabsha station °C

The average, median, maximum and minimum monthly water temperature measured at Kalabsha station

Month	Water temperature (°C)			
	Mean	Median	Maximum	Minimum
January	17.73	17.60	18.71	17.00
February	17.83	18.10	18.36	16.76
March	19.96	19.81	20.78	19.46
April	22.63	22.66	22.91	22.31
May	25.73	25.56	26.50	25.29
June	28.17	28.28	28.44	27.81
July	29.61	29.53	30.08	29.32
August	30.20	29.91	31.62	29.61
September	29.38	28.47	33.72	27.86
October	26.83	26.32	29.26	25.95
November	23.44	24.01	24.67	22.12
December	20.26	20.21	21.28	19.34
Annual	24.72	25.56	33.72	16.76

Aswan Station:

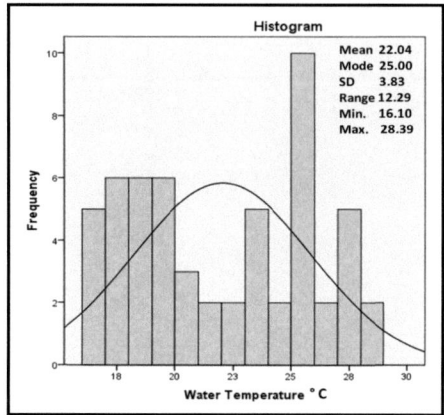

Histogram of the monthly means surface water temperature measured at Aswan station °C

The average, median, maximum and minimum monthly mean surface water temperature measured at Aswan station

Month	Water temperature (°C)			
	Mean	Median	Maximum	Minimum
January	17.31	17.40	18.04	16.41
February	16.83	16.73	17.57	16.10
March	17.71	17.87	18.64	16.53
April	19.85	19.98	20.23	19.29
May	22.16	22.05	23.30	21.26
June	24.90	24.93	25.43	24.31
July	27.86	27.78	28.39	27.50
August	27.00	27.41	28.38	25.87
September	25.81	25.79	26.35	25.19
October	24.62	25.13	25.36	23.52
November	22.07	22.33	23.20	20.68
December	19.16	18.83	19.95	18.70
Annual	22.04	21.89	28.39	16.10

The average, median, maximum and minimum monthly mean surface water temperature measured at Aswan station for different decades

Decade	Water temperature (°C)			
	Mean	Median	Maximum	Minimum
1990s	21.55	21.21	27.50	16.10
2000s	22.17	22.49	28.39	16.41
Average	22.04	21.89	28.39	16.10

El-Alaky Station:

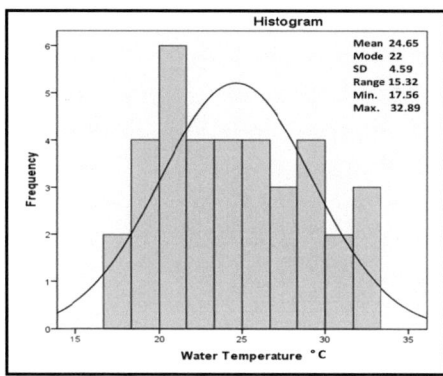

Histogram of monthly means of surface water temperature measured at El-Alaky station

The average, median, maximum and minimum monthly means of surface water temperature measured at El-Alaky station

Month	Water temperature (°C)			
	Mean	Median	Maximum	Minimum
January	18.79	18.78	20.01	17.56
February	18.78	18.78	19.57	17.99
March	19.98	20.17	20.17	19.62
April	22.87	22.54	23.53	22.54
May	25.01	24.87	25.87	24.42
June	29.07	28.96	31.56	26.81
July	31.68	32.41	32.64	29.98
August	30.83	30.27	32.89	29.34
September	28.80	28.80	29.39	28.21
October	24.25	25.75	25.88	21.11
November	23.42	23.02	24.66	22.60
December	20.18	20.29	20.48	19.75
Annual	24.65	24.42	32.89	17.56

The average, median, maximum and minimum monthly means of surface water temperature measured at El-Alaky station for the 1990s and 2000s

Decade	Water temperature (°C)			
	Mean	Median	Maximum	Minimum
1990s	24.96	24.42	32.41	19.57
2000s	24.50	24.42	32.89	17.56
Average	24.65	24.42	32.89	17.56

Abu -Simbel Station:

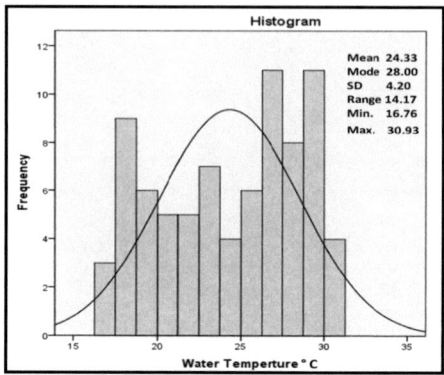

Histogram of the monthly means of surface water temperature measured at Abu -Simbel station °C

The average, median, maximum and minimum monthly means of surface water temperature measured at Abu -Simbel station

Month	Water temperature (°C)			
	Mean	Median	Maximum	Minimum
January	18.07	18.24	19.43	16.76
February	17.92	17.91	18.49	17.12
March	20.24	20.44	21.51	18.83
April	23.53	23.34	25.00	22.40
May	26.14	26.12	26.90	24.96
June	27.77	27.92	29.02	26.73
July	29.60	29.49	30.93	28.75
August	29.36	29.35	30.86	28.35
September	28.31	28.07	30.69	26.35
October	26.35	26.42	27.48	25.30
November	23.33	23.00	24.81	22.45
December	20.69	20.39	21.89	19.84
Annual	24.33	25.00	30.93	16.76

The average, median, maximum and minimum monthly means of surface water temperature measured at Abu -Simbel station for the 1990s and 2000s

Decade	Water temperature (°C)			
	Mean	Median	Maximum	Minimum
1990s	24.35	25.16	30.73	17.12
2000s	24.32	24.96	30.93	16.76
Average	24.33	25.00	30.93	16.76

Toshka Station:

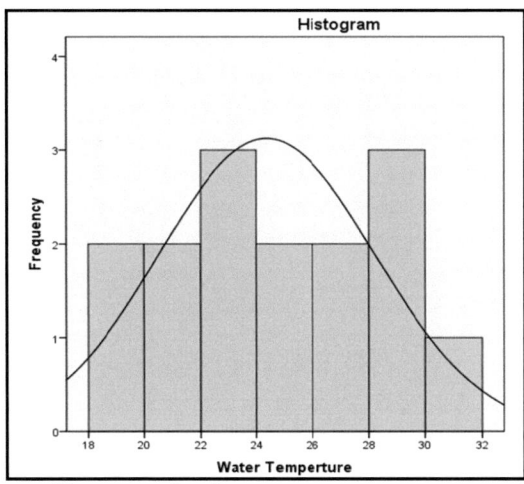

Histogram of the monthly means of surface water temperature measured at Toshka station

The average, median, maximum and minimum monthly surface water temperatures measured at Toshka station

Month	Water temperature (°C)			
	Mean	Median	Maximum	Minimum
January	18.27	18.27	18.27	18.27
February	19.15	19.15	19.15	19.15
March	20.58	20.58	20.58	20.58
April	23.14	23.14	23.14	23.14
May	26.01	26.01	26.01	26.01
June	28.09	28.09	28.09	28.09
July	29.64	29.64	29.64	29.64
August	30.24	30.24	30.24	30.24
September	28.28	28.28	28.28	28.28
October	26.27	26.27	26.61	25.94
November	23.55	23.55	24.33	22.76
December	21.24	21.24	22.10	20.59
Annual	24.37	24.33	30.24	18.27

Amada Station:

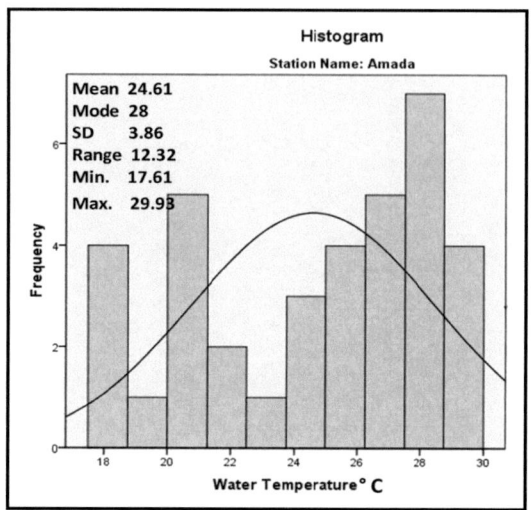

Histogram of monthly means of surface water temperature measured at Amada station °C

The average, median, maximum and minimum monthly surface water temperatures measured at Amada station

Month	Water temperature (°C)			
	Mean	Median	Maximum	Minimum
January	18.38	18.38	18.67	18.09
February	18.37	18.37	18.78	17.95
March	18.98	18.98	20.35	17.61
April	22.69	22.69	23.87	21.51
May	26.82	26.69	28.25	25.51
June	28.01	27.56	29.45	27.01
July	28.70	28.66	29.93	27.55
August	28.13	28.59	29.27	26.05
September	27.36	27.10	28.40	26.84
October	24.65	24.90	26.19	22.64
November	21.79	21.48	23.97	20.22
December	21.01	21.01	21.08	20.93
Annual	24.61	25.78	29.93	17.61

Appendix G 345

Evaporation rates

Kalabsha Station:

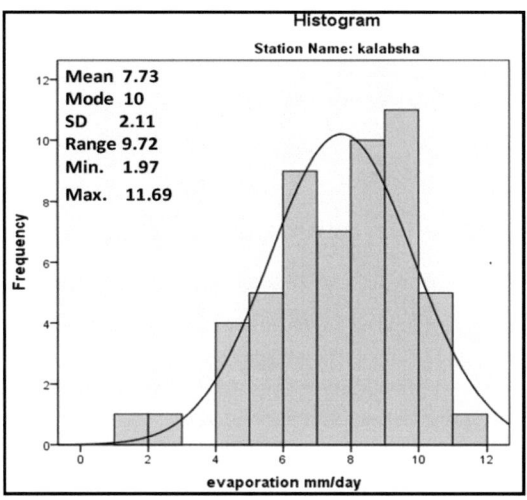

Histogram of the monthly means of evaporation rate measured at Kalabsha station

The average, median, maximum and minimum monthly means of evaporation rates measured at Kalabsha station

Month	Evaporation rate (mm/day)			
	Mean	Median	Maximum	Minimum
January	5.59	5.38	6.44	5.14
February	4.94	4.73	5.85	4.46
March	5.79	6.00	6.32	4.83
April	6.66	6.52	7.36	6.25
May	7.63	7.88	8.30	6.47
June	10.06	10.53	11.69	7.90
July	8.77	8.77	9.31	8.29
August	9.48	9.74	9.91	8.86
September	9.58	9.63	10.22	8.71
October	8.87	8.49	10.79	6.37
November	7.64	8.50	9.64	2.93
December	5.76	6.89	7.29	1.97
Annual	7.73	8.05	11.69	1.97

Aswan Station:

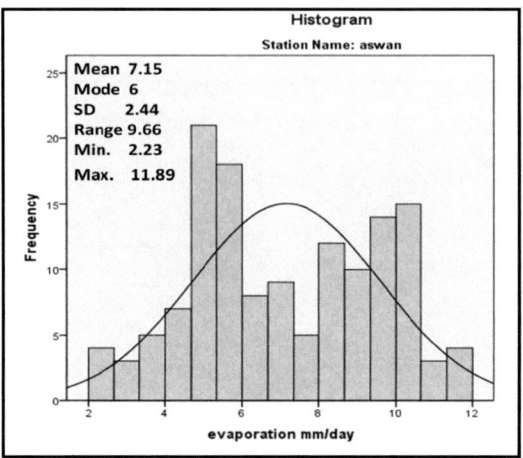

Histogram of the monthly means of evaporation rates measured at Aswan station in mm/day

The mean, median, maximum and minimum monthly means of evaporation rates at Aswan station

Month	Evaporation rate (mm/day)			
	Mean	Median	Maximum	Minimum
January	4.84	4.83	5.79	3.89
February	4.56	4.83	6.24	2.67
March	4.96	5.00	6.47	2.26
April	5.68	6.05	7.17	2.75
May	7.29	7.36	8.49	4.15
June	8.96	8.98	9.79	7.94
July	9.98	9.71	11.89	9.51
August	9.94	10.44	11.58	2.64
September	9.52	10.25	11.40	2.23
October	8.55	8.82	10.32	2.80
November	6.47	6.03	8.09	5.40
December	5.22	5.18	5.98	4.68
Annual	7.15	6.93	11.89	2.23

The average, median, maximum and minimum monthly means of evaporation rates measured at Aswan station for the 1990s and 2000s

Decade	Evaporation rate (mm/day)			
	Mean	Median	Maximum	Minimum
1990s	6.96	7.12	11.89	4.29
2000s	7.58	7.78	11.40	2.23
Average	7.15	7.23	11.89	2.23

El-Alaky Station:

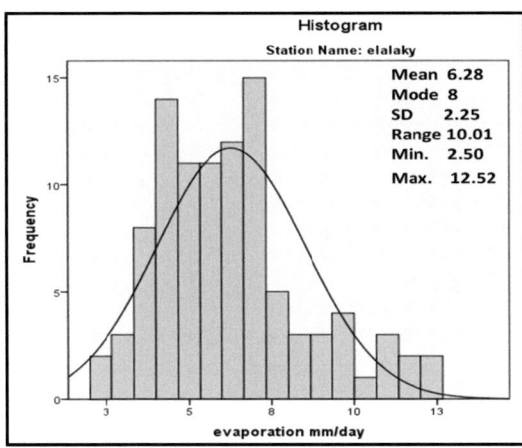

Histogram of the monthly means of evaporation rates measured at El-Alaky station in mm/day

The average, median, maximum and minimum monthly means of evaporation rates measured at El-Alaky station

Month	Evaporation rate (mm/day)			
	Mean	Median	Maximum	Minimum
January	4.11	3.86	6.30	2.64
February	4.28	4.29	5.72	2.50
March	4.69	4.43	6.69	3.05
April	5.17	4.89	7.46	3.84
May	5.96	5.61	9.41	4.57
June	7.21	6.85	11.32	5.15
July	7.85	7.31	11.49	5.20
August	8.54	8.20	12.52	6.29
September	7.87	6.98	11.99	6.03
October	7.85	6.93	12.16	4.54
November	6.72	6.21	11.25	4.30
December	5.13	4.78	6.98	3.99
Annual	6.28	6.03	12.52	2.50

The average, median, maximum and minimum monthly evaporation rate measured at El Alaky station for the last two decades

Decade	Evaporation rate (mm/day)			
	Mean	Median	Maximum	Minimum
1990s	5.64	5.95	7.46	3.84
2000s	6.59	6.06	12.52	2.50
Average	6.28	6.03	12.52	2.50

Abu -Simbel Station:

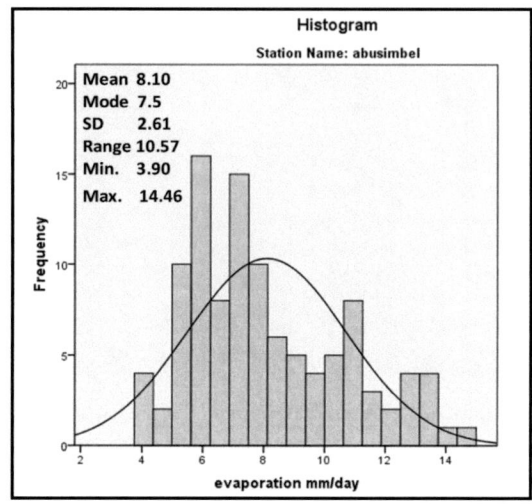

Histogram of the monthly means of evaporation rates measured at Abu -Simbel station in mm/day

The average, median, maximum and minimum monthly evaporation rates measured at Abu–Simbel station in mm/day

Month	Evaporation rate (mm/day)			
	Mean	Median	Maximum	Minimum
January	5.59	5.40	7.79	4.08
February	5.57	5.63	7.86	3.90
March	6.28	6.15	7.00	5.29
April	7.87	7.37	10.13	5.89
May	8.50	7.94	11.37	4.57
June	9.22	9.55	12.99	5.04
July	8.90	8.62	12.79	4.12
August	10.83	10.85	14.46	7.06
September	10.00	9.72	13.66	6.00
October	9.23	9.39	12.57	6.56
November	7.13	7.14	8.89	4.48
December	6.09	6.01	7.04	5.20
Annual	8.10	7.44	14.46	3.90

The average, median, maximum and minimum monthly evaporation rates measured at Abu–Simbel station for different decades

Decade	Evaporation rate (mm/day)			
	Mean	Median	Maximum	Minimum
1990s	7.46	7.06	13.40	4.12
2000s	8.52	7.90	14.46	3.90
Average	8.10	7.44	14.46	3.90

Toshka Station:

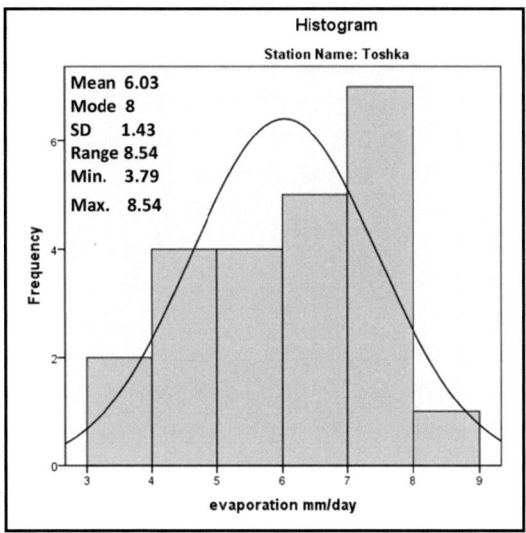

Histogram of the monthly means of evaporation rates measured at Toshka station

The mean, median, maximum and minimum monthly evaporation rate measured at Toshka station

Month	Evaporation rate (mm/day)			
	Mean	Median	Maximum	Minimum
January	4.16	4.16	4.16	4.16
February	3.83	3.83	3.83	3.83
March	3.79	3.79	3.79	3.79
April	4.50	4.50	4.50	4.50
May	5.97	5.97	6.54	5.41
June	7.17	7.17	7.80	6.54
July	7.50	7.50	8.54	6.46
August	6.25	6.25	7.46	5.04
September	7.19	7.19	7.35	7.04
October	6.89	7.35	7.70	5.62
November	6.67	6.70	7.08	6.24
December	4.55	4.29	5.08	4.29
Annual	6.03	6.46	8.54	3.79

Amada Station:

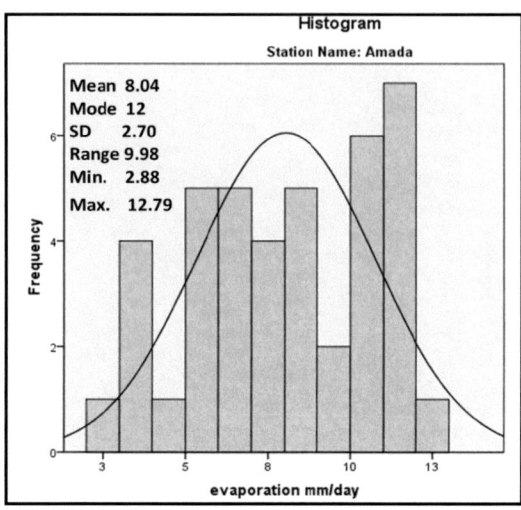

Histogram of the monthly mean evaporation rates measured at Amada station

The mean, median, maximum and minimum monthly evaporation rates measured at Amada station

Month	Evaporation rate (mm/day)			
	Mean	Median	Maximum	Minimum
January	5.56	5.56	5.67	5.45
February	5.46	5.46	5.80	5.12
March	6.49	6.49	6.74	6.24
April	7.73	7.73	8.10	7.35
May	9.85	9.85	10.26	9.44
June	11.02	11.18	11.57	10.31
July	9.52	11.30	11.54	3.93
August	8.87	10.28	11.80	3.57
September	8.99	9.28	12.79	3.91
October	7.85	8.58	11.34	3.97
November	6.62	6.78	8.39	2.88
December	6.69	6.85	7.39	5.69
Annual	8.04	8.10	12.79	2.88

Appendix H

Results of Scenarios for reducing Evaporation losses

Alternatives for eliminating some khors of HADR

Cross section of the alternatives suggested to eliminate Khor Kalabsha:

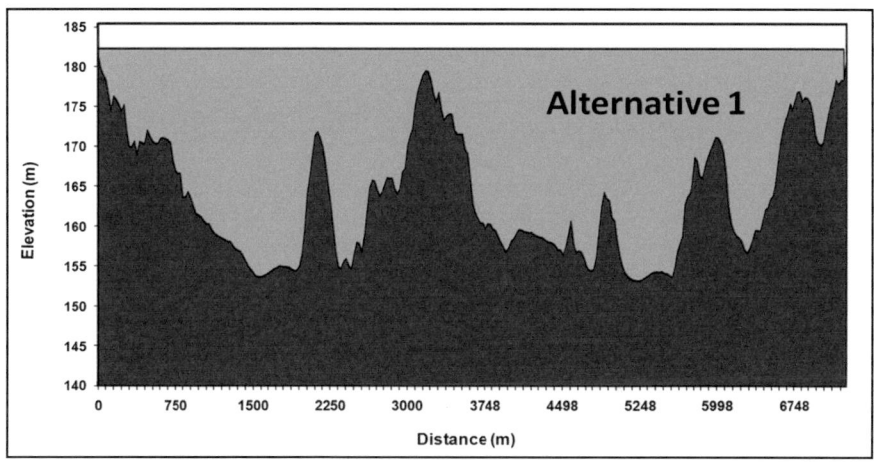

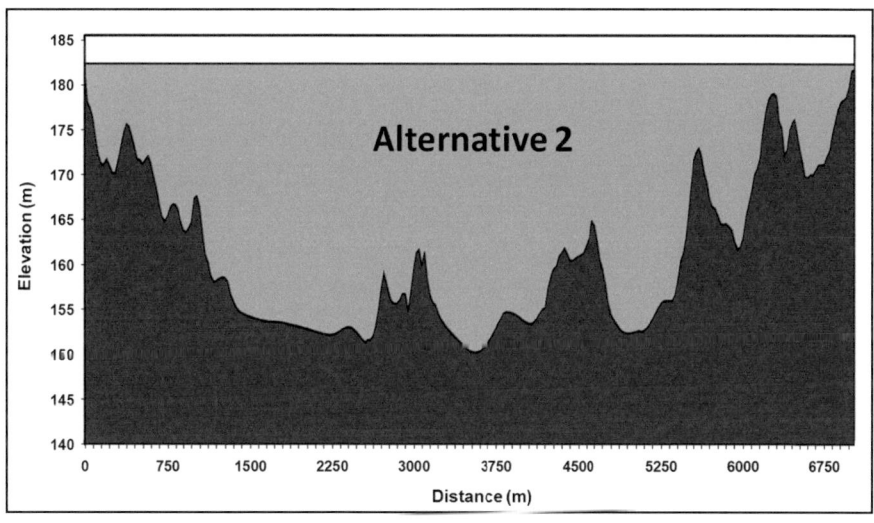

352   Appendix H

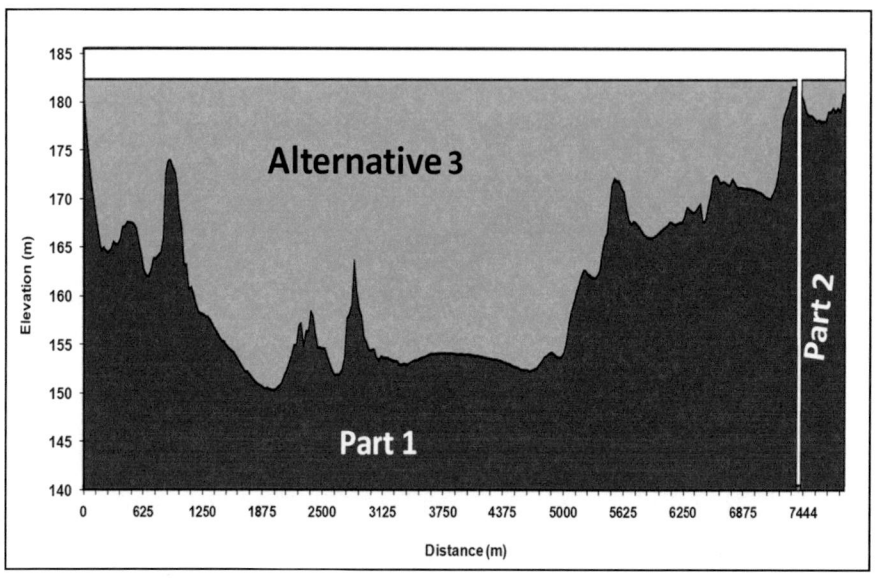

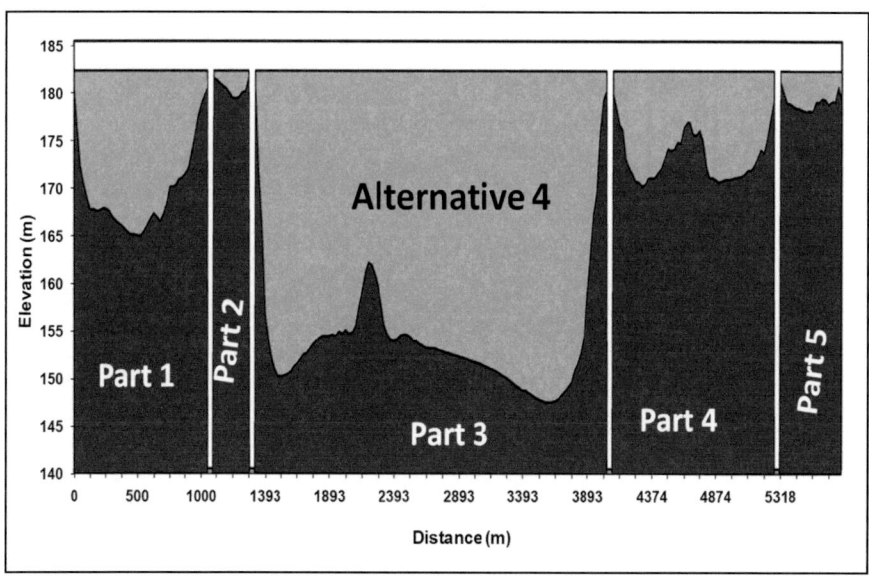

Appendix H 353

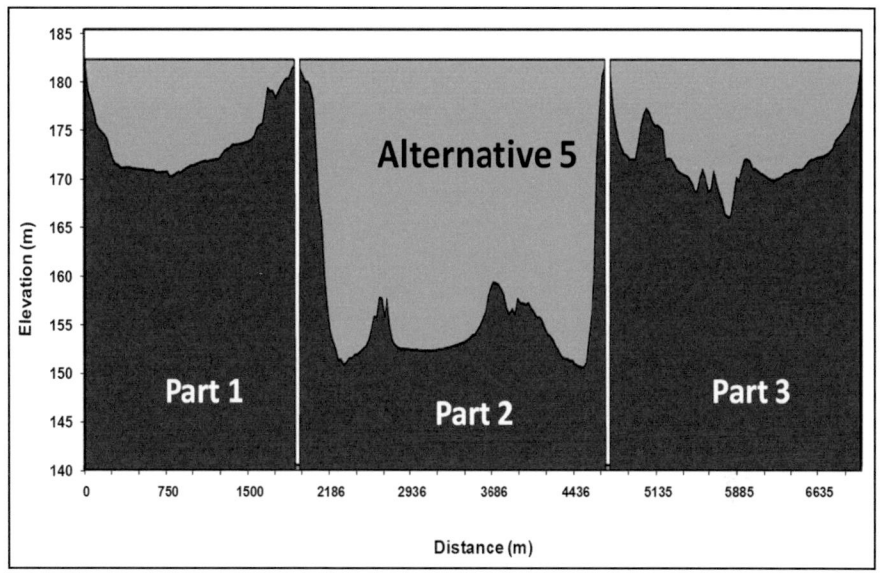

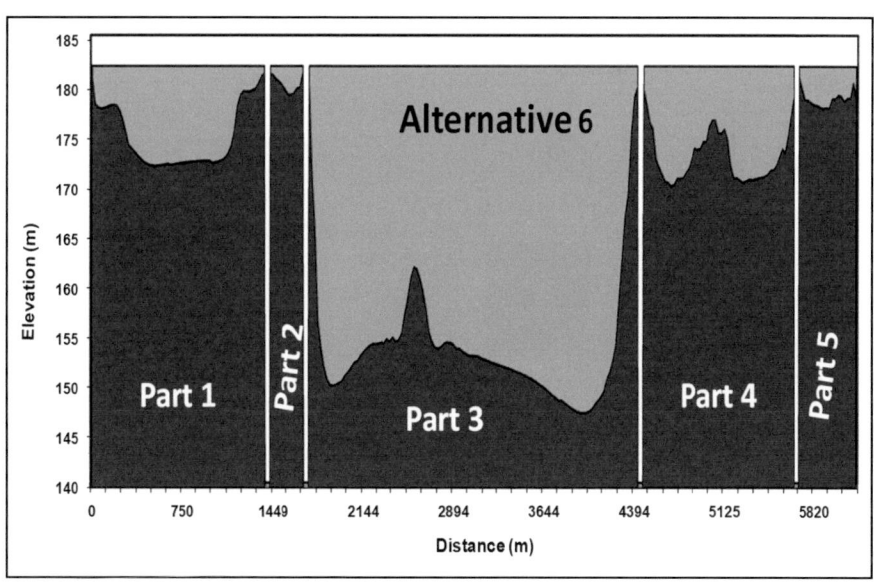

Cross section of alternatives suggested to eliminate Khor El-Alaky

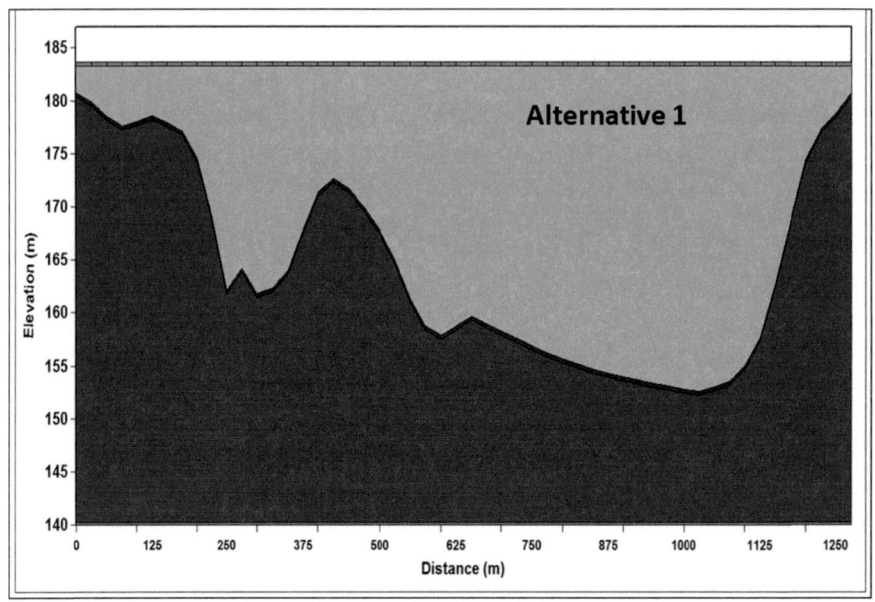

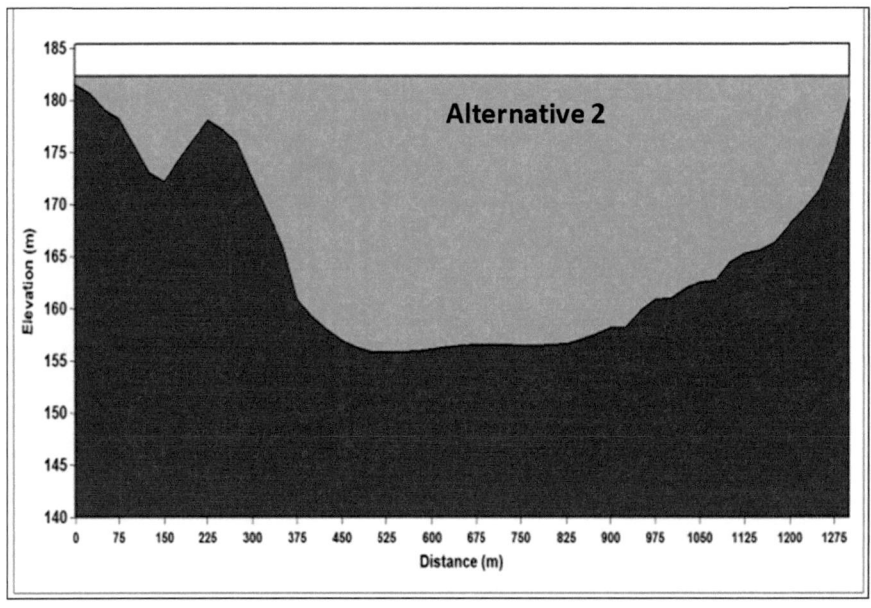

Appendix H

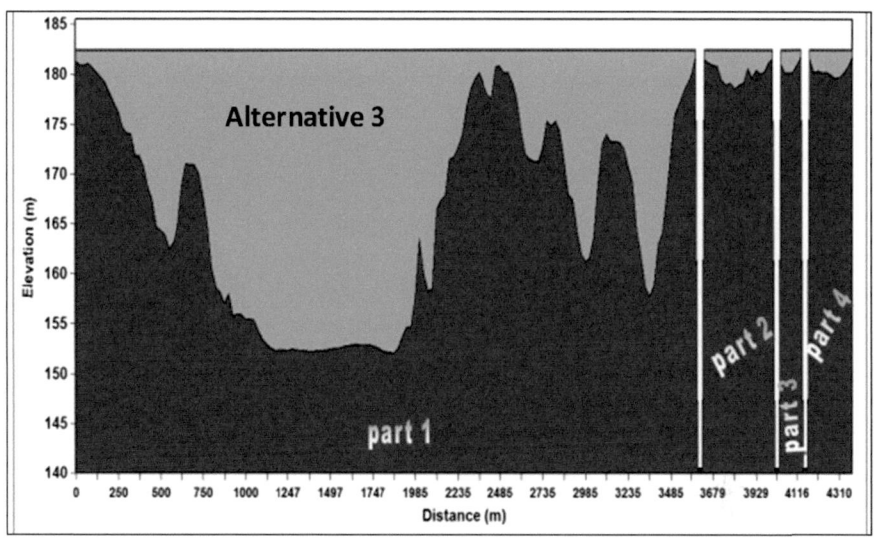

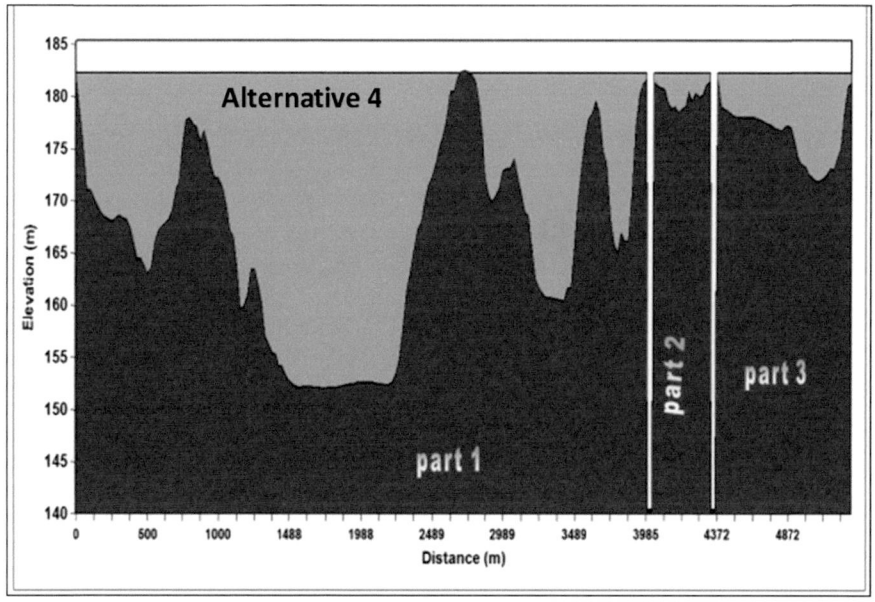

Appendix H

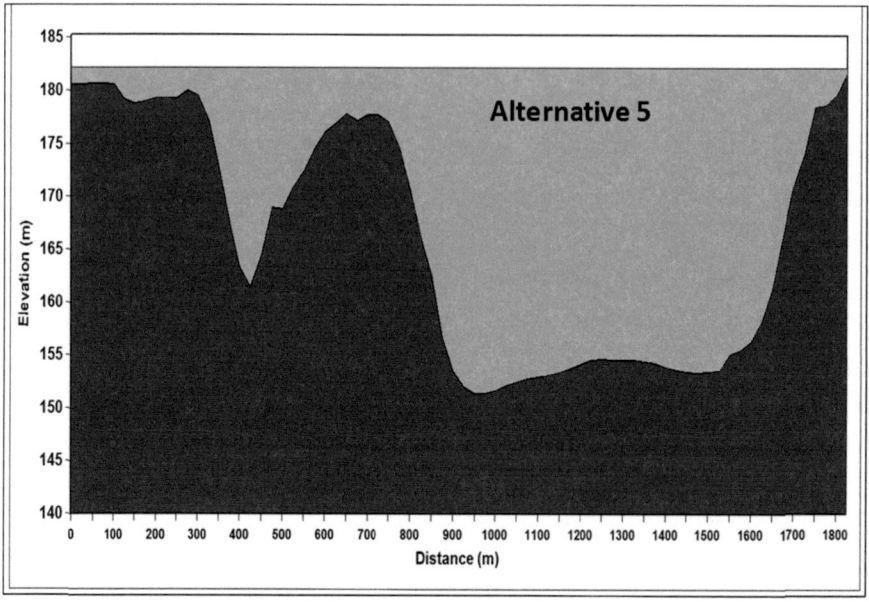

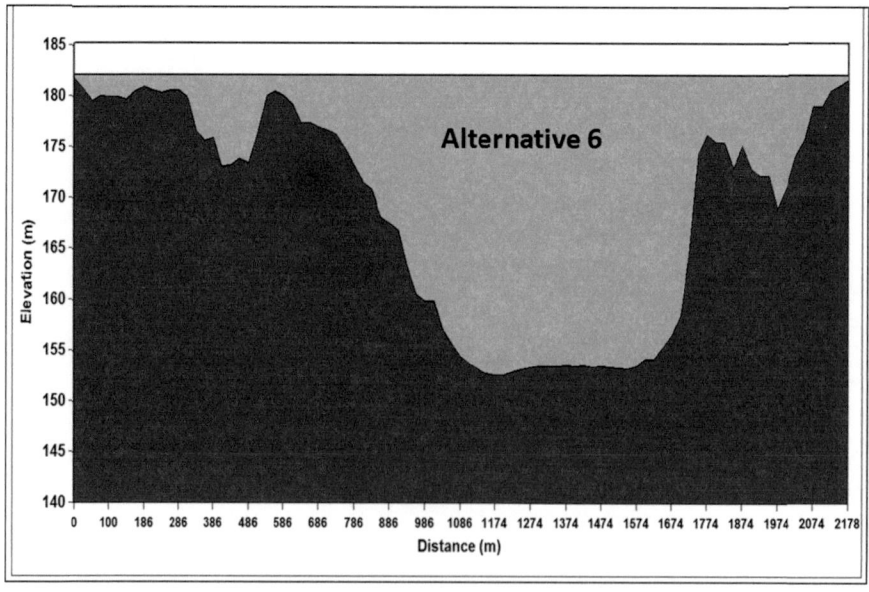

Appendix H 357

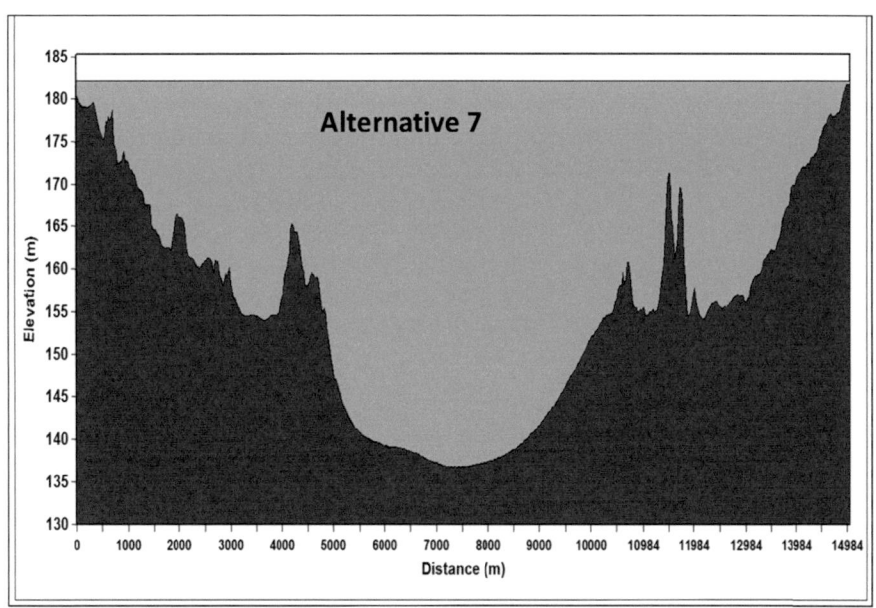

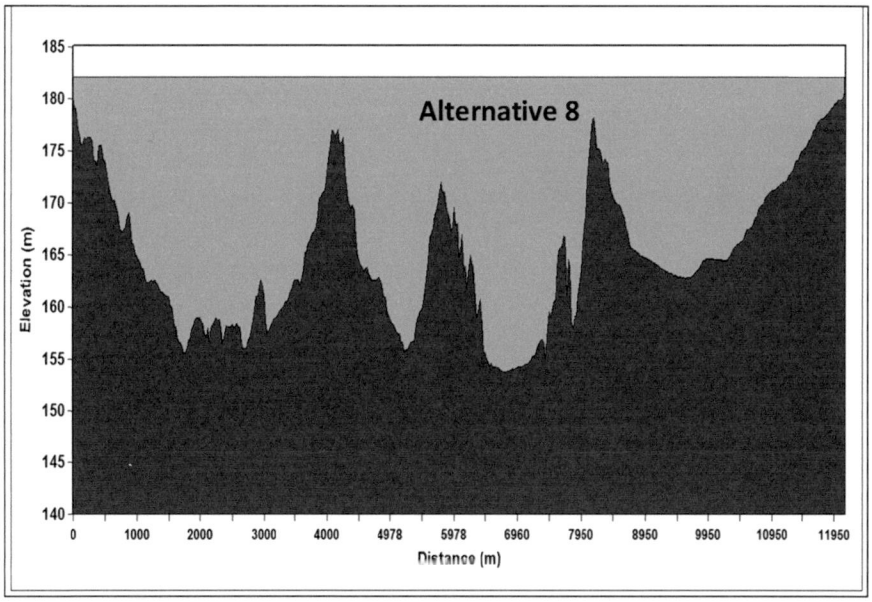

Cross section of the alternatives suggested to eliminate Khor Genina

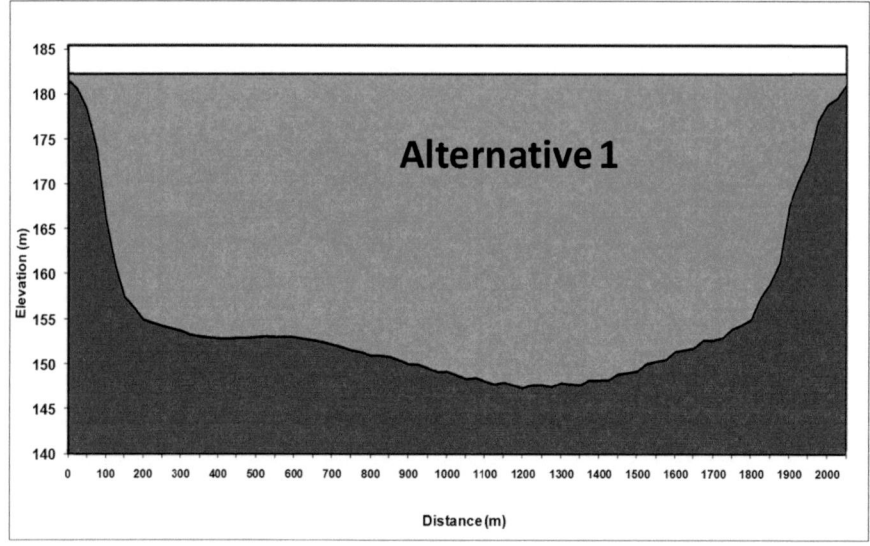

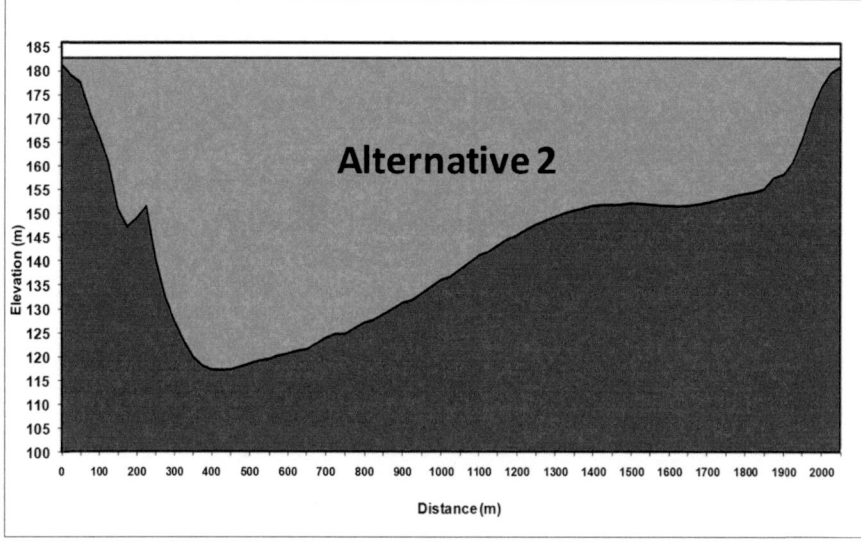

Cross section of the alternatives suggested to eliminate Khor Sara:

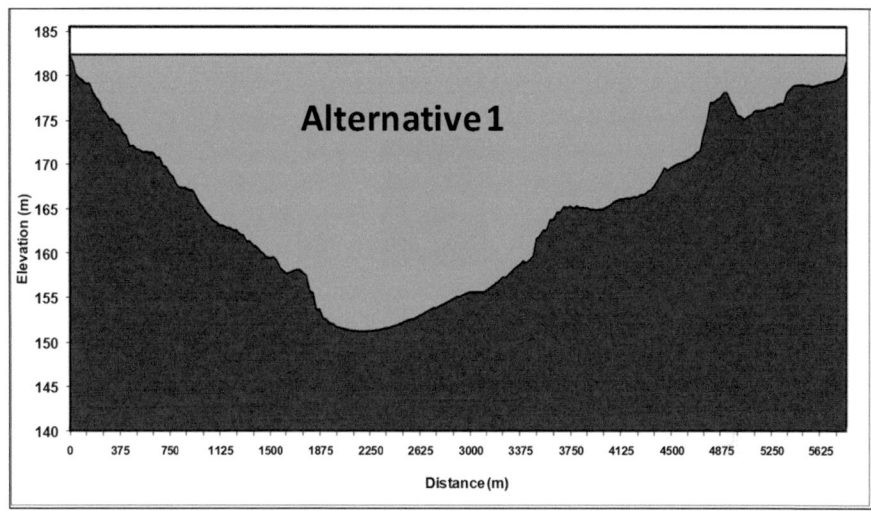

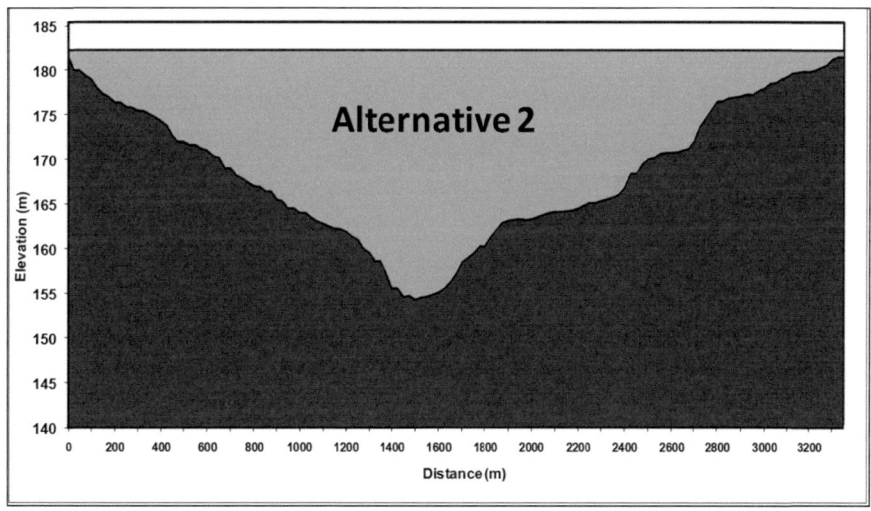

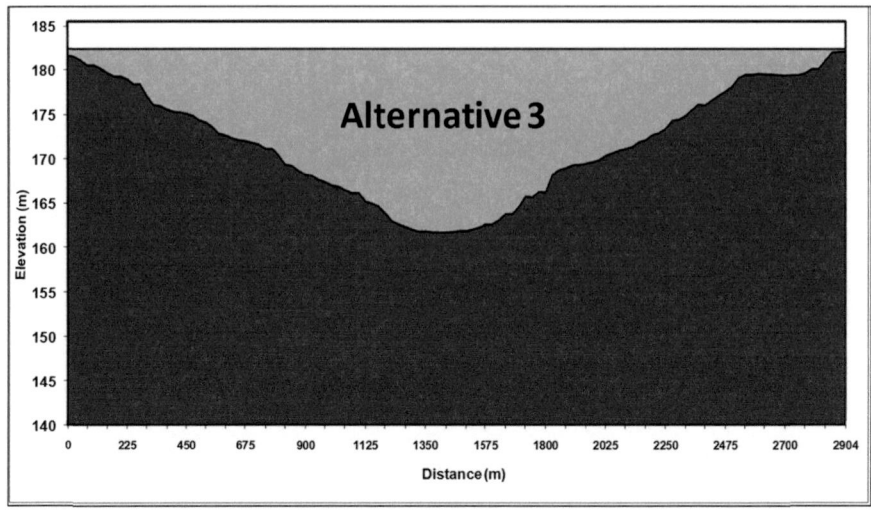

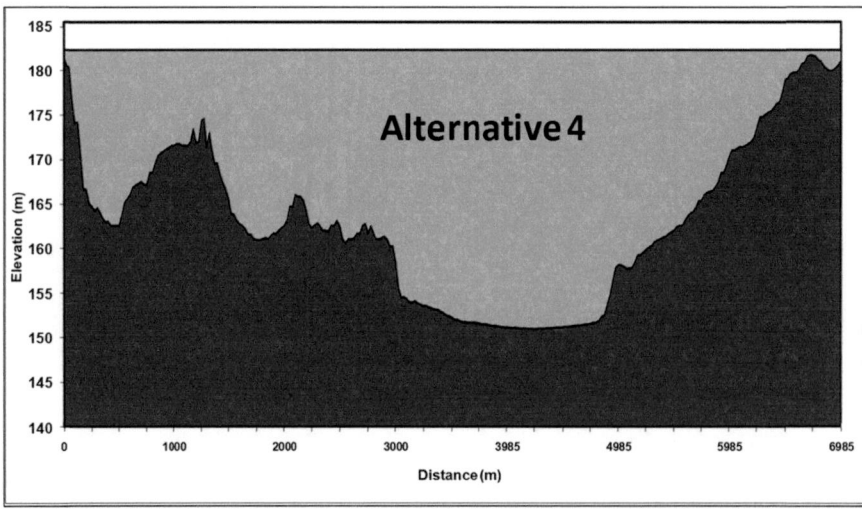

Appendix H 361

Cross section of the alternatives for Establishing a new dam at the El-Madik reach

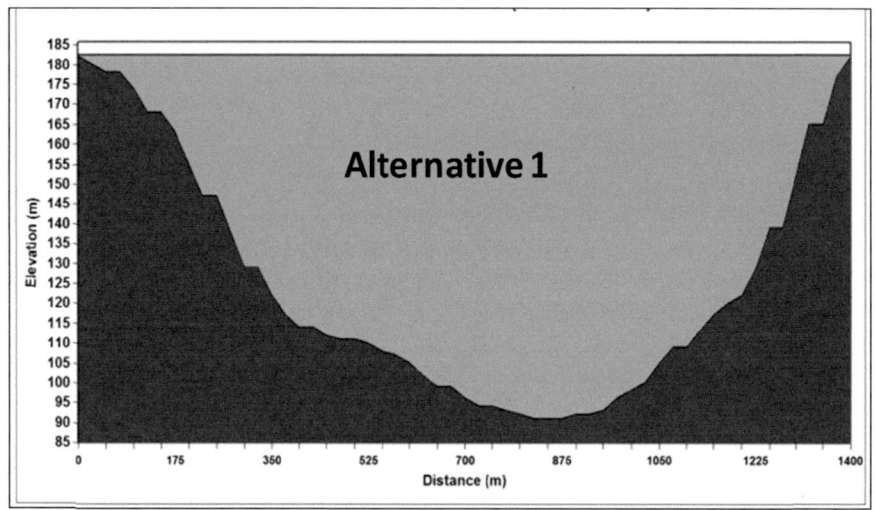

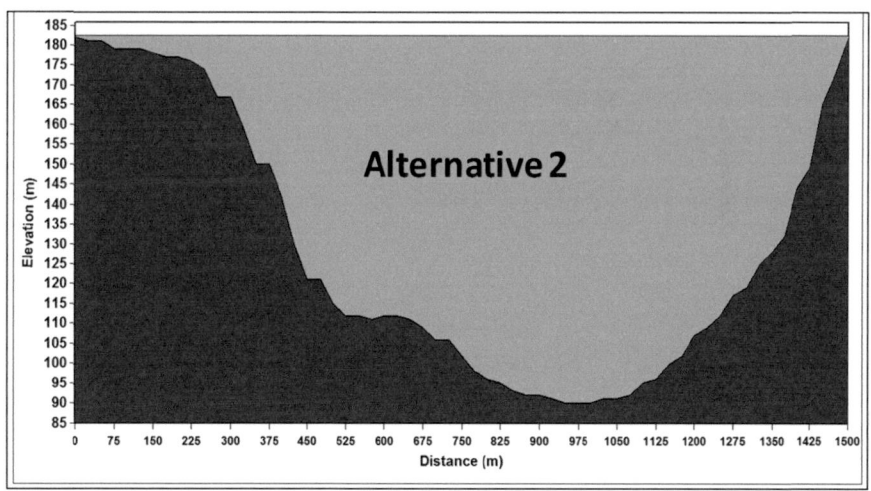

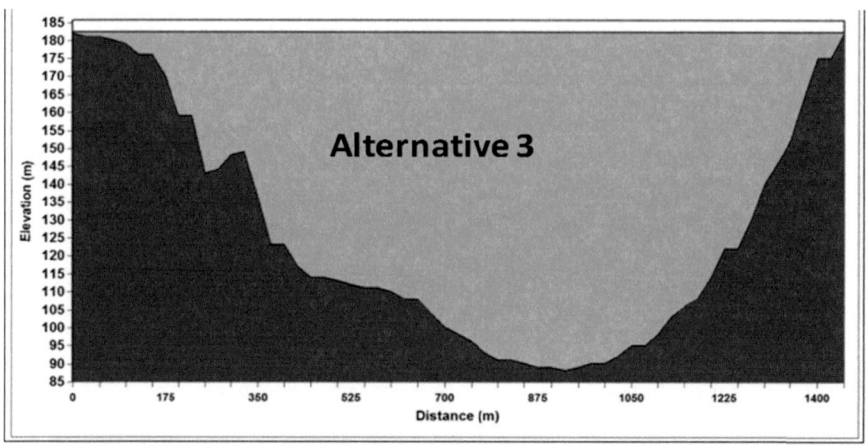
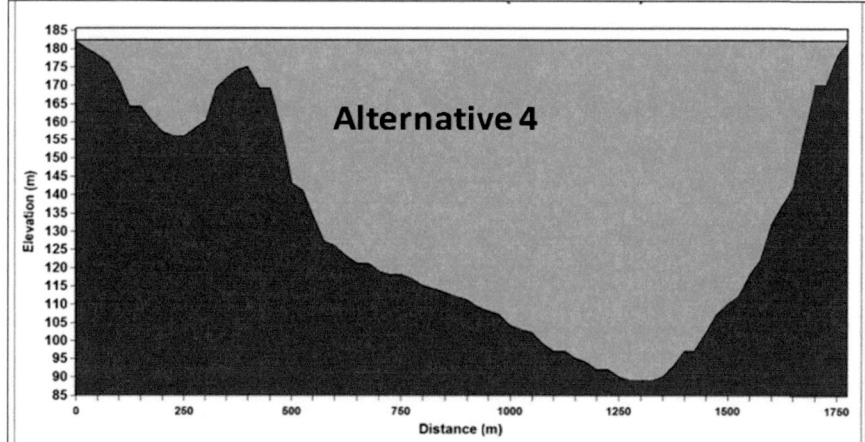
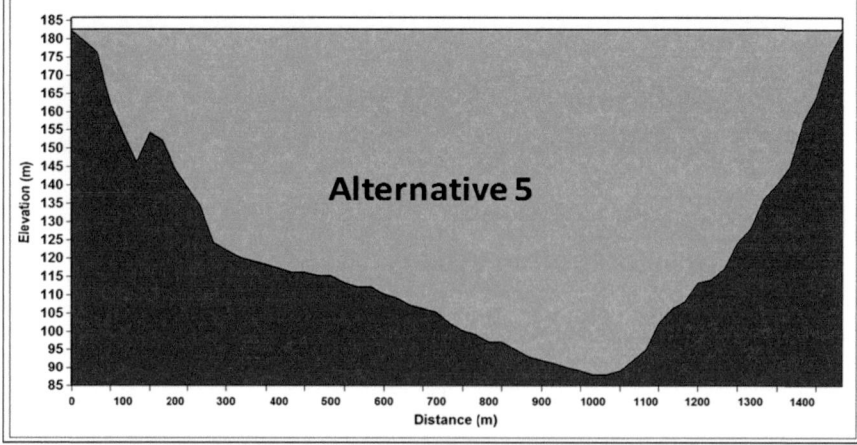

Scenarios for investigating the optimal operational water levels upstream and downstream of the proposed new dam

Scenario (1) results for a reduced water level of one meter in part 2 downstream the proposed dam

Whole HADR			Part 2 Downstream proposed dam			Part 1 upstream proposed dam			Difference	
Water level (m)	Volume (km³)	Surface area (km²)	Water level (m)	Volume (km³)	Surface area (km²)	Volume (km³)	Water level (m)	Surface area (km²)	Increased Water level in part1 (m)	Saved Surface area (km²)
155	43.28	2272.97	154	18.839	1029.45	24.444	155.68	1225.98	0.68	17.54
156	45.61	2379.94	155	19.871	1083.72	25.739	156.73	1289.86	0.73	6.36
157	48.05	2492.20	156	20.952	1136.74	27.093	157.79	1352.68	0.79	2.78
158	50.59	2601.61	157	22.085	1185.31	28.507	158.84	1415.65	0.84	0.64
159	53.25	2720.40	158	23.272	1236.41	29.982	159.89	1483.49	0.89	0.49
160	56.04	2844.82	159	24.514	1294.89	31.522	160.93	1547.04	0.93	2.89
161	58.94	2964.46	160	25.813	1353.83	33.127	161.98	1618.29	0.98	-7.67
162	61.97	3098.85	161	27.173	1413.50	34.799	163.03	1675.48	1.03	9.88
163	65.12	3199.62	162	28.596	1479.25	36.526	164.06	1744.32	1.06	-23.95
164	68.39	3334.65	163	30.084	1525.82	38.305	165.08	1804.14	1.08	4.68
165	71.78	3452.35	164	31.639	1593.73	40.143	166.09	1869.88	1.09	-11.27
166	75.30	3587.20	165	33.264	1653.46	42.038	167.10	1929.87	1.10	3.87
167	78.95	3715.42	166	34.962	1722.99	43.991	168.09	1982.80	1.09	9.63
168	82.72	3813.85	167	36.736	1790.48	45.981	169.06	2053.53	1.06	-30.16
169	86.61	3969.21	168	38.589	1837.50	48.020	170.02	2132.05	1.02	-0.34
170	90.68	4162.59	169	40.523	1920.83	50.152	170.99	2221.46	0.99	20.30
171	94.94	4373.53	170	42.542	2032.68	52.401	171.97	2300.62	0.97	40.24
172	99.41	4552.90	171	44.649	2151.13	54.758	172.96	2367.07	0.96	34.70
173	104.04	4711.98	172	46.847	2250.00	57.192	173.95	2446.68	0.95	15.31
174	108.84	4882.72	173	49.139	2342.36	59.697	174.92	2518.49	0.92	21.87
175	113.80	5040.57	174	51.530	2431.69	62.268	175.89	2608.07	0.89	0.81
176	118.93	5225.60	175	54.022	2516.41	64.910	176.84	2727.77	0.84	-18.58
177	124.30	5502.41	176	56.619	2606.94	67.676	177.81	2853.70	0.81	41.77
178	129.95	5799.07	177	59.325	2754.49	70.621	178.80	3002.56	0.80	42.02
179	135.90	6116.49	178	62.144	2920.59	73.759	179.82	3129.63	0.82	66.25
180	142.14	6357.38	179	65.081	3083.15	77.060	180.85	3224.69	0.85	49.54
181	148.60	6551.88	180	68.138	3206.43	80.457	181.87	3262.02	0.87	83.44
182	155.18	6623.48	181	71.321	3314.09	83.862	182.86		Exceed maximum water level of 182 m AMSL	

Scenario (2) results for reduced water level of two meters in part 2 downstream the proposed dam

Whole HADR			Part 2 Downstream proposed dam			Part 1 upstream proposed dam			Difference	
Water level (m)	Volume (km^3)	Surface area (km^2)	Water level (m)	Volume (km^3)	Surface area (km^2)	Volume (km^3)	Water level (m)	Surface area (km^2)	Increased Water level in part1 (m)	Saved Surface area (km^2)
155	43.28	2272.97	153	17.854	916.16	25.429	156.49	1274.11	1.49	82.70
156	45.61	2379.94	154	18.839	1029.45	26.771	157.54	1338.31	1.54	12.18
157	48.05	2492.20	155	19.871	1083.72	28.175	158.59	1400.99	1.59	7.49
158	50.59	2601.61	156	20.952	1136.74	29.640	159.65	1467.84	1.65	-2.98
159	53.25	2720.40	157	22.085	1185.31	31.168	160.70	1532.81	1.70	2.28
160	56.04	2844.82	158	23.272	1236.41	32.764	161.75	1602.33	1.75	6.08
161	58.94	2964.46	159	24.514	1294.89	34.427	162.80	1662.77	1.80	6.80
162	61.97	3098.85	160	25.813	1353.83	36.159	163.84	1730.37	1.84	14.66
163	65.12	3199.62	161	27.173	1413.50	37.948	164.88	1791.87	1.88	-5.75
164	68.39	3334.65	162	28.596	1479.25	39.793	165.90	1857.90	1.90	-2.51
165	71.78	3452.35	163	30.084	1525.82	41.699	166.92	1920.04	1.92	6.48
166	75.30	3587.20	164	31.639	1593.73	43.663	167.93	1972.52	1.93	20.94
167	78.95	3715.42	165	33.264	1653.46	45.689	168.92	2042.80	1.92	19.16
168	82.72	3813.85	166	34.962	1722.99	47.755	169.90	2121.80	1.90	-30.94
169	86.61	3969.21	167	36.736	1790.48	49.873	170.87	2209.94	1.87	-31.21
170	90.68	4162.59	168	38.589	1837.50	52.086	171.84	2289.73	1.84	35.36
171	94.94	4373.53	169	40.523	1920.83	54.420	172.82	2357.76	1.82	94.95
172	99.41	4552.90	170	42.542	2032.68	56.864	173.82	2436.06	1.82	84.16
173	104.04	4711.98	171	44.649	2151.13	59.390	174.80	2509.88	1.80	50.97
174	108.84	4882.72	172	46.847	2250.00	61.989	175.78	2598.33	1.78	34.40
175	113.80	5040.57	173	49.139	2342.36	64.659	176.75	2716.23	1.75	-18.01
176	118.93	5225.60	174	51.530	2431.69	67.401	177.72	2841.37	1.72	-47.46
177	124.30	5502.41	175	54.022	2516.41	70.274	178.69	2984.76	1.69	1.24
178	129.95	5799.07	176	56.619	2606.94	73.327	179.68	3113.44	1.68	78.69
179	135.90	6116.49	177	59.325	2754.49	76.579	180.70	3211.86	1.70	150.15
180	142.14	6357.38	178	62.144	2920.59	79.996	181.73	3258.23	1.73	178.57
181	148.60	6551.88	179	65.081	3083.15	83.515	182.76	Exceed maximum water level of 182 m AMSL		
182	155.18	6623.48	180	68.138	3206.43	87.045	183.75			

Scenario (3) results for reduced water level of three meters in part 2 downstream the proposed dam

Whole HADR			Part 2 Downstream proposed dam			Part 1 upstream proposed dam			Difference	
Water level (m)	Volume (km³)	Surface area (km²)	Water level (m)	Volume (km³)	Surface area (km²)	Volume (km³)	Water level (m)	Surface area (km²)	Increased Water level in part1 (m)	Saved Surface area (km²)
155	43.28	2272.97	152	16.915	811.01	26.368	157.23	1320.17	2.23	141.79
156	45.61	2379.94	153	17.854	916.16	27.755	158.28	1382.29	2.28	81.50
157	48.05	2492.20	154	18.839	1029.45	29.207	159.34	1447.77	2.34	14.98
158	50.59	2601.61	155	19.871	1083.72	30.722	160.40	1514.67	2.40	3.21
159	53.25	2720.40	156	20.952	1136.74	32.301	161.45	1581.71	2.45	1.95
160	56.04	2844.82	157	22.085	1185.31	33.951	162.50	1646.74	2.50	12.76
161	58.94	2964.46	158	23.272	1236.41	35.669	163.55	1710.81	2.55	17.24
162	61.97	3098.85	159	24.514	1294.89	37.459	164.60	1775.69	2.60	28.28
163	65.12	3199.62	160	25.813	1353.83	39.308	165.64	1840.58	2.64	5.20
164	68.39	3334.65	161	27.173	1413.50	41.215	166.67	1904.63	2.67	16.52
165	71.78	3452.35	162	28.596	1479.25	43.186	167.69	1960.15	2.69	12.94
166	75.30	3587.20	163	30.084	1525.82	45.218	168.69	2026.37	2.69	35.01
167	78.95	3715.42	164	31.639	1593.73	47.314	169.70	2105.04	2.70	16.65
168	82.72	3813.85	165	33.264	1653.46	49.454	170.68	2192.54	2.68	-32.15
169	86.61	3969.21	166	34.962	1722.99	51.647	171.65	2274.45	2.65	-28.23
170	90.68	4162.59	167	36.736	1790.48	53.939	172.62	2344.40	2.62	27.70
171	94.94	4373.53	168	38.589	1837.50	56.354	173.61	2419.43	2.61	116.61
172	99.41	4552.90	169	40.523	1920.83	58.883	174.61	2495.59	2.61	136.48
173	104.04	4711.98	170	42.542	2032.68	61.497	175.60	2581.04	2.60	98.26
174	108.84	4882.72	171	44.649	2151.13	64.187	176.59	2694.42	2.59	37.17
175	113.80	5040.57	172	46.847	2250.00	66.951	177.56	2821.08	2.56	-30.50
176	118.93	5225.60	173	49.139	2342.36	69.792	178.53	2959.93	2.53	-76.68
177	124.30	5502.41	174	51.530	2431.69	72.765	179.50	3092.24	2.50	-21.52
178	129.95	5799.07	175	54.022	2516.41	75.924	180.50	3194.29	2.50	88.36
179	135.90	6110.49	176	56.619	2606.94	79.285	181.52	3252.33	2.52	257.22
180	142.14	6357.38	177	59.325	2754.49	82.816	182.56			
181	148.60	6551.88	178	62.144	2920.59	86.451	183.58	Exceed maximum water level of 182 m AMSL		
182	155.18	6623.48	179	65.081	3083.15	90.102	184.58			

Scenario (4) results for reduced water level of four meters in part 2 downstream the proposed dam

Whole HADR			Part 2 Downstream proposed dam			Part 1 upstream proposed dam			Difference	
Water level (m)	Volume (km³)	Surface area (km²)	Water level (m)	Volume (km³)	Surface area (km²)	Volume (km³)	Water level (m)	Surface area (km²)	Increased Water level in part1 (m)	Saved Surface area (km²)
155	43.28	2272.97	151	16.019	718.62	27.264	157.91	1360.21	2.91	194.13
156	45.61	2379.94	152	16.915	811.01	28.695	158.97	1423.85	2.97	145.08
157	48.05	2492.20	153	17.854	916.16	30.192	160.03	1492.85	3.03	83.20
158	50.59	2601.61	154	18.839	1029.45	31.754	161.09	1557.01	3.09	15.14
159	53.25	2720.40	155	19.871	1083.72	33.383	162.14	1627.37	3.14	9.32
160	56.04	2844.82	156	20.952	1136.74	35.084	163.20	1687.12	3.20	20.95
161	58.94	2964.46	157	22.085	1185.31	36.855	164.25	1755.48	3.25	23.66
162	61.97	3098.85	158	23.272	1236.41	38.701	165.30	1818.62	3.30	43.82
163	65.12	3199.62	159	24.514	1294.89	40.608	166.34	1885.04	3.34	19.68
164	68.39	3334.65	160	25.813	1353.83	42.575	167.37	1944.13	3.37	36.68
165	71.78	3452.35	161	27.173	1413.50	44.609	168.40	2004.86	3.40	33.99
166	75.30	3587.20	162	28.596	1479.25	46.706	169.41	2081.69	3.41	26.26
167	78.95	3715.42	163	30.084	1525.82	48.870	170.41	2168.08	3.41	21.51
168	82.72	3813.85	164	31.639	1593.73	51.079	171.40	2254.53	3.40	-34.41
169	86.61	3969.21	165	33.264	1653.46	53.345	172.37	2327.78	3.37	-12.03
170	90.68	4162.59	166	34.962	1722.99	55.713	173.35	2398.33	3.35	41.27
171	94.94	4373.53	167	36.736	1790.48	58.207	174.35	2476.35	3.35	106.70
172	99.41	4552.90	168	38.589	1837.50	60.817	175.35	2557.00	3.35	158.41
173	104.04	4711.98	169	40.523	1920.83	63.516	176.34	2663.10	3.34	128.05
174	108.84	4882.72	170	42.542	2032.68	66.294	177.33	2791.26	3.33	58.78
175	113.80	5040.57	171	44.649	2151.13	69.149	178.31	2926.56	3.31	-37.12
176	118.93	5225.60	172	46.847	2250.00	72.084	179.28	3066.33	3.28	-90.73
177	124.30	5502.41	173	49.139	2342.36	75.156	180.26	3173.50	3.26	-13.45
178	129.95	5799.07	174	51.530	2431.69	78.416	181.26	3245.07	3.26	122.30
179	135.90	6116.49	175	54.022	2516.41	81.882	182.29			
180	142.14	6357.38	176	56.619	2606.94	85.522	183.33	Exceed maximum water level of 182 m AMSL		
181	148.60	6551.88	177	59.325	2754.49	89.270	184.35			
182	155.18	6623.48	178	62.144	2920.59	93.039	185.35			

Appendix H

Scenario (5) results for reduced water level of five meters in part 2 downstream the proposed dam

Whole HADR			Part 2 Downstream proposed dam			Part 1 upstream proposed dam			Difference	
Water level (m)	Volume (km^3)	Surface area (km^2)	Water level (m)	Volume (km^3)	Surface area (km^2)	Volume (km^3)	Water level (m)	Surface area (km^2)	Increased Water level in part1 (m)	Saved Surface area (km^2)
155	43.28	2272.97	150	15.166	645.41	28.117	158.55	1398.45	3.55	229.11
156	45.61	2379.94	151	16.019	718.62	29.590	159.61	1465.54	3.61	195.78
157	48.05	2492.20	152	16.915	811.01	31.131	160.67	1531.30	3.67	149.89
158	50.59	2601.61	153	17.854	916.16	32.738	161.73	1601.17	3.73	84.28
159	53.25	2720.40	154	18.839	1029.45	34.415	162.79	1662.36	3.79	28.59
160	56.04	2844.82	155	19.871	1083.72	36.165	163.85	1730.61	3.85	30.49
161	58.94	2964.46	156	20.952	1136.74	37.988	164.90	1793.19	3.90	34.53
162	61.97	3098.85	157	22.085	1185.31	39.887	165.95	1861.25	3.95	52.29
163	65.12	3199.62	158	23.272	1236.41	41.850	167.00	1924.84	4.00	38.36
164	68.39	3334.65	159	24.514	1294.89	43.875	168.03	1978.64	4.03	61.12
165	71.78	3452.35	160	25.813	1353.83	45.969	169.06	2053.04	4.06	45.48
166	75.30	3587.20	161	27.173	1413.50	48.129	170.07	2136.68	4.07	37.02
167	78.95	3715.42	162	28.596	1479.25	50.357	171.08	2228.92	4.08	7.24
168	82.72	3813.85	163	30.084	1525.82	52.634	172.07	2307.66	4.07	-19.63
169	86.61	3969.21	164	31.639	1593.73	54.971	173.05	2373.65	4.05	1.83
170	90.68	4162.59	165	33.264	1653.46	57.411	174.03	2453.47	4.03	55.66
171	94.94	4373.53	166	34.962	1722.99	59.981	175.03	2527.07	4.03	123.47
172	99.41	4552.90	167	36.736	1790.48	62.670	176.04	2623.26	4.04	139.16
173	104.04	4711.98	168	38.589	1837.50	65.450	177.04	2752.57	4.04	121.92
174	108.84	4882.72	169	40.523	1920.83	68.313	178.03	2882.76	4.03	79.14
175	113.80	5040.57	170	42.542	2032.68	71.256	179.01	3034.54	4.01	-26.65
176	118.93	5225.60	171	44.649	2151.13	74.282	179.98	3149.16	3.98	-74.69
177	124.30	5502.41	172	46.847	2250.00	77.448	180.97	3235.00	3.97	17.41
178	129.95	5799.07	173	49.139	2342.36	80.807	181.97	3264.89	3.97	191.82
179	135.90	6116.49	174	51.530	2431.69	84.374	183.00			
180	142.14	6357.38	175	54.022	2516.41	88.119	184.04	Exceed maximum water level of 182 m AMSL		
181	148.60	6551.88	176	56.619	2606.94	91.976	185.07			
182	155.18	6623.48	177	59.325	2754.49	95.858	186.08			

Scenario (6) results for reduced water level of six meters in part 2 downstream the proposed dam

Whole HADR			Part 2 Downstream proposed dam			Part 1 upstream proposed dam			Difference	
Water level (m)	Volume (km^3)	Surface area (km^2)	Water level (m)	Volume (km^3)	Surface area (km^2)	Volume (km^3)	Water level (m)	Surface area (km^2)	Increased Water level in part1 (m)	Saved Surface area (km^2)
155	43.28	2272.97	149	14.352	592.99	28.931	159.14	1434.84	4.14	245.13
156	45.61	2379.94	150	15.166	645.41	30.444	160.21	1503.28	4.21	231.25
157	48.05	2492.20	151	16.019	718.62	32.026	161.27	1569.35	4.27	204.23
158	50.59	2601.61	152	16.915	811.01	33.678	162.33	1637.46	4.33	153.13
159	53.25	2720.40	153	17.854	916.16	35.399	163.39	1699.94	4.39	104.31
160	56.04	2844.82	154	18.839	1029.45	37.197	164.45	1766.97	4.45	48.40
161	58.94	2964.46	155	19.871	1083.72	39.070	165.51	1832.00	4.51	48.74
162	61.97	3098.85	156	20.952	1136.74	41.020	166.56	1898.36	4.56	63.75
163	65.12	3199.62	157	22.085	1185.31	43.036	167.61	1956.24	4.61	58.06
164	68.39	3334.65	158	23.272	1236.41	45.117	168.65	2022.81	4.65	75.42
165	71.78	3452.35	159	24.514	1294.89	47.269	169.67	2103.29	4.67	54.17
166	75.30	3587.20	160	25.813	1353.83	49.489	170.69	2193.99	4.69	39.37
167	78.95	3715.42	161	27.173	1413.50	51.780	171.70	2279.09	4.70	22.83
168	82.72	3813.85	162	28.596	1479.25	54.122	172.70	2349.50	4.70	-14.90
169	86.61	3969.21	163	30.084	1525.82	56.526	173.68	2425.04	4.68	18.35
170	90.68	4162.59	164	31.639	1593.73	59.037	174.67	2499.92	4.67	68.93
171	94.94	4373.53	165	33.264	1653.46	61.679	175.67	2587.45	4.67	132.62
172	99.41	4552.90	166	34.962	1722.99	64.444	176.68	2706.31	4.68	123.60
173	104.04	4711.98	167	36.736	1790.48	67.303	177.68	2836.92	4.68	84.58
174	108.84	4882.72	168	38.589	1837.50	70.247	178.68	2983.40	4.68	61.82
175	113.80	5040.57	169	40.523	1920.83	73.275	179.66	3111.46	4.66	8.28
176	118.93	5225.60	170	42.542	2032.68	76.389	180.64	3206.78	4.64	-13.86
177	124.30	5502.41	171	44.649	2151.13	79.646	181.63	3255.33	4.63	95.94
178	129.95	5799.07	172	46.847	2250.00	83.099	182.64			
179	135.90	6116.49	173	49.139	2342.36	86.764	183.67	Exceed maximum water level of 182 m AMSL		
180	142.14	6357.38	174	51.530	2431.69	90.611	184.71			
181	148.60	6551.88	175	54.022	2516.41	94.574	185.75			
182	155.18	6623.48	176	56.619	2606.94	98.564	186.75			

Scenario (7) results for reduced water level of seven meters in part 2 downstream the proposed dam

Whole HADR			Part 2 Downstream proposed dam			Part 1 upstream proposed dam			Difference	
Water level (m)	Volume (km^3)	Surface area (km^2)	Water level (m)	Volume (km^3)	Surface area (km^2)	Volume (km^3)	Water level (m)	Surface area (km^2)	Increased Water level in part1 (m)	Saved Surface area (km^2)
155	43.28	2272.97	148	13.577	545.92	29.705	159.69	1470.85	4.69	256.20
156	45.61	2379.94	149	14.352	592.99	31.257	160.76	1536.39	4.76	250.56
157	48.05	2492.20	150	15.166	645.41	32.880	161.82	1607.42	4.82	239.37
158	50.59	2601.61	151	16.019	718.62	34.573	162.89	1667.65	4.89	215.34
159	53.25	2720.40	152	16.915	811.01	36.339	163.95	1737.48	4.95	171.90
160	56.04	2844.82	153	17.854	916.16	38.182	165.01	1799.62	5.01	129.04
161	58.94	2964.46	154	18.839	1029.45	40.102	166.07	1868.52	5.07	66.48
162	61.97	3098.85	155	19.871	1083.72	42.101	167.13	1931.57	5.13	83.56
163	65.12	3199.62	156	20.952	1136.74	44.169	168.18	1989.20	5.18	73.68
164	68.39	3334.65	157	22.085	1185.31	46.303	169.22	2066.10	5.22	83.24
165	71.78	3452.35	158	23.272	1236.41	48.511	170.25	2152.92	5.25	63.01
166	75.30	3587.20	159	24.514	1294.89	50.788	171.27	2244.25	5.27	48.06
167	78.95	3715.42	160	25.813	1353.83	53.140	172.29	2321.99	5.29	39.60
168	82.72	3813.85	161	27.173	1413.50	55.544	173.28	2392.75	5.28	7.60
169	86.61	3969.21	162	28.596	1479.25	58.013	174.27	2470.81	5.27	19.15
170	90.68	4162.59	163	30.084	1525.82	60.592	175.26	2548.96	5.26	87.80
171	94.94	4373.53	164	31.639	1593.73	63.305	176.27	2653.20	5.27	126.60
172	99.41	4552.90	165	33.264	1653.46	66.142	177.28	2784.34	5.28	115.11
173	104.04	4711.98	166	34.962	1722.99	69.077	178.29	2922.78	5.29	66.21
174	108.84	4882.72	167	36.736	1790.48	72.100	179.29	3066.93	5.29	25.31
175	113.80	5040.57	168	38.589	1837.50	75.209	180.28	3174.94	5.28	28.14
176	118.93	5225.60	169	40.523	1920.83	78.408	181.26	3245.00	5.26	59.78
177	124.30	5502.41	170	42.542	2032.68	81.753	182.25			
178	129.95	5799.07	171	44.649	2151.13	85.297	183.26			
179	135.90	6116.49	172	46.847	2250.00	89.057	184.30	Exceed maximum water level of 182 m AMSL		
180	142.14	6357.38	173	49.139	2342.36	93.001	185.34			
181	148.60	6551.88	174	51.530	2431.69	97.066	186.38			
182	155.18	6623.48	175	54.022	2516.41	101.161	187.39			

Scenario (8) results for reduced water level of eight meters in part 2 downstream the proposed dam

Whole HADR			Part 2 Downstream proposed dam			Part 1 upstream proposed dam			Difference	
Water level (m)	Volume (km^3)	Surface area (km^2)	Water level (m)	Volume (km^3)	Surface area (km^2)	Volume (km^3)	Water level (m)	Surface area (km^2)	Increased Water level in part1 (m)	Saved Surface area (km^2)
155	43.28	2272.97	147	12.840	506.61	30.443	160.20	1503.25	5.20	263.10
156	45.61	2379.94	148	13.577	545.92	32.032	161.27	1569.61	5.27	264.41
157	48.05	2492.20	149	14.352	592.99	33.693	162.34	1637.99	5.34	261.22
158	50.59	2601.61	150	15.166	645.41	35.427	163.41	1701.04	5.41	255.15
159	53.25	2720.40	151	16.019	718.62	37.234	164.47	1768.20	5.47	233.58
160	56.04	2844.82	152	16.915	811.01	39.121	165.54	1833.86	5.54	199.95
161	58.94	2964.46	153	17.854	916.16	41.087	166.60	1900.51	5.60	147.79
162	61.97	3098.85	154	18.839	1029.45	43.134	167.66	1958.78	5.66	110.62
163	65.12	3199.62	155	19.871	1083.72	45.251	168.71	2027.50	5.71	88.40
164	68.39	3334.65	156	20.952	1136.74	47.436	169.75	2109.68	5.75	88.22
165	71.78	3452.35	157	22.085	1185.31	49.697	170.79	2202.66	5.79	64.38
166	75.30	3587.20	158	23.272	1236.41	52.030	171.81	2287.80	5.81	62.99
167	78.95	3715.42	159	24.514	1294.89	54.440	172.83	2358.30	5.83	62.23
168	82.72	3813.85	160	25.813	1353.83	56.904	173.83	2437.36	5.83	22.66
169	86.61	3969.21	161	27.173	1413.50	59.436	174.82	2511.17	5.82	44.54
170	90.68	4162.59	162	28.596	1479.25	62.079	175.82	2601.47	5.82	81.86
171	94.94	4373.53	163	30.084	1525.82	64.860	176.83	2725.48	5.83	122.22
172	99.41	4552.90	164	31.639	1593.73	67.768	177.84	2857.79	5.84	101.38
173	104.04	4711.98	165	33.264	1653.46	70.775	178.85	3010.44	5.85	48.08
174	108.84	4882.72	166	34.962	1722.99	73.874	179.86	3133.94	5.86	25.79
175	113.80	5040.57	167	36.736	1790.48	77.062	180.85	3224.74	5.85	25.35
176	118.93	5225.60	168	38.589	1837.50	80.342	181.84	3261.08	5.84	127.03
177	124.30	5502.41	169	40.523	1920.83	83.772	182.83			
178	129.95	5799.07	170	42.542	2032.68	87.404	183.85			
179	135.90	6116.49	171	44.649	2151.13	91.255	184.88	Exceed maximum water level of 182 m AMSL		
180	142.14	6357.38	172	46.847	2250.00	95.294	185.93			
181	148.60	6551.88	173	49.139	2342.36	99.456	186.97			
182	155.18	6623.48	174	51.530	2431.69	103.653	187.98			

Scenario (9) results for reduced water level of nine meters in part 2 downstream the proposed dam

Whole HADR			Part 2 Downstream proposed dam			Part 1 upstream proposed dam			Difference	
Water level (m)	Volume (km^3)	Surface area (km^2)	Water level (m)	Volume (km^3)	Surface area (km^2)	Volume (km^3)	Water level (m)	Surface area (km^2)	Increased Water level in part1 (m)	Saved Surface area (km^2)
155	43.28	2272.97	146	12.137	481.64	31.146	160.68	1531.90	5.68	259.44
156	45.61	2379.94	147	12.840	506.61	32.770	161.75	1602.56	5.75	270.77
157	48.05	2492.20	148	13.577	545.92	34.468	162.82	1664.14	5.82	282.14
158	50.59	2601.61	149	14.352	592.99	36.240	163.89	1733.58	5.89	275.03
159	53.25	2720.40	150	15.166	645.41	38.088	164.96	1796.45	5.96	278.54
160	56.04	2844.82	151	16.019	718.62	40.017	166.03	1865.73	6.03	260.47
161	58.94	2964.46	152	16.915	811.01	42.026	167.09	1929.56	6.09	223.89
162	61.97	3098.85	153	17.854	916.16	44.118	168.15	1987.37	6.15	195.32
163	65.12	3199.62	154	18.839	1029.45	46.283	169.21	2065.29	6.21	104.87
164	68.39	3334.65	155	19.871	1083.72	48.518	170.25	2153.23	6.25	97.70
165	71.78	3452.35	156	20.952	1136.74	50.830	171.29	2245.72	6.29	69.88
166	75.30	3587.20	157	22.085	1185.31	53.217	172.32	2324.16	6.32	77.72
167	78.95	3715.42	158	23.272	1236.41	55.682	173.34	2397.30	6.34	81.70
168	82.72	3813.85	159	24.514	1294.89	58.204	174.35	2476.28	6.35	42.69
169	86.61	3969.21	160	25.813	1353.83	60.796	175.34	2556.24	6.34	59.15
170	90.68	4162.59	161	27.173	1413.50	63.502	176.34	2662.45	6.34	86.63
171	94.94	4373.53	162	28.596	1479.25	66.347	177.35	2793.69	6.35	100.59
172	99.41	4552.90	163	30.084	1525.82	69.323	178.37	2935.62	6.37	91.46
173	104.04	4711.98	164	31.639	1593.73	72.400	179.38	3078.38	6.38	39.87
174	108.84	4882.72	165	33.264	1653.46	75.572	180.39	3184.79	6.39	44.48
175	113.80	5040.57	166	34.962	1722.99	78.836	181.39	3248.58	6.39	69.00
176	118.93	5225.60	167	36.736	1790.48	82.195	182.38			
177	124.30	5502.41	168	38.589	1837.50	85.706	183.38			
178	129.95	5799.07	169	40.523	1920.83	89.422	184.10			
179	135.90	6116.49	170	42.542	2032.68	93.361	185.44	Exceed maximum water level of 182 m AMSL			
180	142.14	6357.38	171	44.649	2151.13	97.492	186.49			
181	148.60	6551.88	172	46.847	2250.00	101.748	187.53			
182	155.18	6623.48	173	49.139	2342.36	106.044	188.54			

Scenario (10) results for reduced water level of ten meters in part 2 downstream the proposed dam

Whole HADR			Part 2 Downstream proposed dam			Part 1 upstream proposed dam			Difference	
Water level (m)	Volume (km³)	Surface area (km²)	Water level (m)	Volume (km³)	Surface area (km²)	Volume (km³)	Water level (m)	Surface area (km²)	Increased Water level in part1 (m)	Saved Surface area (km²)
155	43.28	2272.97	145	11.469	462.81	31.814	161.13	1559.75	6.13	250.41
156	45.61	2379.94	146	12.137	481.64	33.472	162.20	1630.44	6.20	267.87
157	48.05	2492.20	147	12.840	506.61	35.206	163.27	1692.09	6.27	293.50
158	50.59	2601.61	148	13.577	545.92	37.015	164.34	1760.85	6.34	294.83
159	53.25	2720.40	149	14.352	592.99	38.901	165.41	1825.90	6.41	301.51
160	56.04	2844.82	150	15.166	645.41	40.870	166.48	1893.54	6.48	305.87
161	58.94	2964.46	151	16.019	718.62	42.921	167.55	1953.23	6.55	292.60
162	61.97	3098.85	152	16.915	811.01	45.058	168.62	2020.72	6.62	267.12
163	65.12	3199.62	153	17.854	916.16	47.268	169.67	2103.25	6.67	180.21
164	68.39	3334.65	154	18.839	1029.45	49.550	170.72	2196.55	6.72	108.65
165	71.78	3452.35	155	19.871	1083.72	51.911	171.76	2283.66	6.76	84.97
166	75.30	3587.20	156	20.952	1136.74	54.350	172.79	2355.81	6.79	94.64
167	78.95	3715.42	157	22.085	1185.31	56.868	173.82	2436.18	6.82	93.93
168	82.72	3813.85	158	23.272	1236.41	59.446	174.83	2511.46	6.83	65.97
169	86.61	3969.21	159	24.514	1294.89	62.096	175.82	2602.05	6.82	72.28
170	90.68	4162.59	160	25.813	1353.83	64.862	176.83	2725.58	6.83	83.17
171	94.94	4373.53	161	27.173	1413.50	67.770	177.84	2857.89	6.84	102.15
172	99.41	4552.90	162	28.596	1479.25	70.811	178.86	3012.27	6.86	61.38
173	104.04	4711.98	163	30.084	1525.82	73.955	179.88	3136.98	6.88	49.17
174	108.84	4882.72	164	31.639	1593.73	77.198	180.89	3228.35	6.89	60.64
175	113.80	5040.57	165	33.264	1653.46	80.534	181.89	3262.65	6.89	124.46
176	118.93	5225.60	166	34.962	1722.99	83.969	182.89			
177	124.30	5502.41	167	36.736	1790.48	87.559	183.89			
178	129.95	5799.07	168	38.589	1837.50	91.357	184.91			
179	135.90	6116.49	169	40.523	1920.83	95.380	185.95	Exceed maximum water level of 182 m AMSL			
180	142.14	6357.38	170	42.542	2032.68	99.598	187.01			
181	148.60	6551.88	171	44.649	2151.13	103.946	188.05			
182	155.18	6623.48	172	46.847	2250.00	108.336	189.07			

The Author

Emad El-Din Mohammad Elba, was born 1967 in Egypt. The author got his bachelor's degree in Civil Engineer from Ain Shams University in Egypt. He gained his research experience in the fields of water resources management, sustainable development, preservation and climate change from expert professors of Leuphana University Lueneburg, Germany, where he got both his master's and doctorate degrees. Since 1995, he has been working in Nile water sector at the Egyptian Ministry of Water Resources and Irrigation where he gained his practical experience from the Nile hydrology experts. He worked in the Nile Basin countries and participated in the hydrological studies of the Nile basin of Lake Victoria in Uganda and High Aswan Dam Reservoir (HADR) in Egypt and Sudan. While joining the HADR surveying mission, the author decided to study how to reduce the evaporation losses to optimize Egypt's water resources.